Workbook for Mathematical Economics Part 1

Michael Sampson

Workbook for Mathematical Economics Part 1.

ISBN: 978-0-9947930-1-0

If you find typos or errors, *please tell me!* I value your fresh eyes. Any and all comments can be forwarded to *comments@loglinear.com.*

Published by *Loglinear Publications.* **www.loglinear.com.**

Acknowledgment

I would like to thank my former students for struggling through earlier versions. For pointing out errors and typos, particular thanks go to: Ryan Adams, Mirwais Alamy, Saif Al-Haroun, Kimia Amouzegar, Simon Apter, Viyasan Asokanathan, Audrey Autmezguine, Ramzi Bahsoun, Kahlil Baker, Philippe Baril Lecavalier, Andras A. Birkus, Katherine Boisvert, Jonathan Camarda, Ysa Cao, Patricia Carvajal, Carlo Catino, Yuhsin Chang, Reza Charghi, Alexandru Tudor Ioan Coada, Maxime Comeau, Anthony Delli Colli, Sarkis Ghenberjian, Habibeh Ghaemi, Sabrina Gilbert, Gema Gonzalez, Michael Gross, Heejung Han, Si Si He, Gabriel Hobbs, Sobhi Hodhod, Chris Hotte, Bo Hu, Wan Ting Huang, Wendy Ip, Firas Jahchan, Kåre Johard, Terrel Joseph, Khallad Karime, Amnuai Khongna, Vladimir Khripunov, Mina Kokinos, Mohammad Kureishy, Mercedes Kurowski, Laura King, Alexandre Laquerre, Jonathan Leclerc, Lerona D. Lewis, Ioannis Likakis, Zhou Ling, Alain Lumbroso, Bogdan Malaelea, Curtis McArdle, Diego Mesa, Axel Monsaingeon, Keith Nagy, Andrew Naimi, Ted Nehme, Ha Dao Ngoc, Anh Nguyen, Michael Osvath-Langlais, Igor Paskal, Darren Pellerin, Alexandre Perrault, Benjamin Philion, Luis Miguel Diaz Piccone, Pierre Provencal, Raymond Quach, Jose Antonio Paez Ravelo, Jean-Francois Prud'homme, David Roth, Tiago Rubin, Cyrus Showkati, Chi Ching Tam, Oren Tapiero, Pierre-Luc Vachon, Apira Vigneshwaran, Lise Villeneuve, Richard Yung, Bülent Yurtsever, David Zerkler, and Xiao Chun Zheng.

Michael Sampson is an Associate Professor in the Department of Economics, Concordia University, Montreal.

Cover image taken from John Tenniel's illustrations for Lewis Carroll's *Alice in Wonderland.*

Preface

Practice is the key to learning mathematics, or just about anything for that matter. Here is a large collection of problems for you to practice on. Answers are included, but you should only look at them *as a last resort.* First *struggle* with the problem. A good rule might be to *only look at the answer the next day.* It is the act of struggling that will prepare your mind, not simply reading the answer. The best outcome is to struggle, and get the answer yourself. The next best outcome is to struggle, fail after a long period, and only then look at the answer. Then when you see the answer you're likely to have that 'Aha!' experience, and you'll remember that particular technique for the rest of your life. Otherwise if you simply read the answer each time you get a little frustrated, you'll forget very soon afterwards (typically by the next class).

It is also important to work on the *right kind of problems.* This is depends on where *you are.* There is no point working on A problems if you are currently working at a D level. Work on D and C problems, try to get yourself to a C level, and only then go for the B problems. Assuming you are still ambitious, once you've mastered the B problems, go for the A problems.

Some problems are more important than others. Some are absolutely essential: they deal with skills (for example the chain rule) that you simply *must* master. Other problems are less important but still useful; some are not important at all. Some are there simply because I find them interesting.

With these distinctions in mind I have attempted to categorize the problems according to their level, that is whether they are A, B, C, D problems, and by their importance. The importance is indicated as follows:

*	Not very important.
**	Moderately important.
***	Absolutely essential.

For example *[C***]* would indicate a C level problem that is absolutely essential to master if you want to get a C, while *[A***]* would indicate an A level problem that is absolutely essential to master if you want to get an A . *[B**]* would indicate a moderately important B level problem, and *[A*]* would indicate an unimportant A level problem, although perhaps important for other courses, or simply an interesting or fun problem.

Chapter 1

Mathematical Thinking

1.1 Proofs from Definitions

Problem 1.1 *[C*] Prove that the sum of an odd and even number must be odd.*

Answer: If a is odd then by definition $a = 2m - 1$ where m is an integer. If b is even then by definition $b = 2n$ where n is an integer. Thus

$$a + b = 2m - 1 + 2n = 2(m + n) - 1 = 2r - 1$$

where $r \equiv m + n$ is an integer. Thus $a + b$ is by definition odd.

Problem 1.2 *[B*] Prove that a necessary and sufficient condition for an integer to be the product of two even integers is that it be divisible by 4.*

Answer: Since we are proving a necessary and sufficient condition, two proofs are required. Proof 1: We first prove that if a and b are both even then $a \times b$ is divisible by 4. If a and b are both even then by definition $a = 2m$ and $b = 2n$ where m and n are integers. Then

$$\begin{aligned} a \times b &= 2m \times 2n = 4mn \\ \Longrightarrow \quad \frac{ab}{4} &= \frac{4mn}{4} = mn = r \end{aligned}$$

where $r = m \times n$ is an integer. Thus $a \times b$ is divisible by 4.

Proof 2: Now suppose that $a \times b$ is divisible by 4 so that $\frac{a \times b}{4} = r$ where r is an integer. Then $a \times b$ is the product of two even numbers 2 and s as

$$\begin{aligned} \frac{a \times b}{4} &= r \\ \Longrightarrow \quad a \times b &= 4r = 2(2 \times r) = 2 \times s \end{aligned}$$

where both $2 = 2 \times 1$ and $s = 2 \times r$ are by definition even numbers.

Problem 1.3 *[B*] Prove that two integers are odd if and only if their product is odd.*

Answer: Since we are proving a necessary and sufficient condition, two proofs are required. Proof 1: We first prove that if a and b are both odd then $a \times b$ is odd. If a and b are both odd then by definition $a = 2m-1$ and $b = 2n-1$ where m and n are integers. Then

$$a \times b = (2m-1) \times (2n-1) = 4mn - 2m - 2n + 1 = 2(2mn - m - n + 1) - 1 = 2r - 1$$

where $r = 2mn - m - n + 1$ is an integer. Thus $a \times b$ is by definition odd.

Proof 2: We now prove that if $a \times b$ is odd, then a and b are both odd. Now an integer a is either an odd or an even number. Suppose a is even so that $a = 2 \times m$. Then

$$a \times b = 2m \times b = 2 \times (m \times b) = 2 \times r$$

where $r = m \times b$ is an integer and so $a \times b$ is even. Thus if $a \times b$ is odd then a must be also odd. Similarly suppose b is even so $b = 2 \times n$. Then

$$a \times b = a \times 2 \times n = 2 \times (a \times n) = 2 \times s$$

where $s = a \times n$ is an integer and so $a \times b$ is even. Thus if $a \times b$ is odd then b must also be odd. Since neither a nor b is even, we conclude that both a and b are odd.

1.2 Proof by Counter-Example

Problem 1.4 *[C*] Prove that $\sqrt{x+y} \neq \sqrt{x} + \sqrt{y}$. [A*] Prove that for all $x > 0, y > 0$ that*

$$\sqrt{x} + \sqrt{y} > \sqrt{x+y}.$$

Answer: Any two non-zero numbers will provide a counter-example to $\sqrt{x+y} = \sqrt{x} + \sqrt{y}$. For example suppose $x = 1, y = 1$. Then

$$\sqrt{x+y} = \sqrt{1+1} = \sqrt{2} \approx 1.41 \text{ and } \sqrt{x} + \sqrt{y} = \sqrt{1} + \sqrt{1} = 1 + 1 = 2.$$

To show why $\sqrt{x+y} \neq \sqrt{x} + \sqrt{y}$ for $x \neq 0, y \neq 0$ note that

$$\begin{aligned} \sqrt{x+y} &= \sqrt{x} + \sqrt{y} \\ &\Longrightarrow \left(\sqrt{x+y}\right)^2 = \left(\sqrt{x} + \sqrt{y}\right)^2 \\ &\Longrightarrow x + y = x + y + 2\sqrt{x}\sqrt{y} \\ &\Longrightarrow 2\sqrt{x}\sqrt{y} = 0 \\ &\Longrightarrow x = 0 \text{ or } y = 0. \end{aligned}$$

We have

$$
\begin{aligned}
\sqrt{x}+\sqrt{y}-\sqrt{x+y} &= \left(\sqrt{x}+\sqrt{y}-\sqrt{x+y}\right)\frac{\sqrt{x}+\sqrt{y}+\sqrt{x+y}}{\sqrt{x}+\sqrt{y}+\sqrt{x+y}} \\
&= \frac{\left(\left(\sqrt{x}+\sqrt{y}\right)^2-\left(\sqrt{x+y}\right)^2\right)}{\sqrt{x}+\sqrt{y}+\sqrt{x+y}} \\
&= \frac{x+2\sqrt{x}\sqrt{y}+y-(x+y)}{\sqrt{x}+\sqrt{y}+\sqrt{x+y}} \\
&= \frac{2\sqrt{x}\sqrt{y}}{\sqrt{x}+\sqrt{y}+\sqrt{x+y}} > 0.
\end{aligned}
$$

Problem 1.5 *[A*] A prime number is an integer that has only one and itself as factors. Hence 5 is a prime number since only 1 and 5 divide 5, and 9 is not prime since 3 divides 9. Consider the quadratic*

$$n^2 - n + 41$$

where $n > 0$ is an integer. Show that $n^2 - n + 41$ is prime for $n = 1, 2, 3, 4, 5$. Can you prove that $n^2 - n + 41$ is prime for all n?

Answer: For $n = 1, 2, 3, 4, 5$ we have

$$
\begin{aligned}
1^2 - 1 + 41 &= 41 \\
2^2 - 2 + 41 &= 43 \\
3^2 - 3 + 41 &= 47 \\
4^2 - 4 + 41 &= 53 \\
5^2 - 5 + 41 &= 61
\end{aligned}
$$

and $41, 43, 47, 53, 61$ are all prime numbers. Experiment and you will find that the resulting numbers are all prime a long way up. Proving $n^2 - n + 41$ is prime for all n would be a very difficult, but fortunately it is much easier to prove that it is not true. Take $n = 41$. Then

$$41^2 - 41 + 41 = 41\,(41 - 1 + 1) = 41 \times 41 = 1681$$

and 1681 is not prime since 41 divides it.

Problem 1.6 *[A*] Consider the quadratic $n^2 - 79n + 1601$. Show that $n^2 - 79n + 1601$ is prime for $n = 1, 2, 3, 4, 5$. Can you prove that $n^2 - 79n + 1601$ is prime for all integers $n > 0$?*

Answer: For $n = 1, 2, 3, 4, 5$ we have

$$
\begin{aligned}
1^2 - 79 \times 1 + 1601 &= 1523 \\
2^2 - 79 \times 2 + 1601 &= 1447 \\
3^2 - 79 \times 3 + 1601 &= 1373 \\
4^2 - 79 \times 4 + 1601 &= 1301 \\
5^2 - 79 \times 5 + 1601 &= 1231
\end{aligned}
$$

and $1523, 1447, 1373, 1301, 1231$ are all prime numbers (you might want to verify this). But this formula is not prime for all n since for $n = 80$ we have

$$80^2 - 79 \times 80 + 1601 = 1681 = 41^2.$$

1.3 Proof by Induction

Problem 1.7 *[A*] Define S_n as the sum of the square of the first n integers as*

$$S_n \equiv 1^2 + 2^2 + 3^2 + \cdots + n^2.$$

For example $S_2 = 1^2 + 2^2 = 5$. Use mathematical induction to prove that

$$S_n = \frac{n^3}{3} + \frac{n^2}{2} + \frac{n}{6}.$$

Answer: The statement is true for $n = 1$ since

$$\begin{aligned} S_1 &= 1^2 = 1 = \frac{1^3}{3} + \frac{1^2}{2} + \frac{1}{6} = \frac{1}{3} + \frac{1}{2} + \frac{1}{6} = \frac{2}{6} + \frac{3}{6} + \frac{1}{6} \\ &= \frac{2+3+1}{6} = \frac{6}{6} = 1. \end{aligned}$$

Now assume that the result is true for $n - 1$ as

$$S_{n-1} = \frac{(n-1)^3}{3} + \frac{(n-1)^2}{2} + \frac{n-1}{6}.$$

Then

$$\begin{aligned} S_n &= \underbrace{1^2 + 2^2 + 3^2 + \cdots + (n-1)^2}_{S_{n-1}} + n^2 = \frac{(n-1)^3}{3} + \frac{(n-1)^2}{2} + \frac{n-1}{6} + n^2 \\ &= \frac{n^3 - 3n^2 + 3n - 1}{3} + \frac{n^2 - 2n + 1^2}{2} + \frac{n-1}{6} + n^2 \\ &= \frac{n^3}{3} + \frac{n^2}{2} + \frac{n}{6} + (-n^2 + n^2) + (-n + n) + \left(-\frac{1}{3} + \frac{1}{2} - \frac{1}{6}\right) \\ &= \frac{n^3}{3} + \frac{n^2}{2} + \frac{n}{6}. \end{aligned}$$

Problem 1.8 *[A*] If d_i is the ith odd number then $d_i = 2i - 1$. For example the first three odd numbers are $d_1 = 1, d_2 = 3, d_3 = 5$ as*

$$d_1 = 2 \times 1 - 1 = 1, d_2 = 2 \times 2 - 1 = 3, d_3 = 2 \times 3 - 1 = 5.$$

Using mathematical induction prove that the sum S_n of the first n odd numbers is n^2; that is

$$S_n \equiv d_1 + d_2 + \cdots + d_n = n^2.$$

For example the sum of the first 4 odd numbers is $1+3+5+7=4^2=16$. *Now prove the result using*

$$1+2+\cdots+n=\frac{n(n+1)}{2}.$$

Answer: Clearly $S_1=1=1^2$ so the statement is true for $n=1$. Suppose that it is true for $1,2,\ldots,n-1$; in particular

$$S_{n-1}=d_1+d_2+\cdots+d_{n-1}=(n-1)^2.$$

Then

$$S_n=\underbrace{d_1+d_2+\cdots+d_{n-1}}_{(n-1)^2}+\underbrace{d_n}_{2n-1}=(n-1)^2+2n-1=n^2-2n+1+2n-1=n^2.$$

This completes the proof by induction.

Now using

$$1+2+\cdots+n=\frac{n(n+1)}{2}$$

we have

$$\begin{aligned}S_n &= d_1+d_2+\cdots+d_n=(2\times1-1)+(2\times2-1)+\cdots+(2\times n-1)\\ &= 2(1+2+\cdots+n)-(1+1+\cdots+1)=2\left(\frac{n(n+1)}{2}\right)-n\\ &= n^2+n-n=n^2.\end{aligned}$$

Problem 1.9 *[A*] Show that*

$$S_n\equiv1^3+2^3+3^3+\cdots+n^3=(1+2+3+\cdots+n)^2.$$

Answer: Since

$$1+2+3+\cdots+n=\frac{n(n+1)}{2}$$

the statement is equivalent to

$$S_n=\left(\frac{n(n+1)}{2}\right)^2=\frac{n^2(n+1)^2}{4}.$$

For $n=1$ we have

$$S_1=1^3=\frac{1^2(1+1)^2}{4}=1$$

so the statement is true for $n=1$. Suppose now that it is true for $n-1$ as

$$S_{n-1}=1^3+2^3+3^3+\cdots+(n-1)^3=\frac{(n-1)^2((n-1)+1)^2}{4}=\frac{(n-1)^2n^2}{4}.$$

Then

$$\begin{aligned} S_n &= 1^3+2^3+3^3+\cdots+(n-1)^3+n^3 = \frac{(n-1)^2 n^2}{4}+n^3 = \frac{(n-1)^2 n^2+4n^3}{4} \\ &= \frac{n^4-2n^3+n^2+4n^3}{4} = \frac{n^4+2n^3+n^2}{4} = \frac{n^2\left(n^2+2n+1\right)}{4} = \frac{n^2(n+1)^2}{4}. \end{aligned}$$

Problem 1.10 *[B*] Is it the case that*

$$1^4+2^4+3^4+\cdots+n^4 = \left(1^2+2^2+3^2+\cdots+n^2\right)^2?$$

Answer: This is not true using a proof by counter-example. For $n=2$ we have

$$17 = 1^4+2^4 \neq \left(1^2+2^2\right)^2 = 25.$$

Problem 1.11 *[A*] The binomial coefficients are defined as*

$$\binom{n}{k} \equiv \binom{n-1}{k} + \binom{n-1}{k-1}$$

where $\binom{0}{0} \equiv 1$. *This is illustrated by Pascal's triangle as*

$$\begin{array}{ccccccccccc} & & & & & 1 & & & & & \\ & & & & 1 & & 1 & & & & \\ & & & 1 & & 2 & & 1 & & & \\ & & 1 & & 3 & & 3 & & 1 & & \\ & 1 & & 4 & & 6 & & 4 & & 1 & \\ 1 & & 5 & & 10 & & 10 & & 5 & & 1 \end{array}$$

Use mathematical induction then to prove that $\binom{n}{k} = \frac{n!}{k!(n-k)!}$.

Answer: For $n=0$ and $k=0$ we have $\binom{0}{0} = \frac{0!}{0!0!} = \frac{1}{1\times 1} = 1$. Now assume the result is true for $n-1$. We then have

$$\begin{aligned} \binom{n-1}{k} + \binom{n-1}{k-1} &= \frac{(n-1)!}{k!\,(n-1-k)!} + \frac{(n-1)!}{(k-1)!\,(n-k)!} \\ &= \frac{(n-1)!}{k\times(k-1)!\,(n-1-k)!} + \frac{(n-1)!}{(k-1)!\,(n-k)\,(n-1-k)!} \\ &= \frac{(n-1)!}{(k-1)!\,(n-1-k)!}\left(\frac{1}{k}+\frac{1}{n-k}\right) \\ &= \frac{(n-1)!}{(k-1)!\,(n-1-k)!}\left(\frac{n-k+k}{k\,(n-k)}\right) \\ &= \frac{(n-1)!}{(k-1)!\,(n-1-k)!}\left(\frac{n}{k\,(n-k)}\right) \\ &= \frac{n\times(n-1)!}{(k\times(k-1)!)\,((n-k)\times(n-1-k)!)} \\ &= \frac{n!}{k!\,(n-k)!}. \end{aligned}$$

Problem 1.12 *[A*] Use induction and* $\binom{n}{k} = \frac{n!}{k!(n-k)!}$ *to prove the binomial theorem that*

$$(x+y)^n = \binom{n}{0}x^n y^0 + \binom{n}{1}x^{n-1}y^1 + \binom{n}{2}x^{n-2}y^2 + \cdots + \binom{n}{n}x^0 y^n.$$

Answer: It is clearly true for $n = 1$ since

$$(x+y)^1 = x + y = \binom{1}{0}x^1 y^0 + \binom{1}{1}x^0 y^1.$$

Now assume it is true for $n-1$. We then have

$$\begin{aligned}
(x+y)^n &= (x+y)(x+y)^{n-1} \\
&= (x+y)\left(\begin{array}{c}\binom{n-1}{0}x^{n-1}y^0 + \binom{n-1}{1}x^{n-2}y^1 \\ +\cdots+\binom{n-1}{n-1}x^0 y^{n-1}\end{array}\right) \\
&= \binom{n-1}{0}x^n y^0 + \binom{n-1}{1}x^{n-1}y^1 + \binom{n-1}{2}x^{n-2}y^2 \\
&\quad +\cdots+\binom{n-1}{n-1}x^1 y^{n-1} \\
&\quad +\binom{n-1}{0}x^{n-1}y^1 + \binom{n-1}{1}x^{n-2}y^2 + \binom{n-1}{2}x^{n-3}y^3 \\
&\quad +\cdots+\binom{n-1}{n-1}x^0 y^n \\
&= \binom{n-1}{0}x^n y^0 + \left(\binom{n-1}{1} + \binom{n-1}{0}\right)x^{n-1}y^1 \\
&\quad +\left(\binom{n-1}{2} + \binom{n-1}{1}\right)x^{n-2}y^2 \\
&\quad +\cdots+\left(\binom{n-1}{n-2} + \binom{n-1}{n-1}\right)x^1 y^{n-1} + \binom{n-1}{n-1}x^0 y^n.
\end{aligned}$$

By multiplying the two terms in brackets you will find that the coefficient on $x^{n-j}y^j$ is $\binom{n-1}{j} + \binom{n-1}{j-1} = \binom{n}{j}$. The two terms on $x^n y^0$ $x^n y^0$ have to be treated a little differently but these work as well since

$$\binom{n-1}{0} = 1 = \binom{n}{0}, \; \binom{n-1}{n-1} = 1 = \binom{n}{n}.$$

Thus

$$(x+y)^n = \binom{n}{0}x^n y^0 + \binom{n}{1}x^{n-1}y^1 + \binom{n}{2}x^{n-2}y^2 + \cdots + \binom{n}{n}x^0 y^n.$$

Problem 1.13 *[A*] Consider the Golden Mean* $\phi \equiv \frac{1+\sqrt{5}}{2} \approx 1.618$*. Prove that taking the inverse of* ϕ *is the same as subtracting* 1 *from it as* $\frac{1}{\phi} = \phi - 1$ *or* $\frac{1}{1.618\ldots} = 0.618\ldots$*. Prove squaring* ϕ *is the same as adding one to* ϕ *as* $\phi^2 = \phi + 1$ *or* $(1.618\ldots)^2 = 2.618\ldots$*.*

Answer: We have

$$
\begin{aligned}
\frac{1}{\phi} &= \frac{1}{\frac{1+\sqrt{5}}{2}} = \frac{2}{\sqrt{5}+1} = \frac{2}{\sqrt{5}+1}\frac{\sqrt{5}-1}{\sqrt{5}-1} \\
&= \frac{2\left(\sqrt{5}-1\right)}{\left(\sqrt{5}\right)^2-(1)^2} = \frac{2\left(\sqrt{5}-1\right)}{5-1} = \frac{\sqrt{5}-1}{2} = \frac{1+\sqrt{5}}{2} - \frac{2}{2} = \phi - 1
\end{aligned}
$$

and

$$
\begin{aligned}
\phi^2 &= \left(\frac{1+\sqrt{5}}{2}\right)\left(\frac{1+\sqrt{5}}{2}\right) = \frac{1+2\sqrt{5}+\left(\sqrt{5}\right)^2}{4} = \frac{1+2\sqrt{5}+5}{4} \\
&= \frac{6+2\sqrt{5}}{4} = \frac{3+\sqrt{5}}{2} = \frac{2+1+\sqrt{5}}{2} = \frac{2}{2} + \frac{1+\sqrt{5}}{2} = 1+\phi.
\end{aligned}
$$

Problem 1.14 *[A*] The Fibonacci numbers are defined as* $f_n = f_{n-1} + f_{n-2}$ *with* $f_1 = f_2 = 1$. *For example*

$f_3 = f_2 + f_1 = 1 + 1 = 2$, $f_4 = f_3 + f_2 = 2 + 1 = 3$ *and* $f_5 = f_4 + f_3 = 3 + 2 = 5$.

There is a close link between these numbers and ϕ. *For example* $\frac{f_5}{f_4} = \frac{5}{3} = 1.6 \approx \phi = 1.618\ldots$ *Use proof by induction to show that*

$$
f_n = \frac{\left(\frac{1+\sqrt{5}}{2}\right)^n - \left(\frac{1-\sqrt{5}}{2}\right)^n}{\sqrt{5}}
$$

and from this show that $\lim_{n\to\infty} \frac{f_n}{f_{n-1}} = \phi$.

Answer: For $n = 1$ the result holds as

$$
\frac{\left(\frac{1+\sqrt{5}}{2}\right)^1 - \left(\frac{1-\sqrt{5}}{2}\right)^1}{\sqrt{5}} = \frac{\sqrt{5}}{\sqrt{5}} = 1 = f_1.
$$

Now it is true for $n-1$. Then

$$
\begin{aligned}
f_n &= f_{n-1} + f_{n-2} = \frac{\left(\frac{1+\sqrt{5}}{2}\right)^{n-1} - \left(\frac{1-\sqrt{5}}{2}\right)^{n-1}}{\sqrt{5}} + \frac{\left(\frac{1+\sqrt{5}}{2}\right)^{n-2} - \left(\frac{1-\sqrt{5}}{2}\right)^{n-2}}{\sqrt{5}} \\
&= \frac{\left(\left(\frac{1+\sqrt{5}}{2}\right)^{n-1} + \left(\frac{1+\sqrt{5}}{2}\right)^{n-2}\right) - \left(\left(\frac{1-\sqrt{5}}{2}\right)^{n-1} + \left(\frac{1-\sqrt{5}}{2}\right)^{n-2}\right)}{\sqrt{5}} \\
&= \frac{\left(\frac{1+\sqrt{5}}{2}\right)^{n-2}\left(\frac{1+\sqrt{5}}{2}+1\right) - \left(\frac{1-\sqrt{5}}{2}\right)^{n-2}\left(\frac{1-\sqrt{5}}{2}+1\right)}{\sqrt{5}} \\
&= \frac{\left(\frac{1+\sqrt{5}}{2}\right)^{n-2}\left(\frac{3+\sqrt{5}}{2}\right) - \left(\frac{1-\sqrt{5}}{2}\right)^{n-2}\left(\frac{3-\sqrt{5}}{2}\right)}{\sqrt{5}} \\
&= \frac{\left(\frac{1+\sqrt{5}}{2}\right)^{n-2}\left(\frac{1+\sqrt{5}}{2}\right)^2 - \left(\frac{1-\sqrt{5}}{2}\right)^{n-2}\left(\frac{1-\sqrt{5}}{2}\right)^2}{\sqrt{5}} = \frac{\left(\frac{1+\sqrt{5}}{2}\right)^n - \left(\frac{1-\sqrt{5}}{2}\right)^n}{\sqrt{5}}.
\end{aligned}
$$

Now

$$\frac{f_n}{f_{n-1}} = \frac{\left(\frac{1+\sqrt{5}}{2}\right)^n - \left(\frac{1-\sqrt{5}}{2}\right)^n}{\left(\frac{1+\sqrt{5}}{2}\right)^{n-1} - \left(\frac{1-\sqrt{5}}{2}\right)^{n-1}}.$$

Since $\left|\frac{1-\sqrt{5}}{2}\right| \approx 0.618$ is a fraction, and since $\left(\frac{1-\sqrt{5}}{2}\right)^n \to 0$ as $n \to \infty$ we have

$$\lim_{n\to\infty} \frac{f_n}{f_{n-1}} = \lim_{n\to\infty} \frac{\left(\frac{1+\sqrt{5}}{2}\right)^n - \left(\frac{1-\sqrt{5}}{2}\right)^n}{\left(\frac{1+\sqrt{5}}{2}\right)^{n-1} - \left(\frac{1-\sqrt{5}}{2}\right)^{n-1}} = \lim_{n\to\infty} \frac{\left(\frac{1+\sqrt{5}}{2}\right)^n}{\left(\frac{1+\sqrt{5}}{2}\right)^{n-1}} = \frac{1+\sqrt{5}}{2} \equiv \phi.$$

Problem 1.15 *[A*] Define the* 0th *Fibonacci number as* $f_0 = 0$ *with* $f_n = f_{n-1} + f_{n-2}$ *and* $f_1 = f_2 = 1$. *Using proof by induction show that*

$$\phi^n = f_{n-1} + f_n\phi.$$

Answer: We have

$$\phi^1 = \phi = 0 + 1 \times \phi = f_0 + f_1\phi$$

and so $\phi^n = f_{n-1} + f_n\phi$ is true for $n = 1$. Now suppose it is true for $1, 2, \ldots, n-1$, so that in particular $\phi^{n-1} = f_{n-2} + f_{n-1}\phi$. Then since $\phi^2 = \phi + 1$ and $f_n = f_{n-1} + f_{n-2}$ we have

$$\begin{aligned}\phi^n &= \phi^{n-1}\phi^1 = (f_{n-2} + f_{n-1}\phi)\,\phi = f_{n-2}\phi + f_{n-1}\phi^2 = f_{n-2}\phi + f_{n-1}(\phi + 1) \\ &= f_{n-2}\phi + f_{n-1}\phi + f_{n-1} = f_{n-1} + \underbrace{(f_{n-2} + f_{n-1})}_{f_n}\phi = f_{n-1} + f_n\phi.\end{aligned}$$

1.4 Proof by Contradiction

Problem 1.16 *[A*] Prove that* $\sqrt{p}$ *is irrational where* p *is any prime number (a prime number has only one and itself as factors). This means for example that* $\sqrt{2}, \sqrt{3}, \sqrt{5}$ *are all irrational since* $2, 3, 5$ *are all prime numbers. Hint: it is a fact that any integer* n *can be factored uniquely into* r *prime numbers; for example* $30 = 2 \times 3 \times 5$ *where* $2, 3, 5$ *are prime numbers.*

Answer: Assume that $\sqrt{p}$ is rational so that $\sqrt{p} = \frac{a}{b}$ where a and b are integers. Then

$$\sqrt{p} = \frac{a}{b} \Longrightarrow a^2 = b^2 p.$$

Now if integer a has r prime factors (for example 30 has $r = 3$ factors; i.e., $30 = 2 \times 3 \times 5$) then a^2 has $2r$ prime factors (for example 30^2 has $r = 3 \times 2 = 6$ factors; i.e., $30^2 = 2 \times 2 \times 3 \times 3 \times 5 \times 5$) and so a^2 has an even number of prime factors, say $2r$.

The same argument applies to b^2; that is b^2 must have an even number of prime factors; say $2s$. From this it follows that $b^2 p$ has an odd number of prime factors $2s+1$ since p is prime. But $a^2 = b^2 p$ so a^2 has an odd number of prime factors $2s+1$. Thus a^2 has both an even and an odd number of prime factors, a contradiction. Therefore $\sqrt{p}$ is irrational.

Problem 1.17 *[A*] Prove that there are an infinite number of prime numbers.*

Answer: Under the assumption that there are a finite number of primes: $2, 3, 5, 7, \ldots, p_{\max}$ construct the number

$$n = 2 \times 3 \times 5 \times 7 \times \cdots \times p_{\max} + 1.$$

Clearly n is an integer. Either n is a prime number or it is not a prime number. If n is a prime number then $n > p_{\max}$, which contradicts the assumption that $p_{\max}$ is the largest prime number. If n is not prime then it has at least one prime factor say $\hat{p} > 1$ which must be one of $2, 3, 5, 7, \ldots, p_{\max}$. But this implies that

$$\underbrace{\frac{n}{\hat{p}}}_{\text{integer}} = \underbrace{\frac{2 \times 3 \times 5 \times 7 \times \cdots \times p_{\max}}{\hat{p}}}_{\text{integer}} + \underbrace{\frac{1}{\hat{p}}}_{\text{non-integer}}$$

so that $\frac{n}{\hat{p}}$ is both an integer, and a non-integer, a contradiction. Thus the assumption that there is a largest prime must be false.

Problem 1.18 *[B*] Consider $y = \sqrt{x}$ shown in the figure below. As x increases $\sqrt{x}$ gets flatter and flatter. A natural conjecture then is that $\sqrt{x}$ has some upper limit, that it never gets bigger than some number c so that $\sqrt{x} \leq c$ for all x. Prove that this is not the case; that $\sqrt{x}$ is in fact unbounded.*

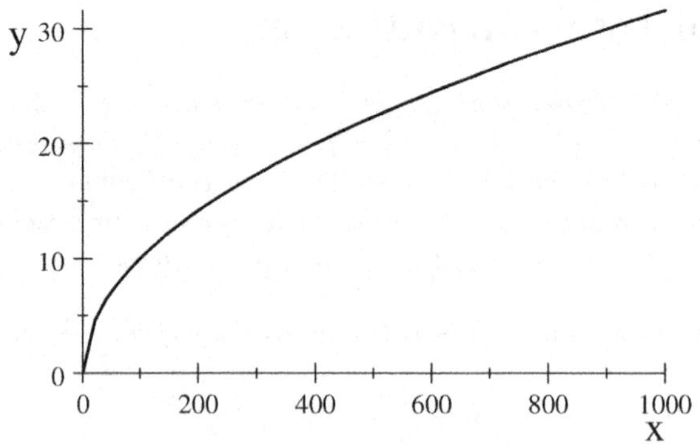

$y = \sqrt{x}$

Answer: Assume to the contrary that $\sqrt{x}$ is bounded so that $\sqrt{x} \leq c$ for all x. Now we are free to choose any x so let $x = (c+1)^2$. We then have

$\sqrt{x} = \sqrt{(c+1)^2} = c+1$. But from $\sqrt{x} \leq c$ we have

$$\begin{aligned}\sqrt{(c+1)^2} &= c+1 \leq c \\ &\Longrightarrow c+1-c \leq c-c \\ &\Longrightarrow 1 \leq 0\end{aligned}$$

which is a contradiction. Thus the original assumption $\sqrt{x} \leq c$ must be false and so $\sqrt{x}$ is unbounded.

Chapter 2

Univariate Calculus

2.1 Exponents

Problem 2.1 *[C***] Simplify the expressions*

$$x^0 - x^3\left(x^{-2}\right)^{\frac{3}{2}}, \frac{\left(x^{\frac{1}{6}}\right)^2}{\sqrt{x^{\frac{2}{3}}}}, \text{ and } \left(\left(\frac{3}{2}\right)^{\frac{3}{5}} x^{\frac{2}{5}}\right)^{\frac{5}{3}}.$$

Answer: Using the properties of exponents we have

$$\begin{aligned}
x^0 - x^3\left(x^{-2}\right)^{\frac{3}{2}} &= 1 - x^3 x^{-2\times\frac{3}{2}} = 1 - x^3 x^{-3} = 1 - x^{3-3} = 1 - x^0 = 1 - 1 = 0 \\
\frac{\left(x^{\frac{1}{6}}\right)^2}{\sqrt{x^{\frac{2}{3}}}} &= \frac{x^{\frac{1}{6}\times 2}}{\left(x^{\frac{2}{3}}\right)^{\frac{1}{2}}} = \frac{x^{\frac{2}{6}}}{x^{\frac{2}{3}\times\frac{1}{2}}} = \frac{x^{\frac{1}{3}}}{x^{\frac{1}{3}}} = x^{\frac{1}{3}-\frac{1}{3}} = x^0 = 1 \\
\left(\left(\frac{3}{2}\right)^{\frac{3}{5}} x^{\frac{2}{5}}\right)^{\frac{5}{3}} &= \left(\frac{3}{2}\right)^{\frac{3}{5}\times\frac{5}{3}} x^{\frac{2}{5}\times\frac{5}{3}} = \left(\frac{3}{2}\right)^1 x^{\frac{2}{3}} = \frac{3}{2} x^{\frac{2}{3}}.
\end{aligned}$$

Problem 2.2 *[B***] Simplify* $\frac{\left(\left(\frac{3}{2}\right)^{\frac{3}{5}} x^{\frac{2}{5}} + x^0 - x^3\left(x^{-2}\right)^{\frac{3}{2}}\right)^{\frac{5}{3}} \left(x^{\frac{1}{6}}\right)^2}{\sqrt{x^{\frac{2}{3}}}}.$

Answer: We have

$$
\begin{aligned}
\frac{\left(\left(\frac{3}{2}\right)^{\frac{3}{5}} x^{\frac{2}{5}} + x^0 - x^3\left(x^{-2}\right)^{\frac{3}{2}}\right)^{\frac{5}{3}}\left(x^{\frac{1}{6}}\right)^2}{\sqrt{x^{\frac{2}{3}}}} &= \frac{\left(\left(\frac{3}{2}\right)^{\frac{3}{5}} x^{\frac{2}{5}} + 1 - x^3 x^{-2\times\frac{3}{2}}\right)^{\frac{5}{3}} x^{\frac{1}{6}\times 2}}{\left(x^{\frac{2}{3}}\right)^{\frac{1}{2}}} \\
&= \frac{\left(\left(\frac{3}{2}\right)^{\frac{3}{5}} x^{\frac{2}{5}} + 1 - x^0\right)^{\frac{5}{3}} x^{\frac{2}{6}}}{x^{\frac{2}{3}\times\frac{1}{2}}} \\
&= \frac{\left(\left(\frac{3}{2}\right)^{\frac{3}{5}} x^{\frac{2}{5}}\right)^{\frac{5}{3}} x^{\frac{1}{3}}}{x^{\frac{1}{3}}} \\
&= \left(\left(\frac{3}{2}\right)^{\frac{3}{5}}\right)^{\frac{5}{3}}\left(x^{\frac{2}{5}}\right)^{\frac{5}{3}} \\
&= \left(\frac{3}{2}\right)^{\frac{3}{5}\times\frac{5}{3}} x^{\frac{2}{5}\times\frac{5}{3}} = \frac{3}{2}x^{\frac{2}{3}}.
\end{aligned}
$$

Problem 2.3 *[B***] Simplify the expressions*

$$
\frac{x^3x^2}{x^5} + x^0 - 2,\ \frac{\left(x^{\frac{1}{10}}\right)^2}{\sqrt{x^{\frac{2}{5}}}},\ \textit{and}\ \left(\left(\frac{4}{3}\right)^{\frac{4}{7}} x^{\frac{3}{7}}\right)^{\frac{7}{4}}.
$$

Answer: We have

$$
\begin{aligned}
\frac{x^3x^2}{x^5} + x^0 - 2 &= \frac{x^{3+2}}{x^5} + 1 - 2 = \frac{x^5}{x^5} + 1 - 2 = 1 + 1 - 2 = 0 \\
\frac{\left(x^{\frac{1}{10}}\right)^2}{\sqrt{x^{\frac{2}{5}}}} &= \frac{x^{\frac{1}{10}\times 2}}{\left(x^{\frac{2}{5}}\right)^{\frac{1}{2}}} = \frac{x^{\frac{2}{10}}}{x^{\frac{2}{5}\times\frac{1}{2}}} = \frac{x^{\frac{1}{5}}}{x^{\frac{1}{5}}} = 1 \\
\left(\left(\frac{4}{3}\right)^{\frac{4}{7}} x^{\frac{3}{7}}\right)^{\frac{7}{4}} &= \left(\frac{4}{3}\right)^{\frac{4}{7}\times\frac{7}{4}} x^{\frac{3}{7}\times\frac{7}{4}} = \left(\frac{4}{3}\right)^1 x^{\frac{3}{4}} = \frac{4}{3}x^{\frac{3}{4}}.
\end{aligned}
$$

Problem 2.4 *[B***] Simplify the expression*

$$
\frac{\left(\left(\frac{4}{3}\right)^{\frac{4}{7}} x^{\frac{3}{7}} + \frac{x^3x^2}{x^5} + x^0 - 2\right)^{\frac{7}{4}}\left(x^{\frac{1}{10}}\right)^2}{\sqrt{x^{\frac{2}{5}}}}.
$$

Answer: The expression simplifies to

$$\frac{\left(\left(\frac{4}{3}\right)^{\frac{4}{7}}x^{\frac{3}{7}}+\frac{x^5}{x^5}+1-2\right)^{\frac{7}{4}}x^{\frac{2}{10}}}{\left(x^{\frac{2}{5}}\right)^{\frac{1}{2}}} = \frac{\left(\left(\frac{4}{3}\right)^{\frac{4}{7}}x^{\frac{3}{7}}+1+1-2\right)^{\frac{7}{4}}x^{\frac{1}{5}}}{x^{\frac{1}{5}}}$$

$$= \left(\left(\frac{4}{3}\right)^{\frac{4}{7}}x^{\frac{3}{7}}\right)^{\frac{7}{4}} = \left(\left(\frac{4}{3}\right)^{\frac{4}{7}}\right)^{\frac{7}{4}}\left(x^{\frac{3}{7}}\right)^{\frac{7}{4}}$$

$$= \frac{4}{3}x^{\frac{3}{4}}.$$

Problem 2.5 *[A*] Prove that $x^m x^n = x^{m+n}$ and $(x^m)^n = x^{m\times n}$ where m and n are positive integers.*

Answer: We have

$$x^m x^n = \Bigg(\underbrace{x\times x\times\cdots\times x}_{m}\Bigg)\Bigg(\underbrace{x\times x\times\cdots\times x}_{n}\Bigg) = \underbrace{x\times x\times\cdots\times x}_{m+n} = x^{m+n}$$

and

$$(x^m)^n = \underbrace{\Bigg(\underbrace{x\times x\times\cdots\times x}_{m}\Bigg)\times\Bigg(\underbrace{x\times x\times\cdots\times x}_{m}\Bigg)\times\cdots\times\Bigg(\underbrace{x\times x\times\cdots\times x}_{m}\Bigg)}_{n}$$

$$= \underbrace{x\times x\times\cdots\times x}_{m\times n} = x^{m\times n}.$$

Problem 2.6 *[B**] For the demand and supply curves*

$$Q_D = 200P^{-\frac{1}{2}},\ Q_S = 2P^{\frac{3}{2}}$$

calculate the equilibrium price P and quantity Q. What is the elasticity of demand and the elasticity of supply? Now suppose the government imposes a tax of $T = \frac{7}{9}P$ on households. What is the new equilibrium price P and quantity Q?

Answer: To find P we use

$$\begin{aligned}
200P^{-\frac{1}{2}} &= Q_D = Q_S = 2P^{\frac{3}{2}}\\
&\implies 200P^{-\frac{1}{2}} = 2P^{\frac{3}{2}}\\
&\implies 200P^{-\frac{1}{2}}P^{\frac{1}{2}} = 2P^{\frac{3}{2}}P^{\frac{1}{2}}\\
&\implies 200P^{-\frac{1}{2}+\frac{1}{2}} = 2P^{\frac{3}{2}+\frac{1}{2}}\\
&\implies 200P^0 = 2P^{\frac{4}{2}}\\
&\implies P^2 = \frac{200}{2} = 100\\
&\implies P = (100)^{\frac{1}{2}} = 10.
\end{aligned}$$

To find the equilibrium quantity we can put $P = 10$ in either the demand curve or supply curve as

$$Q = 200P^{-\frac{1}{2}} = 200 \times 10^{-\frac{1}{2}} \approx 63.25 \text{ or } Q = 2P^{\frac{3}{2}} = 2 \times 10^{\frac{3}{2}} \approx 63.25.$$

Since both the demand and supply curves are of the form Ax^b the elasticities are the exponents on P. The elasticity of demand is the exponent on $P^{-\frac{1}{2}}$ or $-\frac{1}{2}$ and the elasticity of supply is the exponent on $P^{\frac{3}{2}}$ or $\frac{3}{2}$.

With the tax $T = \frac{7}{9}P$ households now have to pay

$$P + T = P + \frac{7}{9}P = \frac{16}{9}P$$

so that the demand curve becomes

$$Q_D = 200\left(\frac{16}{9}P\right)^{-\frac{1}{2}} = 200 \times \frac{3}{4}P^{-\frac{1}{2}} = 150P^{-\frac{1}{2}}.$$

Now to find P use

$$\begin{aligned} 150P^{-\frac{1}{2}} &= Q_D = Q = Q_S = 2P^{\frac{3}{2}} \\ &\Longrightarrow P^2 = \frac{150}{2} = 75 \\ &\Longrightarrow P = 75^{\frac{1}{2}} \approx 8.67. \end{aligned}$$

To find the equilibrium quantity put $P = 75^{\frac{1}{2}}$ in either the demand curve or supply curve as

$$Q = 150P^{-\frac{1}{2}} = 150\left(75^{\frac{1}{2}}\right)^{-\frac{1}{2}} \approx 50.97 \text{ or } Q = 2P^{\frac{3}{2}} = 2\left(75^{\frac{1}{2}}\right)^{\frac{3}{2}} \approx 50.97.$$

2.2 Polynomials

Problem 2.7 *[C***] Find the roots of the quadratic $3x^2 - 4x + 1 = 0$. Use these roots to factor the quadratic.*

Answer: The quadratic can be rewritten as

$$3x^2 - 4x + 1 = (3)\,x^2 + (-4)\,x + (1) = 0$$

so that using

$$\begin{aligned} ax^2 + bx + c &= 0 \\ &\Longrightarrow x = \frac{-b \pm \sqrt{b^2 - 4ac}}{2a} \end{aligned}$$

we have $a = 3, b = -4, c = 1$, and

$$\begin{aligned} (3)\,x^2 + (-4)\,x + (1) &= 0 \\ &\Longrightarrow x = \frac{-(-4) \pm \sqrt{(-4)^2 - 4 \times 3 \times 1}}{2 \times 3} \end{aligned}$$

so that the roots are $x_1 = 1, x_2 = \frac{1}{3}$. The roots can be seen in the figure below where $3x^2 - 4x + 1$ crosses the horizontal axis.

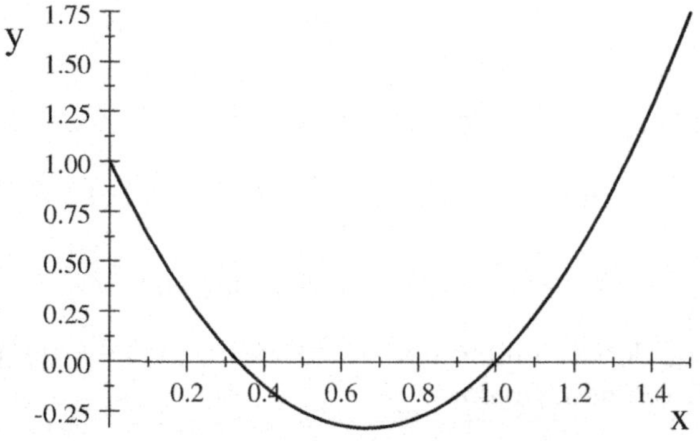

$3x^2 - 4x + 1$

The quadratic is then factored as

$$3x^2 - 4x + 1 = 3(x-1)\left(x - \frac{1}{3}\right).$$

Problem 2.8 *[C***] Find the roots of $2x^2 - 24x + 70 = 0$. Use the roots to factor the polynomial.*

Answer: The quadratic can be rewritten as

$$2x^2 - 24x + 70 = (2)\,x^2 + (-24)\,x + (70) = 0$$

so that using

$$\begin{aligned} ax^2 + bx + c &= 0 \\ \implies x &= \frac{-b \pm \sqrt{b^2 - 4ac}}{2a} \end{aligned}$$

we have $a = 2, b = -24, c = 70$, and

$$\begin{aligned} (2)\,x^2 + (-24)\,x + (70) &= 0 \\ \implies x &= \frac{-(-24) \pm \sqrt{(-24)^2 - 4 \times 2 \times 70}}{2 \times 2} \end{aligned}$$

so that $x_1 = 7, x_2 = 5$. We use the roots to factor the quadratic as

$$2x^2 - 24x + 70 = 2(x-7)(x-5).$$

Problem 2.9 *[B**] If $f(x) = 3x^2 - 6x + 12$ then show that*

$$f(x) = 3(x-1)^2 + 9.$$

Use this result to show that $f(x)$ has a global minimum at $x = 1$; that is $f(x) > f(1)$ for all $x \neq 1$. What is $f(1)$ equal to?

Answer: We have

$$3(x-1)^2+9=3\left(x^2-2x+1\right)+9=3x^2-6x+12=f(x).$$

Since

$$f(1)=3(1-1)^2+9=9$$

and $3(x-1)^2>0$ for $x\neq 1$ it follows that

$$\begin{aligned} f(x) &= \underbrace{3(x-1)^2}_{+}+9>9=f(1) \\ &\implies f(x)>f(1) \text{ for all } x\neq 1 \end{aligned}$$

and so $f(x)$ has a global minimum at $x=1$ with $f(1)=9$ the minimized value. This is shown in the figure below.

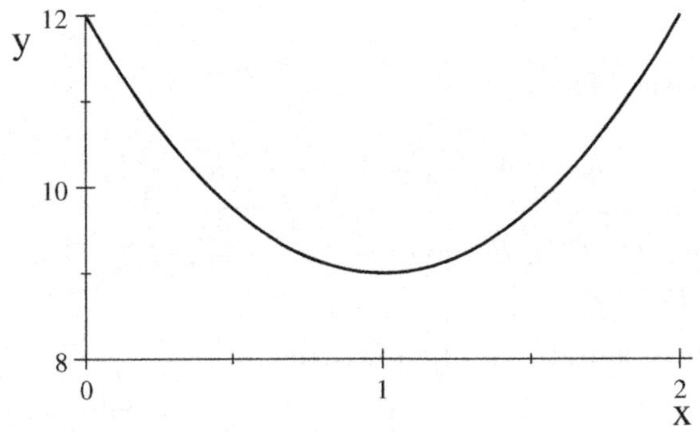

$$f(x)=3x^2-6x+12=3(x-1)^2+9$$

Problem 2.10 *[B*] Show that any quadratic $f(x)$ can be rewritten as*

$$f(x)=ax^2+bx+c=a(x-d)^2+e.$$

Find the values of d and e. This is known as completing the square. Use this result to show that $f(x)$ has a global maximum at $x=d$ if $a<0$, and a global minimum at $x=d$ if $a>0$. Show that $f(d)=e$ is the maximized or minimized value of the function.

Answer: Expanding we have

$$a(x-d)^2+e=a\left(x^2-2dx+d^2\right)+e=ax^2-2adx+ad^2+e=ax^2+bx+c$$

so that equating the coefficients on x and the constant we have

$$\begin{aligned} b &= -2ad \implies d=-\frac{b}{2a} \\ c &= ad^2+e\implies e=c-ad^2=c-a\left(-\frac{b}{2a}\right)^2=c-\frac{b^2}{4a}. \end{aligned}$$

We have

$$f(d) = a(d-d)^2 + e = 0 + e = e = c - \frac{b^2}{4a}.$$

Since $(x-d)^2 > 0$ for $x \neq d$ it follows that if $a < 0$ then

$$f(x) = a(x-d)^2 + e < e = f(d)$$

and so $f(x)$ has a global maximum at d. Similarly if $a > 0$ then $f(x)$ has a global minimum at d since for all $x \neq d$

$$f(x) = a(x-d)^2 + e > e = f(d).$$

Problem 2.11 *[B*] Complete the square for the quadratic*

$$f(x) = -2x^2 + 24x + 5$$

and show that $f(x)$ has a global maximum at $x = 6$.

Answer: Completing the square we have

$$f(x) = -2x^2 + 24x + 5 = -2(x-d)^2 + e = -2x^2 + 4dx - 2d^2 + e$$

so that

$$\begin{aligned} 4d &= 24 \Longrightarrow d = 6 \\ -2d^2 + e &= 5 \Longrightarrow e = 5 + 2d^2 = 5 + 2 \times (6)^2 = 77 \end{aligned}$$

from which it follows that

$$f(x) = -2x^2 + 24x + 5 = -2(x-6)^2 + 77.$$

Since $a = -2 < 0$ it follows that $f(x)$ has a global maximum at $x = 6$ and that $f(6) = 77$.

Problem 2.12 *[A**] Provide a proof for the geometric series; that for $x \neq 1$ the polynomial $S(x)$ given below satisfies*

$$S(x) \equiv 1 + x + x^2 + x^3 + \cdots + x^{n-1} = \frac{1-x^n}{1-x} = \frac{x^n - 1}{x-1}.$$

Answer: We have

$$\begin{aligned} S(x) &\equiv 1 + x + x^2 + x^3 + \cdots + x^{n-1} \\ &\Longrightarrow xS(x) = x + x^2 + x^3 + \cdots + x^n \end{aligned}$$

and so

$$\begin{aligned} S(x) - xS(x) &= (1-x)S(x) \\ &= 1 + x + x^2 + x^3 + \cdots + x^{n-1} - (x + x^2 + x^3 + \cdots + x^n) \\ &= 1 - x^n \end{aligned}$$

and so for $x \neq 1$ we have dividing by $1 - x \neq 0$ that

$$\begin{aligned}(1-x)\,S(x) &= 1-x^n \\ \Longrightarrow\ S(x) &= \frac{1-x^n}{1-x} = \frac{-1}{-1}\frac{1-x^n}{1-x} = \frac{x^n-1}{x-1}.\end{aligned}$$

Problem 2.13 *[B**] Consider a bond that pays \$1000 a year for 30 years with a 5% interest rate or $r = 0.05 = \frac{1}{20}$. What is the price of the bond P_B? What would the price be if the bond paid \$1000 a year forever?*

Answer: We have

$$\begin{aligned}P_B &= \frac{\$1000}{1+\frac{1}{20}} + \frac{\$1000}{\left(1+\frac{1}{20}\right)^2} + \frac{\$1000}{\left(1+\frac{1}{20}\right)^3} + \cdots + \frac{\$1000}{\left(1+\frac{1}{20}\right)^{30}} \\ &= \frac{\$1000}{\frac{1}{20}}\left(1-\left(\frac{1}{1+\frac{1}{20}}\right)^{30}\right) \approx \$15,372.45.\end{aligned}$$

If the bond paid \$1000 a year forever then it would be worth

$$P_B = \frac{\$1000}{\frac{1}{20}} = \$20,000.$$

Problem 2.14 *[A*] Consider integers of the form $a^n - 1$ where $a > 1$ and $n > 1$ are integers. Prove that if $a^n - 1$ is a prime number then $a = 2$. Prove that n must itself be prime for $2^n - 1$ to be prime. Such prime numbers are known as Mersenne primes. Prove that $a = 2$ and n being prime is necessary but not a sufficient condition for $2^n - 1$ to be prime. Usually the largest known prime number is a Mersenne prime. Do a search on the Internet to find the current largest known prime number.*

Answer: From the geometric series we have

$$\begin{aligned}a^n - 1 &= (a-1)\left(1+a+a^2+\cdots+a^{n-1}\right) \\ \Longrightarrow\ \frac{a^n-1}{a-1} &= 1+a+a^2+\cdots+a^{n-1}\end{aligned}$$

and so $a - 1$ divides $a^n - 1$. If $a^n - 1$ is a prime number then only 1 and itself (i.e., $a^n - 1$) can divide $a^n - 1$. Since $n > 1$ we cannot have $a - 1 = a^n - 1$ and so it must be that $a - 1 = 1$ or $a = 2$. (For example $2^3 - 1 = 7$ is prime but $4^3 - 1 = 63$ is not prime.)

Now let's show that if $2^n - 1$ is prime then n must be prime. We do this by showing that if n is not prime then $2^n - 1$ is not prime. Suppose then that n is not prime but factors as $n = r \times s$ for $1 < r < n$ and $1 < s < n$. Then we have

$$2^n - 1 = 2^{r\times s} - 1 = (2^r)^s - 1 = a^s - 1$$

where $a = 2^r > 2$. But from the previous result $a^s - 1$ for $a > 2$ cannot be prime.

Another example of a Mersenne prime is $2^5 - 1 = 31$. But not all numbers of the form $2^n - 1$ with n prime are prime. For example

$$2^{11} - 1 = 2047 = 23 \times 89$$

is not prime, even though 11 is prime. Thus $a = 2$ and n prime is necessary but not sufficient for $2^n - 1$ to be prime.

Currently (July 18, 2005) the largest known prime is a Mersenne prime given by

$$2^{25,964,951} - 1$$

where $25,964,951$ is prime. This (very large) number has 7,816,230 digits!

2.3 Slopes

Problem 2.15 *[D*] For $y = f(x) = x^3$ calculate the slope $\frac{\Delta y}{\Delta x}$. Calculate the slope when $x = 2$ using $\Delta x = 1, \frac{1}{4}, \frac{1}{10}$. What value does the slope appear to approach as Δx becomes smaller? [B**] Show that the derivative is $f'(x) = 3x^2$. [D**] Calculate the slope at $x = 2$ using the derivative.*

Answer: Using the Binomial Theorem we have

$$(x + \Delta x)^3 = x^3 + 3x^2\Delta x + 3x\Delta x^2 + \Delta x^3$$

and so the slope is

$$\begin{aligned}\frac{\Delta y}{\Delta x} &\equiv \frac{f(x+\Delta x) - f(x)}{\Delta x} = \frac{(x+\Delta x)^3 - x^3}{\Delta x} = \frac{x^3 + 3x^2\Delta x + 3x\Delta x^2 + \Delta x^3 - x^3}{\Delta x} \\ &= \frac{3x^2\Delta x + 3x\Delta x^2 + \Delta x^3}{\Delta x} = \frac{3x^2\Delta x}{\Delta x} + \frac{3x\Delta x^2}{\Delta x} + \frac{\Delta x^3}{\Delta x} = 3x^2 + 3x\Delta x + \Delta x^2.\end{aligned}$$

For $x = 2, \Delta x = 1$ we have

$$\frac{\Delta y}{\Delta x} = \frac{(2+1)^3 - 2^3}{1} = 3 \times 2^2 + 3 \times 2 \times 1 + 1^2 = 19.$$

For $x = 2, \Delta x = \frac{1}{4}$ we have

$$\frac{\Delta y}{\Delta x} = \frac{\left(2+\frac{1}{4}\right)^3 - 2^3}{\frac{1}{4}} = 3 \times 2^2 + 3 \times 2 \times \frac{1}{4} + \left(\frac{1}{4}\right)^2 = \frac{217}{16} \approx 13.6.$$

For $x = 2, \Delta x = \frac{1}{10}$ we have

$$\frac{\Delta y}{\Delta x} = \frac{\left(2+\frac{1}{10}\right)^3 - 2^3}{\frac{1}{10}} = 3 \times 2^2 + 3 \times 2 \times \frac{1}{10} + \left(\frac{1}{10}\right)^2 = \frac{1261}{100} = 12.61.$$

As $\Delta x \to 0$ the slope at $x = 2$ appears to be approaching 12.

As $\Delta x \to 0$ the slope approaches the derivative as

$$f'(x) \equiv \frac{dy}{dx} = \lim_{\Delta x \to 0} \left(3x^2 + 3x\Delta x + \Delta x^2\right) = 3x^2.$$

When $x = 2$ the derivative formula $f'(x) = 3x^2$ states that the slope is indeed 12 as $f'(2) = 3 \times 2^2 = 12$.

Problem 2.16 *[D**] For $y = f(x) = \frac{1}{x}$ calculate the slope $\frac{\Delta y}{\Delta x}$. Calculate the slope when $x = 2$ using $\Delta x = 1, \Delta x = \frac{1}{4}, \Delta x = \frac{1}{10}$. What value does the slope appear to approach as Δx becomes smaller? [B**] Show that the derivative is $f'(x) = -\frac{1}{x^2}$. Calculate the slope at $x = 2$ using the derivative.*

Answer: We have

$$\begin{aligned}
\frac{\Delta y}{\Delta x} &\equiv \frac{f(x+\Delta x) - f(x)}{\Delta x} = \frac{\frac{1}{x+\Delta x} - \frac{1}{x}}{\Delta x} \\
&= \frac{\frac{x}{(x+\Delta x)x} - \frac{x+\Delta x}{x(x+\Delta x)}}{\Delta x} = \frac{\frac{x-(x+\Delta x)}{(x+\Delta x)x}}{\Delta x} = \frac{\frac{-\Delta x}{(x+\Delta x)x}}{\Delta x} \\
&= \frac{-1}{(x+\Delta x)\,x}.
\end{aligned}$$

For $x = 2, \Delta x = 1$ we have

$$\frac{\Delta y}{\Delta x} = \frac{\frac{1}{2+1} - \frac{1}{2}}{1} = \frac{-1}{(2+1) \times 2} = -\frac{1}{6} = -0.166\cdots.$$

For $x = 2, \Delta x = \frac{1}{4}$ we have

$$\frac{\Delta y}{\Delta x} = \frac{\frac{1}{2+\frac{1}{4}} - \frac{1}{2}}{\frac{1}{4}} = \frac{-1}{\left(2+\frac{1}{4}\right) \times 2} = -\frac{2}{9} = -0.222\cdots.$$

For $x = 2, \Delta x = \frac{1}{10}$ we have

$$\frac{\Delta y}{\Delta x} = \frac{\frac{1}{2+\frac{1}{10}} - \frac{1}{2}}{\frac{1}{10}} = \frac{-1}{\left(2+\frac{1}{10}\right) \times 2} = -\frac{5}{21} \approx -0.238.$$

As $\Delta x \to 0$ the slope at $x = 2$ appears to be approaching $-\frac{1}{4} = -0.25$.

As $\Delta x \to 0$ the slope approaches the derivative as

$$f'(x) \equiv \frac{dy}{dx} = \lim_{\Delta x \to 0} \frac{-1}{(x+\Delta x)\,x} = -\frac{1}{x^2}.$$

When $x = 2$ the derivative formula $f'(x) = -\frac{1}{x^2}$ states that the slope is indeed $-\frac{1}{4}$ as

$$f'(2) = -\frac{1}{2^2} = -\frac{1}{4} = -0.25.$$

Problem 2.17 *[B**] Given* $f(x) = x^n$ *for* n *a positive integer, prove that* $f'(x) = nx^{n-1}$.

Answer: Using the Binomial Theorem

$$(x + \Delta x)^n = x^n + nx^{n-1}\Delta x + \binom{n}{2}x^{n-2}\Delta x^2 + \cdots + \binom{n}{n}\Delta x^n$$

we have

$$(x + \Delta x)^n - x^n = nx^{n-1}\Delta x + \Delta x^2 Q$$

where Q is defined as

$$Q \equiv \binom{n}{2}x^{n-2} + \binom{n}{3}x^{n-3}\Delta x + \cdots + \binom{n}{n}\Delta x^{n-2}.$$

Note that as $\Delta x \to 0$ we have $Q \to \binom{n}{2}x^{n-2}$. The slope is

$$\frac{\Delta y}{\Delta x} = \frac{f(x + \Delta x) - f(x)}{\Delta x} = \frac{nx^{n-1}\Delta x + \Delta x^2 Q}{\Delta x} = nx^{n-1} + \Delta x Q.$$

To find the derivative let $\Delta x \to 0$ so that $\Delta x Q \to 0$ and so

$$\frac{dy}{dx} = \lim_{\Delta x \to 0}\left(nx^{n-1} + \Delta x Q\right) = nx^{n-1}.$$

2.4 Elasticities

Problem 2.18 *[D***] Suppose the demand curve is* $Q = P^{-\frac{1}{2}}$ *where* P *is the price. Using derivatives show that the demand curve slopes downwards (is globally decreasing) and that it is globally convex. Using the definition of elasticity show that the elasticity of demand is the exponent on* P *or* $\eta(P) = -\frac{1}{2}$. *Show that at* $P = 3$ *the slope and elasticity are different. If the slope equals the elasticity, what must* P *be equal to?*

Answer: The demand curve slopes downwards (is globally decreasing) since for all P we have

$$\begin{aligned} Q &= P^{-\frac{1}{2}} \\ \Longrightarrow \frac{dQ}{dP} &= -\frac{1}{2}\underbrace{P^{-\frac{3}{2}}}_{+} < 0 \end{aligned}$$

and the demand curve is globally convex since

$$\begin{aligned} \frac{dQ}{dP} &= -\frac{1}{2}P^{-\frac{3}{2}} \\ \Longrightarrow \frac{d^2Q}{dP^2} &= \frac{3}{4}\underbrace{P^{-\frac{5}{2}}}_{+} > 0. \end{aligned}$$

The elasticity of demand is

$$\eta(P) \equiv \frac{dQ}{dP}\frac{P}{Q} = -\frac{1}{2}P^{-\frac{3}{2}}\frac{P}{P^{-\frac{1}{2}}} = -\frac{1}{2}\frac{P^{-\frac{1}{2}}}{P^{-\frac{1}{2}}} = -\frac{1}{2}.$$

At $P = 3$ the elasticity is $-\frac{1}{2}$ but the slope is

$$\frac{dQ}{dP} = -\frac{1}{2}3^{-\frac{3}{2}} \approx -0.1.$$

If the slope and elasticity are the same then it must be that $P = 1$ as

$$\begin{aligned} \frac{dQ}{dP} &= -\frac{1}{2}P^{-\frac{3}{2}} = -\frac{1}{2} \\ &\Longrightarrow P^{-\frac{3}{2}} = 1 \Longrightarrow P = 1. \end{aligned}$$

Problem 2.19 *[D***] Suppose the demand curve is $Q = P^{-\frac{1}{2}} - 1$ for $P < 1$. Show that the elasticity $\eta(P)$ is not the exponent $-\frac{1}{2}$. Show that $\eta(P) < -\frac{1}{2}$. What is the elasticity if $P = \frac{1}{4}$? When will demand be elastic?*

Answer: The elasticity is

$$\eta(P) \equiv \frac{dQ}{dP}\frac{P}{Q} = -\frac{1}{2}P^{-\frac{3}{2}}\frac{P}{P^{-\frac{1}{2}} - 1} = -\frac{1}{2}\frac{P^{-\frac{1}{2}}}{P^{-\frac{1}{2}} - 1} = -\frac{1}{2}\frac{1}{1 - P^{\frac{1}{2}}}.$$

The elasticity is always less than $-\frac{1}{2}$ since

$$\begin{aligned} P &< 1 \\ &\Longrightarrow \frac{1}{1 - P^{\frac{1}{2}}} > 1 \\ &\Longrightarrow \eta(P) = -\frac{1}{2}\frac{1}{1 - P^{\frac{1}{2}}} < -\frac{1}{2}. \end{aligned}$$

If $P = \frac{1}{4}$ then

$$\eta\left(\frac{1}{4}\right) = -\frac{1}{2}\frac{1}{1 - \left(\frac{1}{4}\right)^{\frac{1}{2}}} = -1.$$

Demand is elastic (i.e., $\eta(P) < -1$) for $P > \frac{1}{4}$ since

$$\begin{aligned} \eta(P) &< -1 \\ &\Longrightarrow -\frac{1}{2}\frac{1}{1 - P^{\frac{1}{2}}} < -1 \\ &\Longrightarrow \frac{1}{1 - P^{\frac{1}{2}}} > 2 \Longrightarrow 1 - P^{\frac{1}{2}} < \frac{1}{2} \\ &\Longrightarrow P^{\frac{1}{2}} > \frac{1}{2} \Longrightarrow \left(P^{\frac{1}{2}}\right)^2 > \left(\frac{1}{2}\right)^2 \Longrightarrow P > \frac{1}{4}. \end{aligned}$$

Problem 2.20 *[D***]Given the demand curve* $Q = \frac{10}{P^5}$ *what is the elasticity of demand? Is demand elastic or inelastic? [C***]If* $Q(P) = \frac{10}{P^5} + 5$ *then what is the elasticity of demand when* $P = 1$*?*

Answer: Since $Q = \frac{10}{P^5} = 10P^{-5}$ the elasticity of demand is the exponent on P or $\eta = -5$. Demand is elastic since $\eta < -1$. If $Q(P) = \frac{10}{P^5} + 5$ then $Q'(P) = -\frac{50}{P^6}$ so that at $P = 1$ we have

$$\begin{aligned} \eta(P) &= \frac{Q'(P)P}{Q(P)} = \frac{-50P^{-5}}{10P^{-5}+5} \\ &\Longrightarrow \eta(1) = \frac{-50}{10+5} \approx -3.33. \end{aligned}$$

2.5 Product Rule

Problem 2.21 *[B**]Use* $\frac{d}{dx}x^n = nx^{n-1}$ *for* $n > 0$ *and the product rule to prove*

$$\frac{d}{dx}x^{-n} = -nx^{-n-1}.$$

Answer: Using the product rule on $x^n x^{-n} = 1$ we have

$$\begin{aligned} x^n x^{-n} &= 1 \\ &\Longrightarrow \frac{d}{dx}\left(x^n x^{-n}\right) = 0 \\ &\Longrightarrow \left(\frac{d}{dx}x^n\right)x^{-n} + x^n\left(\frac{d}{dx}x^{-n}\right) = 0 \\ &\Longrightarrow \left(nx^{n-1}\right)x^{-n} + x^n\left(\frac{d}{dx}x^{-n}\right) = 0 \\ &\Longrightarrow x^n\left(\frac{d}{dx}x^{-n}\right) = -nx^{n-1}x^{-n} = -nx^{-1} \\ &\Longrightarrow \frac{d}{dx}x^{-n} = -nx^{-1}x^{-n} = -nx^{-n-1}. \end{aligned}$$

Problem 2.22 *[A***] If* $h(x) = f(x) + g(x)$ *where the elasticities of* $h(x), f(x), g(x)$ *are denoted as* $\eta_h(x), \eta_f(x), \eta_g(x)$*, show that the elasticity of* $h(x)$ *is not the sum of the elasticities of* $f(x)$ *and* $g(x)$*, but a weighted sum*

$$\eta_h(x) = \alpha(x)\eta_f(x) + (1-\alpha(x))\eta_g(x)$$

where $\alpha(x) = \frac{f(x)}{f(x)+g(x)}$*. But show that if* $h(x) = f(x)g(x)$ *then* $\eta_h(x) = \eta_f(x) + \eta_g(x)$*.*

Answer: If $h(x) = f(x) + g(x)$ then the elasticity of $h(x)$ is

$$\begin{aligned}
\eta_h(x) &= \frac{h'(x)x}{h(x)} = \frac{(f'(x) + g'(x))x}{f(x) + g(x)} = \frac{f'(x)x + g'(x)x}{f(x) + g(x)} \\
&= \frac{f'(x)x}{f(x) + g(x)} + \frac{g'(x)x}{f(x) + g(x)} \\
&= \frac{f(x)}{f(x) + g(x)} \frac{f'(x)x}{f(x)} + \frac{g(x)}{f(x) + g(x)} \frac{g'(x)x}{g(x)} \\
&= \alpha(x)\eta_f(x) + \frac{g(x)}{f(x) + g(x)}\eta_g(x) \\
&= \alpha(x)\eta_f(x) + (1 - \alpha(x))\eta_g(x)
\end{aligned}$$

since

$$1 - \alpha(x) = 1 - \frac{f(x)}{f(x) + g(x)} = \frac{f(x) + g(x)}{f(x) + g(x)} - \frac{f(x)}{f(x) + g(x)} = \frac{g(x)}{f(x) + g(x)}.$$

If $h(x) = f(x)g(x)$ then the elasticity of $h(x)$ is

$$\begin{aligned}
\eta_h(x) &= \frac{h'(x)x}{h(x)} = \frac{(f'(x)g(x) + g'(x)f(x))x}{f(x)g(x)} \\
&= \frac{f'(x)g(x)x + g'(x)f(x)x}{f(x)g(x)} \\
&= \frac{f'(x)g(x)x}{f(x)g(x)} + \frac{g'(x)f(x)x}{f(x)g(x)} \\
&= \frac{f'(x)x}{f(x)} + \frac{g'(x)x}{g(x)} = \eta_f(x) + \eta_g(x).
\end{aligned}$$

Problem 2.23 *[B***] If $h(x) = f(x)g(x)$ then what is $h''(x)$?*

Answer: If $h(x) = f(x)g(x)$ then the product rules gives

$$h'(x) = f'(x)g(x) + f(x)g'(x).$$

Applying the product rule one more time yields

$$\begin{aligned}
h''(x) &= \frac{d}{dx}(f'(x)g(x) + f(x)g'(x)) = \frac{d}{dx}(f'(x)g(x)) + \frac{d}{dx}(f(x)g'(x)) \\
&= f''(x)g(x) + f'(x)g'(x) + f'(x)g'(x) + f(x)g''(x) \\
&= f''(x)g(x) + 2f'(x)g'(x) + f(x)g''(x).
\end{aligned}$$

Problem 2.24 *[B***]Given a demand curve $Q = Q(P)$ and revenue given by $R(P) = P \times Q(P)$, show that*

$$R'(P) = Q(P)(1 + \eta(P))$$

where $\eta(P)$ is the elasticity of demand. If the firm maximizes revenue instead of profits at P^*, then what will the elasticity of demand be?

Answer: Using the product rule we have

$$R'(P) = Q(P) + PQ'(P) = Q(P)\left(1 + \frac{PQ'(P)}{Q(P)}\right) = Q(P)(1 + \eta(P))$$

where $\eta(P) \equiv \frac{PQ'(P)}{Q(P)}$ is the elasticity of demand. If a firm maximizes revenue PQ at P^*, then $R'(P^*) = 0$ and so

$$\begin{aligned} R'(P^*) &= Q(P^*)(1 + \eta(P^*)) = 0 \\ &\Longrightarrow 1 + \eta(P^*) = 0 \Longrightarrow \eta(P^*) = -1. \end{aligned}$$

Problem 2.25 *[A***] Suppose that $f(x)$ and $g(x)$ are two globally positive, increasing, and convex functions. Show that $h(x) = f(x)g(x)$ is globally increasing and convex.*

Answer: By the product rule

$$h'(x) = f(x)g'(x) + f'(x)g(x) > 0$$

since $f(x) > 0, g(x) > 0$ and $f'(x) > 0, g'(x) > 0$. Furthermore

$$\begin{aligned} h''(x) &= f'(x)g'(x) + f(x)g''(x) + f'(x)g'(x) + f''(x)g(x) \\ &= f(x)g''(x) + 2f'(x)g'(x) + f''(x)g(x) > 0 \end{aligned}$$

and so $h(x)$ is also globally convex.

2.6 Chain Rule

Problem 2.26 *[D***] If $h(x) = \left(\frac{1}{1+\frac{1}{x^2}}\right)^{\frac{3}{2}}$ then what is $h'(x)$?*

Answer: We have

$$h(x) = \left(\frac{1}{1+\frac{1}{x^2}}\right)^{\frac{3}{2}} = \left(\frac{1}{1+x^{-2}}\right)^{\frac{3}{2}} = \left(\left(1+x^{-2}\right)^{-1}\right)^{\frac{3}{2}} = \left(1+x^{-2}\right)^{-\frac{3}{2}}.$$

Using the chain rule with $h(x) = f(g(x))$ where $f(x) = x^{-\frac{3}{2}}$ and $g(x) = 1 + x^{-2}$ we have

$$h'(x) = -\frac{3}{2}\left(1+x^{-2}\right)^{-\frac{3}{2}-1}\left(-2x^{-3}\right) = 3\left(1+x^{-2}\right)^{-\frac{5}{2}}x^{-3}.$$

Problem 2.27 *[D***]For each of*

$$h(x) = \left(1+x^2\right)^{\frac{1}{2}}, h(x) = \frac{1}{1+x^{-2}} \text{ and } h(x) = \sqrt{1+\sqrt{\sqrt{x}}}$$

find the inside function $g(x)$ and the outside function $f(x)$ such that $h(x) = f(g(x))$. Verify that $h(x) = f(g(x))$ and show that $h(x) \neq g(f(x))$. Use the chain rule to calculate $h'(x)$.

Answer: For $h(x) = \left(1+x^2\right)^{\frac{1}{2}}$ we have inside function $g(x) = 1 + x^2$ and outside function $f(x) = x^{\frac{1}{2}}$ since

$$h(x) = f(g(x)) = (g(x))^{\frac{1}{2}} = \left(1+x^2\right)^{\frac{1}{2}}.$$

We have $h(x) \neq g(f(x))$ since

$$g(f(x)) = 1 + (f(x))^2 = 1 + \left(x^{\frac{1}{2}}\right)^2 = 1 + x \neq \left(1+x^2\right)^{\frac{1}{2}}.$$

Now using the chain rule we have $f'(x) = \frac{1}{2}x^{-\frac{1}{2}}$ so that replacing x with the inside function $1+x^2$ yields $\frac{1}{2}\left(1+x^2\right)^{-\frac{1}{2}}$. Multiplying this with the derivative of the inside function $g'(x) = 2x$ yields

$$h'(x) = \frac{1}{2}\left(1+x^2\right)^{-\frac{1}{2}} \times 2x = \left(1+x^2\right)^{-\frac{1}{2}} x.$$

For

$$h(x) = \frac{1}{1+x^{-2}} = \left(1+x^{-2}\right)^{-1}$$

we have inside function $g(x) = 1 + x^{-2}$ and outside function $f(x) = x^{-1}$ since

$$h(x) = f(g(x)) = (g(x))^{-1} = \left(1+x^{-2}\right)^{-1} = \frac{1}{1+x^{-2}}.$$

We have $h(x) \neq g(f(x))$ since

$$g(f(x)) = 1 + (f(x))^{-2} = 1 + \left(x^{-1}\right)^{-2} = 1 + x^2 \neq \frac{1}{1+x^{-2}}.$$

Now using the chain rule we have $f'(x) = -x^{-2}$ so that replacing x with the inside function $1+x^{-2}$ yields $-\left(1+x^{-2}\right)^{-2}$. Multiplying this by the derivative of the inside function $g'(x) = -2x^{-3}$ yields

$$h'(x) = -\left(1+x^{-2}\right)^{-2}\left(-2x^{-3}\right) = 2\left(1+x^{-2}\right)^{-2} x^{-3}.$$

For the last function we have

$$h(x) = \sqrt{1+\sqrt{\sqrt{x}}} = \left(1+\sqrt{\sqrt{x}}\right)^{\frac{1}{2}} = \left(1+\left((x)^{\frac{1}{2}}\right)^{\frac{1}{2}}\right)^{\frac{1}{2}} = \left(1+x^{\frac{1}{4}}\right)^{\frac{1}{2}}.$$

We have inside function $g(x) = 1 + x^{\frac{1}{4}}$ and outside function $f(x) = x^{\frac{1}{2}}$ since

$$h(x) = f(g(x)) = (g(x))^{\frac{1}{2}} = \left(1+x^{\frac{1}{4}}\right)^{\frac{1}{2}}.$$

We have $h(x) \neq g(f(x))$ since

$$g(f(x)) = 1 + (f(x))^{\frac{1}{4}} = 1 + \left(x^{\frac{1}{2}}\right)^{\frac{1}{4}} = 1 + x^{\frac{1}{8}} \neq \left(1+x^{\frac{1}{4}}\right)^{\frac{1}{2}}.$$

Now using the chain rule we have $f'(x) = \frac{1}{2}x^{-\frac{1}{2}}$ so that replacing x with the inside function $1+x^{\frac{1}{4}}$ yields $\frac{1}{2}\left(1+x^{\frac{1}{4}}\right)^{-\frac{1}{2}}$. Multiplying this with the derivative of the inside function $g'(x) = \frac{1}{4}x^{-\frac{3}{4}}$ yields

$$h'(x) = \frac{1}{2}\left(1+x^{\frac{1}{4}}\right)^{-\frac{1}{2}}\frac{1}{4}x^{-\frac{3}{4}} = \frac{1}{8}\left(1+x^{\frac{1}{4}}\right)^{-\frac{1}{2}}x^{-\frac{3}{4}}.$$

Problem 2.28 *[D***] If $h(x) = \left(3+\frac{1}{x^2}\right)^{-3}$ then what is $h'(x)$?*

Answer: Using the chain rule with $f(x) = x^{-3}$ and $g(x) = 3 + x^{-2}$ we have

$$h'(x) = (-3 \times -2)\left(3+x^{-2}\right)^{-4}x^{-3} = 6\left(3+x^{-2}\right)^{-4}x^{-3}.$$

Problem 2.29 *[B***] Suppose that $f(x)$ and $g(x)$ are two globally increasing, and convex functions. Show that $h(x) = g(f(x))$ is globally increasing and convex.*

Answer: We have from the chain rule that

$$h'(x) = g'(f(x))\,f'(x) > 0$$

where $g'(f(x)) > 0$ and $f'(x) > 0$ by the assumption that $f(x)$ and $g(x)$ are globally increasing. Differentiating yet again we find that

$$h''(x) = g''(f(x))\left(f'(x)\right)^2 + g'(f(x))\,f''(x) > 0$$

where $g''(f(x)) > 0$ and $f''(x) > 0$ by the assumption that $f(x)$ and $g(x)$ are globally convex.

Problem 2.30 *[A***] Suppose for all x that $f'(x) > 0$, $f''(x) < 0$, $g'(x) > 0$ and $g''(x) < 0$. If $h(x) = g(f(x))$ then show that $h'(x) > 0$ and determine if $h(x)$ is globally concave or convex.*

Answer: We have using the chain rule that $h(x)$ is globally increasing as

$$h'(x) = \overbrace{g'(f(x))}^{+}\overbrace{f'(x)}^{+} > 0.$$

Differentiating again with respect to x we obtain

$$h''(x) = \overbrace{g''(f(x))}^{-}\overbrace{f'(x)^2}^{+} + \overbrace{g'(f(x))}^{+}\overbrace{f''(x)}^{-} < 0$$

and so $h(x)$ is globally concave.

2.7 Inverse Functions

Problem 2.31 *[D***] If $f(x) = \left(\frac{1}{\sqrt{x}}\right)^3$ then what is the inverse function $g(x)$? What is the elasticity of the inverse function?*

Answer: We have

$$f(x) = \left(\frac{1}{\sqrt{x}}\right)^3 = \left(x^{-\frac{1}{2}}\right)^3 = x^{-\frac{3}{2}} \Longrightarrow g(x) = x^{-\frac{2}{3}}.$$

Here $g(x) = x^{-\frac{2}{3}}$ satisfies the definition of an inverse function as

$$\begin{aligned} g(f(x)) &= \left(x^{-\frac{3}{2}}\right)^{-\frac{2}{3}} = x^{-\frac{3}{2}\times-\frac{2}{3}} = x^1 = x \\ f(g(x)) &= \left(x^{-\frac{2}{3}}\right)^{-\frac{3}{2}} = x^{-\frac{2}{3}\times-\frac{3}{2}} = x^1 = x. \end{aligned}$$

The elasticity of $g(x)$ is the exponent on x or $-\frac{2}{3}$.

Problem 2.32 *[D***] If the demand curve takes the form $Q = 8P^{-3}$ then what is the inverse demand curve? What is the elasticity of the demand curve, and what is the elasticity of the inverse demand curve?*

Answer: The inverse demand curve is found as

$$Q = 8P^{-3} \Longrightarrow P^{-3} = 8^{-1}Q \Longrightarrow P = \left(8^{-1}Q\right)^{-\frac{1}{3}} = 2Q^{-\frac{1}{3}}.$$

Both the demand curve $Q = 8P^{-3}$ and the inverse demand curve $P = 2Q^{-\frac{1}{3}}$ have the form Ax^b and so the elasticity is the exponent on the independent variable. The elasticity of the demand curve is -3, and the elasticity of the inverse demand curve is $-\frac{1}{3}$.

Problem 2.33 *[B***] If the demand curve takes the form $Q = AP^{-\beta}$ then what is the inverse demand curve? What is the elasticity of the demand curve, and what is the elasticity of the inverse demand curve?*

Answer: The inverse demand curve is found as

$$\begin{aligned} Q &= AP^{-\beta} \Longrightarrow P^{-\beta} = A^{-1}Q \\ &\Longrightarrow P = \left(A^{-1}Q\right)^{-\frac{1}{\beta}} = A^{\frac{1}{\beta}}Q^{-\frac{1}{\beta}}. \end{aligned}$$

Both the demand curve $Q = AP^{-\beta}$ and the inverse demand curve $P = A^{\frac{1}{\beta}}Q^{-\frac{1}{\beta}}$ have the form Ax^b and so the elasticity is the exponent on the independent variable. The elasticity of the demand curve is $-\beta$, and the elasticity of the inverse demand curve is $-\frac{1}{\beta}$.

Problem 2.34 *[B**] Suppose that for $x > 0$ that $f(x)$ is given by*

$$y = f(x) = x^2 + 2x + 1.$$

Show that $y = f(x)$ has an inverse function $g(x)$ and find $g(x)$. Verify that $g(x)$ and $f(x)$ satisfy the definition for inverse functions.

Answer: Since $x > 0$ and so for all x we have

$$f'(x) = 2x + 2 > 0.$$

It follows that $f(x)$ has an inverse function $g(x)$. To find $g(x)$ solve for $y = g(x)$ as

$$\begin{aligned} x &= y^2 + 2y + 1 \\ &\Longrightarrow (y+1)^2 = x \\ &\Longrightarrow y + 1 = x^{\frac{1}{2}} \\ &\Longrightarrow y = g(x) = x^{\frac{1}{2}} - 1. \end{aligned}$$

We then have

$$\begin{aligned} f(g(x)) &= \left(x^{\frac{1}{2}} - 1\right)^2 + 2\left(x^{\frac{1}{2}} - 1\right) + 1 = x - 2x^{\frac{1}{2}} + 1 + 2x^{\frac{1}{2}} - 2 + 1 = x \\ g(f(x)) &= \left(x^2 + 2x + 1\right)^{\frac{1}{2}} - 1 = \left((x+1)^2\right)^{\frac{1}{2}} - 1 = x + 1 - 1 = x. \end{aligned}$$

Problem 2.35 *[B**] Suppose that for $x > 0$ that $f(x)$ is given by*

$$f(x) = x^2 - 2x + 1.$$

Show that $f(x)$ does not have an inverse function. Show how by restricting the domain of $f(x)$ it is possible to give $f(x)$ an inverse function.

Answer: Since

$$f'(x) = 2x - 2 = 2(x - 1)$$

we have $f'(x) < 0$ for $x < 1$ and $f'(x) > 0$ for $x > 1$. It follows that $f(x)$ is neither globally increasing nor globally decreasing and hence has no inverse function.

If we restrict the domain of $f(x)$ so that $x > 1$ (or $0 < x < 1$) then $f'(x) > 0$ and so $f(x)$ is globally increasing. The inverse function $y = g(x)$ is found as

$$\begin{aligned} y^2 - 2y + 1 &= x \\ &\Longrightarrow (y-1)^2 = x \\ &\Longrightarrow y - 1 = x^{\frac{1}{2}} \\ &\Longrightarrow y = g(x) = x^{\frac{1}{2}} + 1. \end{aligned}$$

Problem 2.36 *[B**] Consider*

$$f(x) = \frac{(x-5)^3}{3} + 7x + 8.$$

Show that $f(x)$ has an inverse function. What would you need to do to find the inverse function?

Answer: We have for all x that

$$f'(x) = (x-5)^2 + 7 > 0$$

since $(x-5)^2 \geq 0$. Since $f'(x) > 0$ for all x it follows that $f(x)$ has an inverse function. To find this inverse function we need to solve

$$\begin{aligned} \frac{(y-5)^3}{3} + 7y + 8 &= x \\ &\Longrightarrow \frac{(y-5)^3}{3} + 7y + 8 - x = 0. \end{aligned}$$

No normal human being knows the formula for the roots of a cubic, but the computer will find it as

$$\begin{aligned} y &= \frac{1}{2}\left(-516 + 12x + 4\sqrt{18013 - 774x + 9x^2}\right)^{\frac{1}{3}} \\ &\quad - \frac{14}{\left(-516 + 12x + 4\sqrt{18013 - 774x + 9x^2}\right)^{\frac{1}{3}}} + 5. \end{aligned}$$

Problem 2.37 *[A**] When will the general cubic*

$$f(x) = \frac{a}{3}x^3 + \frac{b}{2}x^2 + cx + d$$

have an inverse function?

Answer: We require that for

$$f'(x) = ax^2 + bx + c$$

that either $f'(x) > 0$ for all x, or that $f'(x) < 0$ for all x. This is equivalent to requiring that $f'(x) = 0$ has no solution. Since

$$\begin{aligned} f'(x) &= 0 \Longrightarrow ax^2 + bx + c = 0 \\ &\Longrightarrow x = \frac{-b \pm \sqrt{b^2 - 4ac}}{2a} \end{aligned}$$

this will occur only when the term under the square root sign is negative (since you are not allowed to take the square root of a negative number), or when $b^2 - 4ac < 0$.

Problem 2.38 *[A***] If $f(x)$ is globally increasing and concave, show that its inverse function $g(x)$ is globally increasing and convex.*

Answer: Differentiating both sides of $f(g(x)) = x$ with respect to x and using the chain rule leads to

$$f'(g(x))\,g'(x) = 1 \Longrightarrow g'(x) = \frac{1}{f'(g(x))} > 0$$

since $f'(g(x)) > 0$. Differentiating both sides of $f'(g(x))\,g'(x) = 1$ with respect to x and using the chain and product rules leads to

$$\begin{aligned} f''(g(x))\,g'(x)^2 + f'(g(x))\,g''(x) &= 0 \\ \Longrightarrow\quad g''(x) &= -\frac{\overbrace{f''(g(x))}^{-}\overbrace{g'(x)^2}^{+}}{\underbrace{f'(g(x))}_{+}} > 0. \end{aligned}$$

2.8 Optimization

Problem 2.39 *[D***] Consider*

$$f(x) = 5x^{\frac{4}{5}} - 2x$$

*for $x > 0$. Find the first-order condition and solve. Show that your solution is a global maximizer. [B**] If $f(x) = 0$ then what is x?*

Answer: We have $f'(x) = 4x^{-\frac{1}{5}} - 2$. Solving the first-order condition $f'(x^*) = 0$ yields

$$\begin{aligned} 4(x^*)^{-\frac{1}{5}} - 2 &= 0 \\ \Longrightarrow\quad (x^*)^{-\frac{1}{5}} &= \frac{1}{2} \\ \Longrightarrow\quad x^* &= \left(\frac{1}{2}\right)^{-5} = 32. \end{aligned}$$

The function $f(x) = 5x^{\frac{4}{5}} - 2x$ is globally concave since $x^{-\frac{6}{5}} > 0$ for all x and so

$$f''(x) = -\frac{4}{5}x^{-\frac{6}{5}} < 0.$$

Thus $x^* = 32$ is a global maximizer. (It is not enough to say that $f''(32) < 0$ since this only proves that $x^* = 32$ is a local maximizer.)

Let x be the root which satisfies $f(x) = 0$ for $x > 0$. Since $x > 0$ we can divide by x as

$$\begin{aligned} f(x) &= 5x^{\frac{4}{5}} - 2x = 0 \\ &\implies x^{\frac{4}{5}} = \frac{2}{5}x \implies \frac{x^{\frac{4}{5}}}{x} = \frac{2}{5}\frac{x}{x} \\ &\implies x^{-\frac{1}{5}} = \frac{2}{5} \implies x = \left(\frac{2}{5}\right)^{-5} \approx 97.66. \end{aligned}$$

Problem 2.40 *Consider*

$$f(x) = 3x^{\frac{2}{3}} - 5x$$

for $x > 0$. Find the first-order condition and solve. Show that your solution is a global maximizer. Is the function globally increasing or globally decreasing?

Answer: Since

$$f'(x) = 2x^{-\frac{1}{3}} - 5$$

the first-order condition $f'(x^*) = 0$ yields

$$\begin{aligned} 2(x^*)^{-\frac{1}{3}} - 5 &= 0 \\ &\implies (x^*)^{-\frac{1}{3}} = \frac{5}{2} \\ &\implies x^* = \left(\frac{5}{2}\right)^{-3} = \frac{8}{125}. \end{aligned}$$

We have $f''(x) = -\frac{2}{3}x^{-\frac{4}{3}} < 0$ since $x^{-\frac{4}{3}} > 0$. Thus $f(x)$ is globally concave and $x^* = \frac{8}{125}$ is a global maximizer. The function $f(x)$ is neither globally increasing nor decreasing since $f'(x) > 0$ for $x < \frac{8}{125}$ and $f'(x) < 0$ for $x > \frac{8}{125}$.

Problem 2.41 *[D***] Suppose $f(x)$ is given by $f(x) = x^{\frac{1}{2}} - x$ for $x > 0$. Find the first-order condition and solve. Show that your solution is a global maximizer. [C**] Is $f(x)$ globally increasing or globally decreasing?*

Answer: We have $f'(x) = \frac{1}{2}x^{-\frac{1}{2}} - 1$. The first-order condition $f'(x^*) = 0$ yields

$$\begin{aligned} \frac{1}{2}(x^*)^{-\frac{1}{2}} - 1 &= 0 \\ &\implies (x^*)^{-\frac{1}{2}} = 2 \\ &\implies x^* = (2)^{-2} = \frac{1}{4}. \end{aligned}$$

The function $f(x)$ is globally concave since for all x

$$f''(x) = -\frac{1}{4}x^{-\frac{3}{2}} < 0$$

since $x^{-\frac{3}{2}} > 0$. So $x^* = \frac{1}{4}$ is a global maximizer. (Note it is not enough to say that $f''\left(\frac{1}{4}\right) < 0$ since this only proves that $x^* = \frac{1}{4}$ is a local maximizer.) The function $f(x)$ is neither globally increasing nor globally decreasing since $f'(x) > 0$ for $x < \frac{1}{4}$ while $f'(x) < 0$ for $x > \frac{1}{4}$.

Problem 2.42 *[D***] Consider*

$$f(x) = \frac{x^3}{3} - 2x^2 + 3x + 7$$

*for $x \geq 0$. Find the solutions to the first-order condition and determine if the solutions is a local maximizer or minimizer. [B**] Where is the global maximum and where is the global minimum?*

Answer: Since

$$f'(x) = x^2 - 4x + 3$$

the first-order condition $f'(x^*) = 0$ yields

$$\begin{aligned} (x^*)^2 + (-4)x^* + (3) &= 0 \\ \Longrightarrow x^* &= \frac{-(-4) \pm \sqrt{(-4)^2 - 4 \times (3)}}{2} \\ \Longrightarrow x_1^* &= 1, x_2^* = 3. \end{aligned}$$

Now $f''(x) = 2x - 4$ so that $x_1^* = 1$ is a local maximizer since

$$f''(x_1^*) = 2(x_1^*) - 4 = (2 \times 1) - 4 = -2 < 0$$

and $x_2^* = 3$ is a local minimizer since

$$f''(x_2^*) = 2(x_2^*) - 4 = (2 \times 3) - 4 = 2 > 0.$$

The function $f(x)$ has no global maximum since $f(x) \longrightarrow \infty$ as $x \longrightarrow \infty$. It turns out that $x_2^* = 3$ is a global minimizer, but is not a unique global minimizer since $f(0) = f(3) = 7$. The other global minimum at $x_3^* = 0$ does not solve the first-order condition $f'(x^*) = 0$, but is instead a corner solution on the boundary of the domain of $f(x)$. This is illustrated in the figure below.

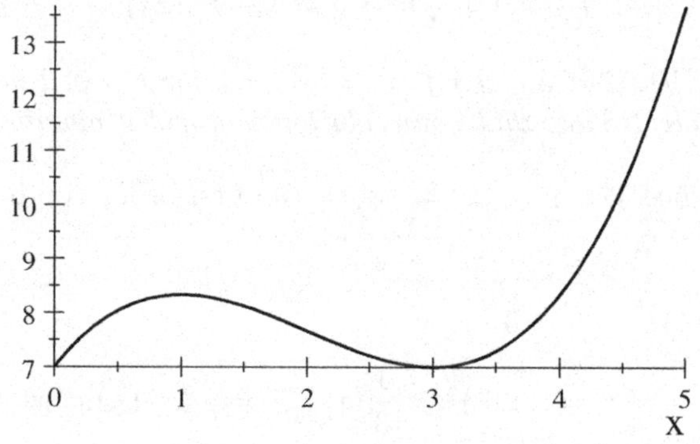

$f(x) = \frac{x^3}{3} - 2x^2 + 3x + 7$

Problem 2.43 *[B**] Given the general cubic*

$$f(x) = \frac{a}{3}x^3 + \frac{b}{2}x^2 + cx + d$$

show that two solutions x_1^*, x_2^* to the first-order condition exist as long as $b^2 - 4ac > 0$. Show that one solution is a local maximizer and the other is a local minimizer with $f''(x_1^*) = -f''(x_2^*)$.

Answer: Since

$$f'(x) = ax^2 + bx + c$$

the first-order condition $f'(x^*) = 0$ yields two solutions

$$\begin{aligned} a(x^*)^2 + bx^* + c &= 0 \\ \implies x^* &= \frac{-b \pm \sqrt{b^2 - 4ac}}{2a} \end{aligned}$$

or

$$x_1^* = \frac{-b + \sqrt{b^2 - 4ac}}{2a}, x_2^* = \frac{-b - \sqrt{b^2 - 4ac}}{2a}$$

as long as $b^2 - 4ac > 0$ (since we cannot take the square root of a negative number). Now $f''(x) = 2ax + b$ so that

$$\begin{aligned} f''(x_1^*) &= 2a\left(\frac{-b + \sqrt{b^2 - 4ac}}{2a}\right) + b = \sqrt{b^2 - 4ac} > 0 \\ f''(x_2^*) &= 2a\left(\frac{-b - \sqrt{b^2 - 4ac}}{2a}\right) + b = -\sqrt{b^2 - 4ac} < 0 \end{aligned}$$

so that x_1^* is a local minimizer and x_2^* is a local maximizer with

$$f''(x_1^*) = \sqrt{b^2 - 4ac} = -f''(x_2^*).$$

Problem 2.44 *[D***] Consider $f(x) = x^{\frac{1}{2}} - x^{\frac{3}{2}}$ for $x > 0$. Find the first-order condition and solve. Show that your solution is a global maximizer.*

Answer: Since $f'(x) = \frac{1}{2}x^{-\frac{1}{2}} - \frac{3}{2}x^{\frac{1}{2}}$ the first-order condition $f'(x^*) = 0$ yields

$$\begin{aligned} \frac{1}{2}(x^*)^{-\frac{1}{2}} - \frac{3}{2}(x^*)^{\frac{1}{2}} &= 0 \\ &\implies (x^*)^{\frac{1}{2}} = \frac{1}{3}(x^*)^{-\frac{1}{2}} \implies (x^*)^{\frac{1}{2}}(x^*)^{\frac{1}{2}} = \frac{1}{3}(x^*)^{\frac{1}{2}}(x^*)^{-\frac{1}{2}} \\ &\implies (x^*)^{\frac{1}{2}+\frac{1}{2}} = \frac{1}{3}(x^*)^{\frac{1}{2}-\frac{1}{2}} \implies (x^*)^1 = \frac{1}{3}(x^*)^0 \implies x^* = \frac{1}{3}. \end{aligned}$$

Furthermore for all x we have

$$f''(x) = -\frac{1}{4}x^{-\frac{3}{2}} - \frac{3}{4}x^{-\frac{1}{2}} < 0$$

(since $x^{-\frac{3}{2}} > 0$ and $x^{-\frac{1}{2}} > 0$) so that $f(x)$ is globally concave. This means that $x^* = \frac{1}{3}$ is a global maximizer.

Problem 2.45 *[D***] Consider the function*

$$f(x) = x^{\frac{4}{3}} - x^{\frac{2}{3}} - 1$$

*for $x > 0$. Find the first-order condition and solve. Show that $f(x)$ is globally concave and that your solution is a global maximizer. [B**] For what value of x is $f(x) = 0$?*

Answer: We have $f'(x) = \frac{4}{3}x^{\frac{1}{3}} - \frac{2}{3}x^{-\frac{1}{3}}$ and so the first-order condition $f'(x^*) = 0$ yields

$$\begin{aligned}
\frac{4}{3}(x^*)^{\frac{1}{3}} - \frac{2}{3}(x^*)^{-\frac{1}{3}} &= 0 \\
&\Longrightarrow \frac{4}{3}(x^*)^{\frac{1}{3}} = \frac{2}{3}(x^*)^{-\frac{1}{3}} \Longrightarrow (x^*)^{\frac{1}{3}} = \frac{1}{2}(x^*)^{-\frac{1}{3}} \\
&\Longrightarrow (x^*)^{\frac{1}{3}}(x^*)^{\frac{1}{3}} = \frac{1}{2}(x^*)^{-\frac{1}{3}}(x^*)^{\frac{1}{3}} \Longrightarrow (x^*)^{\frac{2}{3}} = \frac{1}{2} \\
&\Longrightarrow x^* = \left(\frac{1}{2}\right)^{\frac{3}{2}} \approx 0.35.
\end{aligned}$$

Calculating $f''(x)$ we have for all x that

$$f''(x) = \frac{4}{9}\underbrace{x^{-\frac{2}{3}}}_{+} + \frac{2}{9}\underbrace{x^{-\frac{4}{3}}}_{+} > 0$$

so $f(x)$ is globally convex (**not concave** as the question states: don't be intimidated by the question!), and so $x^* = \left(\frac{1}{2}\right)^{\frac{3}{2}}$ is a global minimizer.

To solve

$$f(x) = x^{\frac{4}{3}} - x^{\frac{2}{3}} - 1 = 0$$

set $y = x^{\frac{2}{3}}$ and $y^2 = x^{\frac{4}{3}}$ so that $f(x)$ becomes a quadratic as

$$\begin{aligned}
x^{\frac{4}{3}} - x^{\frac{2}{3}} - 1 &= 0 \\
&\Longrightarrow y^2 - y - 1 = 0 \\
&\Longrightarrow y = x^{\frac{2}{3}} = \frac{1+\sqrt{1+4}}{2} = \frac{1+\sqrt{5}}{2} \\
&\Longrightarrow x = \left(\frac{1+\sqrt{5}}{2}\right)^{\frac{3}{2}} \approx 2.06.
\end{aligned}$$

We ignore the negative root since $x > 0 \Longrightarrow y > 0$ and $\frac{1-\sqrt{5}}{2} \approx -0.62 < 0$.

Problem 2.46 *Consider the function*

$$f(x) = 3x^{\frac{1}{3}} - 2x^{\frac{3}{2}}$$

*for $x > 0$. [D***] Calculate $f'(x)$ and $f''(x)$. Is $f(x)$ globally decreasing or globally increasing? [C***] Show that $f(x)$ is globally concave. [D***] Find the first-order condition and solve. Show that the solution to the first-order condition is a global maximizer.*

Answer: We have $f'(x) = x^{-\frac{2}{3}} - 3x^{\frac{1}{2}}$ and so the first-order condition $f'(x^*) = 0$ yields

$$\begin{aligned} (x^*)^{-\frac{2}{3}} - 3(x^*)^{\frac{1}{2}} &= 0 \\ &\Longrightarrow (x^*)^{-\frac{2}{3}} = 3(x^*)^{\frac{1}{2}} \Longrightarrow (x^*)^{-\frac{2}{3}-\frac{1}{2}} = 3 \Longrightarrow (x^*)^{-\frac{7}{6}} = 3 \\ &\Longrightarrow x^* = 3^{-\frac{6}{7}} \approx 0.39. \end{aligned}$$

Since

$$f''(x) = -\frac{2}{3}x^{-\frac{5}{3}} - \frac{3}{2}x^{-\frac{1}{2}} < 0$$

it follows that $f(x)$ is globally concave and $x^* = 3^{-\frac{6}{7}}$ is a global maximizer.

The function $f(x)$ is neither globally increasing nor decreasing. To show this just pick any two points to the left and right of $x^* = 3^{-\frac{6}{7}}$. For example

$$f'(0.1) = (0.1)^{-\frac{2}{3}} - 3\times(0.1)^{\frac{1}{2}} \approx 3.7 > 0 \text{ and } f'(1) = (1)^{-\frac{2}{3}} - 3\times(1)^{\frac{1}{2}} = -2 < 0.$$

Problem 2.47 *[D***] Consider the function*

$$f(x) = -\frac{1}{5}x^5 - \frac{2}{x^2}$$

for $x > 0$. Calculate $f'(x)$ and $f''(x)$. Is $f(x)$ is globally decreasing? Show that $f(x)$ is globally concave. Find the first-order condition and solve.

Answer: We have $f'(x) = -x^4 + 4x^{-3}$ and so the first-order condition $f'(x^*) = 0$ yields

$$\begin{aligned} -(x^*)^4 + 4(x^*)^{-3} &= 0 \\ &\Longrightarrow (x^*)^4 = 4(x^*)^{-3} \Longrightarrow (x^*)^7 = 4 \\ &\Longrightarrow x^* = 4^{\frac{1}{7}}. \end{aligned}$$

Since

$$f''(x) = -4x^3 - 12x^{-4} < 0$$

it follows that $f(x)$ is globally concave and $x^* = 4^{\frac{1}{7}}$ is a global maximizer.

The function $f(x)$ is neither globally increasing nor decreasing. To show this just pick any two points to the left and right of $x^* = 4^{\frac{1}{7}} \approx 1.219$. For example

$$f'(1) = -(1)^4 + 4\times(1)^{-3} = 3 > 0 \text{ and } f'(2) = -(2)^4 + 4\times(2)^{-3} \approx -15.5 < 0.$$

Problem 2.48 *[A***] Consider the function*

$$f(x) = x^a - x^b \text{ for } x > 0$$

where $0 < a < 1$ *and* $b > 1$. *Solve the first-order condition and show that this is a global maximizer.*

Answer: We have

$$f'(x) = ax^{a-1} - bx^{b-1}$$

and so the first-order condition $f'(x^*) = 0$ yields

$$\begin{aligned} a(x^*)^{a-1} - b(x^*)^{b-1} &= 0 \\ &\Longrightarrow a(x^*)^{a-1} = b(x^*)^{b-1} \\ &\Longrightarrow \frac{(x^*)^{a-1}}{(x^*)^{b-1}} = \frac{b}{a} \Longrightarrow (x^*)^{a-b} = \frac{b}{a} \\ &\Longrightarrow x^* = \left(\frac{b}{a}\right)^{\frac{1}{a-b}}. \end{aligned}$$

To show that $f(x)$ has a global maximum at x^* note that

$$f''(x) = \overbrace{a}^{+}\overbrace{(a-1)}^{-}x^{a-2} - \overbrace{b}^{+}\overbrace{(b-1)}^{+}x^{b-2} < 0$$

where $a - 1 < 0$ since $a < 1$ and $b - 1 > 0$ since $b > 1$. Therefore $f(x)$ is globally concave so that $x^* = \left(\frac{b}{a}\right)^{\frac{1}{a-b}}$ is a global maximizer.

2.8.1 Profit Maximization

Problem 2.49 *[D***] Suppose a competitive firm's production function is* $Q = \frac{3}{2}L^{\frac{2}{3}}$. *Show that there is a positive and diminishing marginal product of labour. Find profits as a function of L. From the first-order condition find the firm's demand for labour* L^* *and determine the elasticity of demand. Show that* L^* *maximizes profits. [C***] Find the supply curve and the elasticity of supply.*

Answer: We have a positive and diminishing marginal product of labour since

$$MP_L(L) = f'(L) = \overbrace{L^{-\frac{1}{3}}}^{+} > 0, MP_L'(L) = f''(L) = -\overbrace{\frac{1}{3}L^{-\frac{4}{3}}}^{+} < 0.$$

Profits as a function of L are found by replacing Q in $\pi \equiv PQ - WL$ with $Q = \frac{3}{2}L^{\frac{2}{3}}$ as $\pi(L) = P\frac{3}{2}L^{\frac{2}{3}} - WL$ so that $\pi'(L) = PL^{-\frac{1}{3}} - W$. From the first-order condition $\pi'(L^*) = 0$ we have for real wage $w \equiv \frac{W}{P}$ that the demand for labour is

$$\begin{aligned} P\left(L^{*}\right)^{-\frac{1}{3}}-W &= 0 \\ &\Longrightarrow \left(L^{*}\right)^{-\frac{1}{3}}=\frac{W}{P}=w \\ &\Longrightarrow L^{*}=w^{-3}. \end{aligned}$$

Since L^* has the form $y = Ax^b$ the elasticity of demand is the exponent on w or -3.

Since for all L we have

$$\pi''(L) = -\frac{1}{3}\overbrace{PL^{-\frac{4}{3}}}^{+} < 0$$

it follows that $\pi(L) = P\frac{3}{2}L^{\frac{2}{3}} - WL$ is globally concave and so $L^* = w^{-3}$ is a global profit maximizer.

The supply curve is found as

$$Q^* = f(L^*) = \frac{3}{2}(L^*)^{\frac{2}{3}} = \frac{3}{2}\left(w^{-3}\right)^{\frac{2}{3}} = \frac{3}{2}w^{-2} = \frac{3}{2}\left(w^{-1}\right)^2 = \frac{3}{2}p^2$$

where $p \equiv \frac{P}{W} = w^{-1}$ is the real price of Q. The elasticity of supply is the exponent on p or 2.

Problem 2.50 *[D***] Suppose a competitive firm's production function is $Q = 9L^{\frac{1}{3}}$. Show that there is a positive and diminishing marginal product of labour. Find profits as a function of L. From the first-order condition find the firm's demand for labour L^* and determine the elasticity of demand. Show that L^* maximizes profits. [C***] Find the supply curve and the elasticity of supply.*

Answer: The marginal product of labour is positive and diminishing as

$$\begin{aligned} MP_L(L) &= f'(L) = \overbrace{3L^{-\frac{2}{3}}}^{+} > 0 \\ MP_L'(L) &= f''(L) = -\overbrace{2L^{-\frac{5}{3}}}^{+} < 0. \end{aligned}$$

Profits as a function of L are found by replacing Q in $\pi \equiv PQ - WL$ with $Q = 9L^{\frac{1}{3}}$ as $\pi(L) = P \times 9L^{\frac{1}{3}} - WL$ so that $\pi'(L) = 3P \times L^{-\frac{2}{3}} - W$. From the first-order condition for profit maximization $\pi'(L^*) = 0$ we have for real wage $w \equiv \frac{W}{P}$ that the demand for labour is

$$\begin{aligned} 3P\left(L^{*}\right)^{-\frac{2}{3}}-W &= 0 \\ &\Longrightarrow \left(L^{*}\right)^{-\frac{2}{3}}=\frac{1}{3}\frac{W}{P}=\frac{1}{3}w \\ &\Longrightarrow L^{*}=\left(\frac{1}{3}w\right)^{-\frac{3}{2}}=3^{\frac{3}{2}}w^{-\frac{3}{2}}. \end{aligned}$$

Since this is of the form $y = Ax^b$ the elasticity of demand is $-\frac{3}{2}$.

Now for all L we have $\pi''(L) = -2PL^{-\frac{5}{3}} < 0$ so that $\pi(L) = P \times 9L^{\frac{1}{3}} - WL$ is globally concave and hence $L^* = 3^{\frac{3}{2}} w^{-\frac{3}{2}}$ is a global profit maximizer.

The supply curve is found as

$$Q^* = f(L^*) = 9(L^*)^{\frac{1}{3}} = 9\left(3^{\frac{3}{2}} w^{-\frac{3}{2}}\right)^{\frac{1}{3}} = 3^{\frac{5}{2}} w^{-\frac{1}{2}} = 3^{\frac{5}{2}}\left(w^{-1}\right)^{\frac{1}{2}} = 3^{\frac{5}{2}} p^{\frac{1}{2}}$$

where $p \equiv \frac{P}{W} = w^{-1}$ is the real price of Q. The elasticity of supply is the exponent on p or $\frac{1}{2}$.

Problem 2.51 *[B***] Suppose that the firm has a production function $Q = AL^{\alpha}$ where $A > 0$ and $0 < \alpha < 1$. Find the firm's demand curve for labour L^* and show from the elasticity of demand that labour demand slopes downwards. Show that L^* maximizes profits. Find the elasticity of supply and use this to show that the supply curve slopes upwards.*

Answer: We have

$$\pi(L) = PAL^{\alpha} - WL$$

so that

$$\pi'(L) = \alpha APL^{\alpha-1} - W.$$

The first-order condition for profit maximization $\pi'(L^*) = 0$ then yields

$$\begin{aligned} \alpha PA(L^*)^{\alpha-1} - W &= 0 \\ &\Longrightarrow (L^*)^{\alpha-1} = \alpha^{-1}A^{-1}\frac{W}{P} = \alpha^{-1}A^{-1}w \\ &\Longrightarrow L^* = \alpha^{\frac{1}{1-\alpha}} A^{\frac{1}{1-\alpha}} w^{\frac{1}{\alpha-1}}. \end{aligned}$$

This corresponds to a global profit maximum since

$$\pi''(L) = \alpha(\alpha-1)APL^{\alpha-2} < 0$$

since $0 < \alpha < 1$ implies $\alpha - 1 < 0$.

The elasticity of labour demand is then the exponent on the real wage or $\eta = \frac{1}{\alpha-1}$ which is negative since

$$0 < \alpha < 1 \Longrightarrow \eta = \frac{1}{\alpha-1} < 0.$$

The supply curve is found as

$$\begin{aligned} Q^* &= f(L^*) = A(L^*)^{\alpha} = A\left(\alpha^{\frac{1}{1-\alpha}} A^{\frac{1}{1-\alpha}} w^{\frac{1}{\alpha-1}}\right)^{\alpha} = \alpha^{\frac{\alpha}{1-\alpha}} AA^{\frac{\alpha}{1-\alpha}} w^{\frac{\alpha}{\alpha-1}} \\ &= \alpha^{\frac{\alpha}{1-\alpha}} A^{\frac{1}{1-\alpha}} \left(w^{-1}\right)^{-\frac{\alpha}{\alpha-1}} = \alpha^{\frac{\alpha}{1-\alpha}} A^{\frac{1}{1-\alpha}} p^{\frac{\alpha}{1-\alpha}} \end{aligned}$$

where $p \equiv \frac{P}{W} \equiv w^{-1}$. The elasticity of supply, the exponent on p, is $\eta = \frac{\alpha}{1-\alpha} > 0$ (since $0 < \alpha < 1$) and so the supply curve slopes upwards.

Problem 2.52 *[C***] Suppose a competitive firm has a production function*

$$Q = \left(1 + L^{-1}\right)^{-1}.$$

Show that this production function is globally increasing and concave. Show that the marginal product of labour $MP_L(L)$ is positive and diminishing. Find the firm's labour demand curve and prove that the labour demand curve is downward sloping. Is the elasticity constant?

Answer: We have

$$\begin{aligned} MP_L(L) &\equiv f'(L) = \left(1 + L^{-1}\right)^{-2} L^{-2} = \left(\left(1 + L^{-1}\right) L\right)^{-2} = (1 + L)^{-2} > 0 \\ MP_L'(L) &\equiv f''(L) = -2(1 + L)^{-3} < 0. \end{aligned}$$

Profits are

$$\pi(L) = P\left(1 + L^{-1}\right)^{-1} - WL$$

so that $\pi'(L) = P(1 + L)^{-2} - W$ and the first-order condition $\pi'(L^*) = 0$ yields

$$\begin{aligned} P(1 + L^*)^{-2} - W &= 0 \Longrightarrow (1 + L^*)^{-2} = \frac{W}{P} = w \\ &\Longrightarrow (1 + L^*) = w^{-\frac{1}{2}} \\ &\Longrightarrow L^* = w^{-\frac{1}{2}} - 1. \end{aligned}$$

(We require $w < 1$ or else $L^* = 0$, a corner solution.) Here $L^* = w^{-\frac{1}{2}} - 1$ corresponds to a global profit maximum since for all L we have

$$\pi''(L) = -2P(1 + L)^{-3} < 0.$$

We have

$$\frac{dL^*}{dw} = -\frac{1}{2} w^{-\frac{3}{2}} < 0$$

so the labour demand curve is downward sloping. The elasticity is not constant since

$$\eta(\omega) = \frac{dL^*}{dw}\frac{w}{L^*} = -\frac{1}{2} w^{-\frac{3}{2}} \frac{w}{w^{-\frac{1}{2}} - 1} = -\frac{1}{2} \frac{w^{-\frac{1}{2}}}{w^{-\frac{1}{2}} - 1}.$$

Problem 2.53 *[C***] Suppose a competitive firm's production function is*

$$Q = f(L) = \left(1 + \frac{L^{\frac{3}{2}}}{L^{\frac{1}{2}} L^{\frac{5}{2}}}\right)^{-\frac{2}{3}}.$$

Find the marginal product of labour and show that there is a diminishing marginal product of labour. Find the firm's labour demand curve.

Answer: Simplifying first we have

$$f(L) = \left(1 + \frac{L^{\frac{3}{2}}}{L^{\frac{1}{2}} L^{\frac{5}{2}}}\right)^{-\frac{2}{3}} = \left(1 + L^{\frac{3}{2}-\frac{1}{2}-\frac{5}{2}}\right)^{-\frac{2}{3}} = \left(1 + L^{-\frac{3}{2}}\right)^{-\frac{2}{3}}.$$

Using the chain rule the marginal product of labour is

$$MP_L(L) = f'(L) = -\frac{2}{3}\left(1 + L^{-\frac{3}{2}}\right)^{-\frac{2}{3}-1}\left(-\frac{3}{2}L^{-\frac{3}{2}-1}\right) = \left(1 + L^{-\frac{3}{2}}\right)^{-\frac{5}{3}} L^{-\frac{5}{2}}$$

which can be simplified as

$$\begin{aligned} MP_L(L) &= \left(1 + L^{-\frac{3}{2}}\right)^{-\frac{5}{3}} L^{-\frac{5}{2}} = \left(1 + L^{-\frac{3}{2}}\right)^{-\frac{5}{3}} \left(L^{\frac{3}{2}}\right)^{-\frac{5}{3}} \\ &= \left(\left(1 + L^{-\frac{3}{2}}\right)\left(L^{\frac{3}{2}}\right)\right)^{-\frac{5}{3}} = \left(1 + L^{\frac{3}{2}}\right)^{-\frac{5}{3}}. \end{aligned}$$

There is a diminishing marginal product of labour since

$$MP_L'(L) = -\frac{5}{3}\left(1 + L^{\frac{3}{2}}\right)^{-\frac{5}{3}-1}\frac{3}{2}L^{\frac{3}{2}-1} = -\frac{5}{2}\left(1 + L^{\frac{3}{2}}\right)^{-\frac{8}{3}} L^{\frac{1}{2}} < 0$$

since $L^{\frac{1}{2}} > 0$ and $\left(1 + L^{\frac{3}{2}}\right)^{-\frac{8}{3}} > 0$.

Profits are

$$\pi(L) = P\left(1 + L^{-\frac{3}{2}}\right)^{-\frac{2}{3}} - WL$$

so that $\pi'(L) = P\left(1 + L^{\frac{3}{2}}\right)^{-\frac{5}{3}} - W$ and the first-order condition $\pi'(L^*) = 0$ yields

$$\begin{aligned} P\left(1 + (L^*)^{\frac{3}{2}}\right)^{-\frac{5}{3}} - W &= 0 \\ &\implies \left(1 + (L^*)^{\frac{3}{2}}\right)^{-\frac{5}{3}} = w \\ &\implies 1 + (L^*)^{\frac{3}{2}} = w^{-\frac{3}{5}} \\ &\implies (L^*)^{\frac{3}{2}} = w^{-\frac{3}{5}} - 1 \\ &\implies L^* = \left(w^{-\frac{3}{5}} - 1\right)^{\frac{2}{3}}. \end{aligned}$$

(We require $w < 1$ or else $L^* = 0$, a corner solution.) This corresponds to a global profit maximum since for all L we have

$$\pi''(L) = -\frac{5}{2}P\left(1 + L^{\frac{3}{2}}\right)^{-\frac{8}{3}} L^{\frac{1}{2}} < 0.$$

Problem 2.54 *[A ***] A version of the CES (Constant Elasticity of Substitution) production function is given by*

$$Q = f(L) = \frac{1}{\beta}(\alpha + \beta L^{\rho})^{\frac{1}{\rho}}$$

where $\alpha > 0$, $\beta > 0$ *and* $\rho < 1$. *Note* $\rho < 0$ *is allowed! Show that marginal product of labour is globally positive, concave, and diminishing. Find the firm's demand for labour.*

Answer: We have using the chain rule that

$$\begin{aligned} MP_L(L) &= f'(L) = \frac{1}{\beta\rho}(\alpha + \beta L^{\rho})^{\frac{1}{\rho}-1} \times \rho\beta L^{\rho-1} \\ &= \underbrace{(\alpha + \beta L^{\rho})^{\frac{1}{\rho}-1}}_{+}\underbrace{L^{\rho-1}}_{+} > 0. \end{aligned}$$

The presence of L inside and outside the brackets makes this expression difficult to manipulate. We can however get the second term back inside the brackets and this will make our life easier

$$\begin{aligned} MP_L(L) &= (\alpha + \beta L^{\rho})^{\frac{1}{\rho}-1} L^{\rho-1} = (\alpha + \beta L^{\rho})^{\frac{1}{\rho}-\frac{\rho}{\rho}} L^{-(1-\rho)} \\ &= (\alpha + \beta L^{\rho})^{\frac{1-\rho}{\rho}} \left(L^{-\rho}\right)^{\frac{(1-\rho)}{\rho}}. \end{aligned}$$

Note that $L^{-\rho}$ and $\alpha + \beta L^{\rho}$ now have the same exponent $\frac{1-\rho}{\rho}$ so that using the property of exponents $(xy)^a = x^a y^a$ we have

$$\begin{aligned} MP_L(L) &= (\alpha + \beta L^{\rho})^{\frac{1-\rho}{\rho}} \left(L^{-\rho}\right)^{\frac{(1-\rho)}{\rho}} = \left((\alpha + \beta L^{\rho})\left(L^{-\rho}\right)\right)^{\frac{1-\rho}{\rho}} \\ &= \left(\alpha L^{-\rho} + \beta L^{\rho} L^{-\rho}\right)^{\frac{1-\rho}{\rho}} = \left(\alpha L^{-\rho} + \beta\right)^{\frac{1-\rho}{\rho}}. \end{aligned}$$

Now to show concavity and a diminishing marginal product of labour we again use the chain rule as

$$\begin{aligned} f''(L) &= MP_L'(L) = \frac{1-\rho}{\rho}\left(\alpha L^{-\rho} + \beta\right)^{\frac{1-\rho}{\rho}-1} \alpha(-\rho) L^{-\rho-1} \\ &= -\left(\frac{1-\rho}{\rho}\right)\left(\alpha L^{-\rho} + \beta\right)^{\frac{1-\rho}{\rho}-1} \alpha\rho L^{-\rho-1} \\ &= -\underbrace{(1-\rho)\alpha}_{+}\underbrace{\left(\alpha L^{-\rho} + \beta\right)^{\frac{1-\rho}{\rho}-1} L^{-\rho-1}}_{+} < 0 \end{aligned}$$

where $1-\rho > 0$ since $\rho < 1$. Thus $f''(L) = MP_L'(L) < 0$ and so $f(L)$ is concave and there is a diminishing marginal product of labour.

From profits we have

$$\begin{aligned} \pi(L) &= P\frac{1}{\beta}(\alpha + \beta L^{\rho})^{\frac{1}{\rho}} - WL \\ &\implies \pi'(L) = P\left(\alpha L^{-\rho} + \beta\right)^{\frac{1-\rho}{\rho}} - W \end{aligned}$$

so that the first-order condition $\pi'(L^*) = 0$ is

$$P\left(\alpha (L^*)^{-\rho} + \beta\right)^{\frac{1-\rho}{\rho}} - W = 0$$

which yields

$$\begin{aligned}\left(\alpha (L^*)^{-\rho} + \beta\right)^{\frac{1-\rho}{\rho}} &= w \\ &\implies \alpha (L^*)^{-\rho} + \beta = w^{\frac{\rho}{1-\rho}} \implies (L^*)^{-\rho} = \frac{w^{\frac{\rho}{1-\rho}} - \beta}{\alpha} \\ &\implies L^* = \left(\frac{w^{\frac{\rho}{1-\rho}} - \beta}{\alpha}\right)^{-\frac{1}{\rho}}.\end{aligned}$$

This corresponds to a global profit maximum since for all L we have

$$\pi''(L) = -P(1-\rho)\alpha\left(\alpha L^{-\rho} + \beta\right)^{\frac{1-\rho}{\rho}-1} L^{-\rho-1} < 0.$$

Problem 2.55 *[D***] Suppose a firm has a production function given by*

$$Q = f(L) = 2L^{\frac{1}{2}} + \frac{4}{3}L^{\frac{3}{4}}.$$

*Show that the marginal product of labour is positive and diminishing. Show that the production function is concave. [B***] Find the demand for labour L^* and the supply curve Q^*. Verify that L^* maximizes profits.*

Answer: We have

$$MP_L(L) = f'(L) = L^{-\frac{1}{2}} + L^{-\frac{1}{4}} > 0 \text{ and } MP_L'(L) = f''(L) = -\frac{1}{2}L^{-\frac{3}{2}} - \frac{1}{4}L^{-\frac{5}{4}} < 0.$$

For profits $\pi(L)$ we have

$$\begin{aligned}\pi(L) &= P\left(2L^{\frac{1}{2}} + \frac{4}{3}L^{\frac{3}{4}}\right) - WL \\ &\implies \pi'(L) = P\left(L^{-\frac{1}{2}} + L^{-\frac{1}{4}}\right) - W\end{aligned}$$

so that the first-order condition $\pi'(L^*) = 0$ yields

$$\begin{aligned}P\left((L^*)^{-\frac{1}{2}} + (L^*)^{-\frac{1}{4}}\right) - W &= 0 \\ &\implies (L^*)^{-\frac{1}{2}} + (L^*)^{-\frac{1}{4}} = \frac{W}{P} = w \\ &\implies \left((L^*)^{-\frac{1}{4}}\right)^2 + (L^*)^{-\frac{1}{4}} - w = 0.\end{aligned}$$

Using $x = (L^*)^{-\frac{1}{4}}$ this becomes $x^2 + x - w = 0$ which is a quadratic. We can solve for L^* as

$$\begin{aligned} x &= \frac{-1 \pm \sqrt{1^2 - 4 \times 1 \times (-w)}}{2} \\ &\implies x = (L^*)^{-\frac{1}{4}} = \frac{\sqrt{1+4w} - 1}{2} \\ &\implies L^* = \left(\frac{\sqrt{1+4w}-1}{2}\right)^{-4} \end{aligned}$$

where we ignore the negative root. The labour demand curve is shown below.

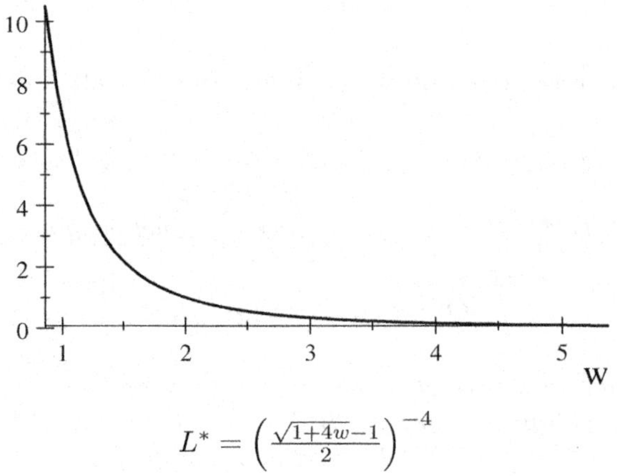

We now find the supply curve as

$$\begin{aligned} Q^* &= f(L^*) = 2(L^*)^{\frac{1}{2}} + \frac{4}{3}(L^*)^{\frac{3}{4}} \\ &= 2\left(\left(\frac{\sqrt{1+4w}-1}{2}\right)^{-4}\right)^{\frac{1}{2}} + \frac{4}{3}\left(\left(\frac{\sqrt{1+4w}-1}{2}\right)^{-4}\right)^{\frac{3}{4}} \\ &= 2\left(\frac{\sqrt{1+4w}-1}{2}\right)^{-2} + \frac{4}{3}\left(\frac{\sqrt{1+4w}-1}{2}\right)^{-3} \\ &= \frac{8}{(\sqrt{1+4w}-1)^2} + \frac{32}{3}\frac{1}{(\sqrt{1+4w}-1)^3} \\ &= \frac{8}{\left(\sqrt{1+4p^{-1}}-1\right)^2} + \frac{32}{3}\frac{1}{\left(\sqrt{1+4p^{-1}}-1\right)^3} \end{aligned}$$

using $w = p^{-1}$. The supply curve is shown below.

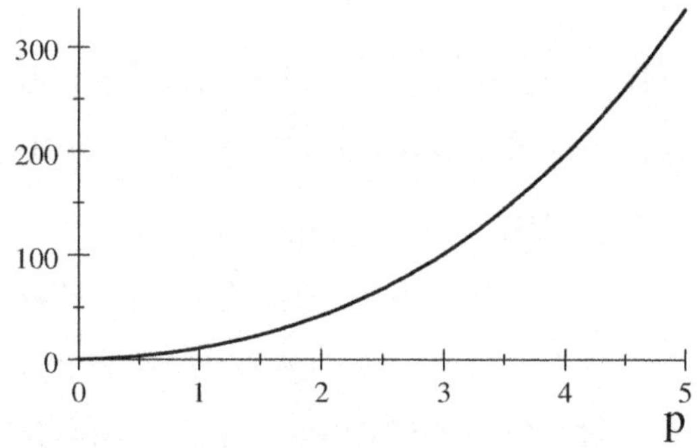

Supply Curve

The second-order condition for a profit maximum is satisfied since

$$\pi''(L) = P\left(-\frac{1}{2}L^{-\frac{3}{2}} - \frac{1}{4}L^{-\frac{5}{4}}\right) < 0.$$

2.8.2 The Cost Function

Problem 2.56 *[D***] Suppose a firm faces a cost function $C(Q) = 5Q^4 + 120$. Find the marginal and average cost functions. Show that the cost function is globally convex. Show that the average cost function $AC(Q)$ is also globally convex. [B**] Find the Q that minimizes $AC(Q)$. If $P = 540$ what will the profit maximizing Q^* be?*

Answer: Marginal $MC(Q)$ and average cost $AC(Q)$ are

$$MC(Q) \equiv C'(Q) = 20Q^3,\ AC(Q) \equiv \frac{C(Q)}{Q} = \frac{5Q^4 + 120}{Q} = 5Q^3 + \frac{120}{Q}$$

and for all Q

$$C''(Q) = MC'(Q) = 60Q^2 > 0$$

so that $C(Q)$ is globally convex.

Now

$$AC'(Q) = 15Q^2 - \frac{120}{Q^2} \Longrightarrow AC''(Q) = 30Q + \frac{240}{Q^3} > 0 \text{ for all } Q$$

so that $AC(Q)$ is globally convex.

To find the minimum of $AC(Q)$ use the first-order condition $AC'(Q^*) = 0$ as

$$\begin{aligned} 15(Q^*)^2 - \frac{120}{(Q^*)^2} &= 0 \\ &\Longrightarrow (Q^*)^4 = \frac{120}{15} = 8 \\ &\Longrightarrow Q^* = 8^{\frac{1}{4}} \approx 1.68. \end{aligned}$$

Now if $P = 540$

$$\pi(Q) = PQ - C(Q) = 540Q - \left(5Q^4 + 120\right).$$

To maximize profits use the first-order condition $\pi'(Q^*) = 0$ as

$$\begin{aligned} 540 - 20(Q^*)^3 &= 0 \\ &\Longrightarrow (Q^*)^3 = \frac{540}{20} = 27 \\ &\Longrightarrow Q^* = 27^{\frac{1}{3}} = 3. \end{aligned}$$

(Alternatively set $P = MC(Q^*)$.) Now use $\pi''(Q) = -60Q^2 < 0$ so that $\pi(Q)$ is globally concave and $Q^* = 3$ is a global maximizer.

Problem 2.57 *[C***] Find the cost function when the production function is* $Q = L^{\frac{2}{3}}$. *Show that the cost function is convex. [B***] Find profits as a function of* Q *and the firm's supply curve. What is the elasticity of supply?*

Answer: By solving for L from $Q = L^{\frac{2}{3}}$ we have

$$Q = L^{\frac{2}{3}} \Longrightarrow L = Q^{\frac{3}{2}}.$$

The cost function is then found by replacing L with $Q^{\frac{3}{2}}$ to obtain costs as a function of Q as

$$C(Q) = WL = WQ^{\frac{3}{2}}.$$

Differentiation of $C(Q)$ then yields

$$\begin{aligned} MC(Q) &= C'(Q) = \frac{3}{2}WQ^{\frac{1}{2}} \\ &\Longrightarrow MC'(Q) = C''(Q) = \frac{3}{4}\overbrace{WQ^{-\frac{1}{2}}}^{+} > 0 \end{aligned}$$

and so $C(Q)$ is globally convex.

Profits are found by replacing $C \equiv WL$ with $C(Q)$ as

$$\pi(Q) = PQ - WL = PQ - C(Q) = PQ - WQ^{\frac{3}{2}}$$

so that letting $p = \frac{P}{W}$ be the real price of Q we have from the first-order condition $\pi'(Q^*) = 0$ that

$$\begin{aligned} P - \frac{3}{2}W(Q^*)^{\frac{1}{2}} &= 0 \\ &\Longrightarrow (Q^*)^{\frac{1}{2}} = \frac{2}{3}\frac{P}{W} = \frac{2}{3}p \\ &\Longrightarrow Q^* = \left(\frac{2}{3}p\right)^2 = \frac{4}{9}p^2. \end{aligned}$$

Since Q^* has the form $y = Ax^b$, the elasticity of supply is the exponent on p or 2.

Problem 2.58 *[B***] Suppose the firm's production function is given by*

$$Q = (L^{\frac{1}{2}} + 4)^2.$$

Find the firm's cost function $C(Q)$, marginal cost and show that $C''(Q) > 0$. Find the firm's supply curve.

Answer: Solving $Q = (L^{\frac{1}{2}} + 4)^2$ for L we find that

$$\begin{aligned} Q &= (L^{\frac{1}{2}} + 4)^2 \\ &\Longrightarrow L = \left(Q^{\frac{1}{2}} - 4\right)^2 \\ &\Longrightarrow L = Q - 8Q^{\frac{1}{2}} + 16. \end{aligned}$$

Replacing L in $C \equiv WL$ with $Q - 8Q^{\frac{1}{2}} + 16$ we obtain

$$\begin{aligned} C(Q) &= W\left(Q - 8Q^{\frac{1}{2}} + 16\right) \\ &\Longrightarrow MC(Q) = C'(Q) = W\left(1 - 4Q^{-\frac{1}{2}}\right) \\ &\Longrightarrow MC'(Q) = C''(Q) = 2WQ^{-\frac{3}{2}} > 0. \end{aligned}$$

Profits are found by replacing $C \equiv WL$ with $C(Q)$ as

$$\pi(Q) = PQ - WL = PQ - C(Q) = PQ - W\left(Q - 8Q^{\frac{1}{2}} + 16\right)$$

so that letting $p = \frac{P}{W}$ be the real price of Q we have from the first-order condition $\pi'(Q^*) = 0$ that

$$\begin{aligned} P - W\left(1 - 4(Q^*)^{-\frac{1}{2}}\right) &= 0 \\ &\Longrightarrow 1 - 4(Q^*)^{-\frac{1}{2}} = \frac{P}{W} = p \Longrightarrow Q^* = 16(1-p)^{-2}. \end{aligned}$$

Problem 2.59 *[A***] Show that a firm with a globally increasing concave production function $Q = f(L)$ will have a globally increasing convex cost function $C(Q)$. Show that $MC(Q) = \frac{W}{MP_L}$. Show that a competitive firm maximizes profits by setting price P equal to marginal cost MC. Show that this is equivalent to setting the real wage w equal to the marginal product of labour MP_L.*

Answer: If $Q = f(L)$ with $f'(L) > 0, f''(L) < 0$ then the inverse function $L = g(Q)$ exists with $f(g(Q)) = Q$. From the chain rule we have

$$\begin{aligned} f(g(Q)) &= Q \Longrightarrow f'(g(Q))\, g'(Q) = 1 \\ &\Longrightarrow g'(Q) = \frac{1}{f'(g(Q))} = \frac{1}{MP_L(g(Q))} > 0. \end{aligned}$$

One more application of the chain rule yields

$$\begin{aligned} f''(g(Q))\,g'(Q)^2 + f'(g(Q))\,g''(Q) &= 0 \\ &\Longrightarrow g''(Q) = -\frac{f''(g(Q))\,g'(Q)^2}{f'(g(Q))} > 0 \end{aligned}$$

since $f'(g(Q)) > 0, f''(g(Q)) < 0$. Now if costs are $C \equiv WL$ then the cost function is $C(Q) = Wg(Q)$ and so

$$MC(Q) \equiv C'(Q) = Wg'(Q) > 0 \text{ and } MC'(Q) \equiv C''(Q) = Wg''(Q) > 0.$$

Since $L = g(Q)$ and $g'(Q) = \frac{1}{MP_L(g(Q))} = \frac{1}{MP_L(L)}$ we have

$$MC(Q) \equiv Wg'(Q) = \frac{W}{MP_L(g(Q))} = \frac{W}{MP_L(L)}.$$

Profit are $\pi \equiv PQ - WL$. Replace WL with costs $C(Q)$ we have

$$\begin{aligned} \pi(Q) &= PQ - C(Q) \\ &\Longrightarrow \pi'(Q) = P - C'(Q) = P - MC(Q) \end{aligned}$$

so that the first-order condition $\pi'(Q^*) = 0$ yields

$$P - MC(Q^*) = 0 \Longrightarrow P = MC(Q^*).$$

Since $MC(Q^*) = \frac{W}{MP_L(L^*)}$ we have

$$P = MC(Q^*) = \frac{W}{MP_L(L^*)} \Longrightarrow MP_L(L^*) = \frac{W}{P} = w.$$

This is a global maximizer since $C''(Q) > 0$ and so for all Q

$$\begin{aligned} \pi'(Q) &= P - C'(Q) \\ &\Longrightarrow \pi''(Q) = -C''(Q) < 0. \end{aligned}$$

Problem 2.60 *[B**] Suppose that a firm has a Cobb-Douglas production function $Q = L^{\frac{1}{2}}K^{\frac{1}{2}}$ with costs $C \equiv WL + RK$. Find the cost minimizing L, K for producing a given Q units of output. Find the firm's cost function $C^* \equiv WL^* + RK^*$.*

Answer: Once L is chosen so too is K since

$$\begin{aligned} Q &= L^{\frac{1}{2}}K^{\frac{1}{2}} \\ &\Longrightarrow K^{\frac{1}{2}} = QL^{-\frac{1}{2}} \\ &\Longrightarrow K = Q^2L^{-1} \end{aligned}$$

where $K = Q^2L^{-1}$ is the isoquant corresponding to output Q. Therefore we can write cost solely as a function of L as

$$C(L) = WL + RK = WL + RQ^2L^{-1}$$

where Q is given. The first-order condition $C'(L^*) = 0$ yields

$$\begin{aligned} W - RQ^2 (L^*)^{-2} &= 0 \\ &\Longrightarrow (L^*)^{-2} = Q^{-2} W R^{-1} \\ &\Longrightarrow L^* = Q W^{-\frac{1}{2}} R^{\frac{1}{2}}. \end{aligned}$$

This gives the cost minimizing amount of labour L^*. To find the optimal K^* calculate the required K^* need to produce Q as

$$K^* = Q^2 (L^*)^{-1} = Q^2 \left(Q W^{-\frac{1}{2}} R^{\frac{1}{2}} \right)^{-1} = Q W^{\frac{1}{2}} R^{-\frac{1}{2}}.$$

The firm's cost function is then

$$\begin{aligned} C^* &= WL^* + RK^* = WQW^{-\frac{1}{2}} R^{\frac{1}{2}} + RQW^{\frac{1}{2}} R^{-\frac{1}{2}} = QW^{\frac{1}{2}} R^{\frac{1}{2}} + QW^{\frac{1}{2}} R^{\frac{1}{2}} \\ &= 2QW^{\frac{1}{2}} R^{\frac{1}{2}}. \end{aligned}$$

Problem 2.61 *[A ***] Given a Cobb-Douglas production function $Q = L^\alpha K^\beta$, find the cost minimizing L, K. Find the firm's cost function $C^*(Q, W, R) \equiv WL^* + RK^*$. Using $W = R = 1$, show that there are decreasing returns to scale if $\alpha + \beta < 1$, constant returns to scale if $\alpha + \beta = 1$, and increasing returns to scale if $\alpha + \beta > 1$.*

Answer: We have

$$\begin{aligned} Q &= L^\alpha K^\beta \\ &\Longrightarrow K^\beta = Q L^{-\alpha} \\ &\Longrightarrow K = Q^{\frac{1}{\beta}} L^{-\frac{\alpha}{\beta}} \end{aligned}$$

and so costs can be written as a function of L as

$$C(L) = WL + RK = WL + RQ^{\frac{1}{\beta}} L^{-\frac{\alpha}{\beta}}.$$

The first-order condition $C'(L^*) = 0$ then yields

$$\begin{aligned} W - \frac{\alpha}{\beta} R Q^{\frac{1}{\beta}} (L^*)^{-\frac{\alpha}{\beta} - 1} &= 0 \\ &\Longrightarrow (L^*)^{-\frac{\alpha}{\beta} - 1} = \frac{\beta}{\alpha} Q^{-\frac{1}{\beta}} \frac{W}{R} \\ &\Longrightarrow (L^*)^{-\frac{(\alpha+\beta)}{\beta}} = \frac{\beta}{\alpha} Q^{-\frac{1}{\beta}} \frac{W}{R} \\ &\Longrightarrow L^* = \left(\frac{\beta}{\alpha} \right)^{-\frac{\beta}{\alpha+\beta}} \left(\frac{W}{R} \right)^{-\frac{\beta}{\alpha+\beta}} Q^{\frac{1}{\alpha+\beta}}. \end{aligned}$$

The optimal K^* is then found as

$$\begin{aligned} K^* &= Q^{\frac{1}{\beta}}(L^*)^{-\frac{\alpha}{\beta}} = Q^{\frac{1}{\beta}}\left(\left(\frac{\beta}{\alpha}\right)^{-\frac{\beta}{\alpha+\beta}}\left(\frac{W}{R}\right)^{-\frac{\beta}{\alpha+\beta}}Q^{\frac{1}{\alpha+\beta}}\right)^{-\frac{\alpha}{\beta}} \\ &= \left(\frac{\beta}{\alpha}\right)^{\frac{\alpha}{\alpha+\beta}}\left(\frac{W}{R}\right)^{\frac{\alpha}{\alpha+\beta}}Q^{\frac{1}{\beta}-\frac{\alpha}{\beta}\frac{1}{\alpha+\beta}} = \left(\frac{\beta}{\alpha}\right)^{\frac{\alpha}{\alpha+\beta}}\left(\frac{W}{R}\right)^{\frac{\alpha}{\alpha+\beta}}Q^{\frac{1}{\beta}\left(1-\frac{\alpha}{\alpha+\beta}\right)} \\ &= \left(\frac{\beta}{\alpha}\right)^{\frac{\alpha}{\alpha+\beta}}\left(\frac{W}{R}\right)^{\frac{\alpha}{\alpha+\beta}}Q^{\frac{1}{\beta}\left(\frac{\beta}{\alpha+\beta}\right)} = \left(\frac{\beta}{\alpha}\right)^{\frac{\alpha}{\alpha+\beta}}\left(\frac{W}{R}\right)^{\frac{\alpha}{\alpha+\beta}}Q^{\frac{1}{\alpha+\beta}}. \end{aligned}$$

The cost function is then

$$\begin{aligned} C^*(Q,W,R) &\equiv WL^* + RK^* \\ &= W\left(\frac{\beta}{\alpha}\right)^{-\frac{\beta}{\alpha+\beta}}\left(\frac{W}{R}\right)^{-\frac{\beta}{\alpha+\beta}}Q^{\frac{1}{\alpha+\beta}} + R\left(\frac{\beta}{\alpha}\right)^{\frac{\alpha}{\alpha+\beta}}\left(\frac{W}{R}\right)^{\frac{\alpha}{\alpha+\beta}}Q^{\frac{1}{\alpha+\beta}} \\ &= Q^{\frac{1}{\alpha+\beta}}\left(\left(\frac{\beta}{\alpha}\right)^{-\frac{\beta}{\alpha+\beta}}W\left(\frac{W}{R}\right)^{-\frac{\beta}{\alpha+\beta}} + \left(\frac{\beta}{\alpha}\right)^{\frac{\alpha}{\alpha+\beta}}R\left(\frac{W}{R}\right)^{\frac{\alpha}{\alpha+\beta}}\right) \\ &= Q^{\frac{1}{\alpha+\beta}}\left(\left(\frac{\beta}{\alpha}\right)^{-\frac{\beta}{\alpha+\beta}}W^{1-\frac{\beta}{\alpha+\beta}}R^{\frac{\beta}{\alpha+\beta}} + \left(\frac{\beta}{\alpha}\right)^{\frac{\alpha}{\alpha+\beta}}R^{1-\frac{\alpha}{\alpha+\beta}}W^{\frac{\alpha}{\alpha+\beta}}\right) \\ &= Q^{\frac{1}{\alpha+\beta}}W^{\frac{\alpha}{\alpha+\beta}}R^{\frac{\beta}{\alpha+\beta}}\left(\left(\frac{\beta}{\alpha}\right)^{-\frac{\beta}{\alpha+\beta}} + \left(\frac{\beta}{\alpha}\right)^{\frac{\alpha}{\alpha+\beta}}\right) \\ &= AQ^{\frac{1}{\alpha+\beta}}W^{\frac{\alpha}{\alpha+\beta}}R^{\frac{\beta}{\alpha+\beta}} \end{aligned}$$

where

$$A = \left(\frac{\beta}{\alpha}\right)^{-\frac{\beta}{\alpha+\beta}} + \left(\frac{\beta}{\alpha}\right)^{\frac{\alpha}{\alpha+\beta}}.$$

If $W = R = 1$ then

$$C^*(Q) = AQ^{\frac{1}{\alpha+\beta}}.$$

For average costs $AC^*(Q)$ then

$$\begin{aligned} AC^*(Q) &\equiv \frac{C^*(Q)}{Q} = AQ^{\frac{1}{\alpha+\beta}-1} = AQ^{\frac{1-(\alpha+\beta)}{\alpha+\beta}} \\ &\Longrightarrow \frac{dAC^*(Q)}{dQ} = \left(\frac{1}{\alpha+\beta}-1\right)AQ^{\frac{1-(\alpha+\beta)}{\alpha+\beta}-1}. \end{aligned}$$

So there are decreasing returns to scale $\frac{dAC^*(Q)}{dQ} > 0$ if $\alpha+\beta < 1$ since

$$\alpha+\beta < 1 \Longrightarrow \frac{1}{\alpha+\beta} - 1 > 0$$

constant returns to scale $\frac{dAC^*(Q)}{dQ} = 0$ if $\alpha+\beta = 1$, and increasing returns to scale $\frac{dAC^*(Q)}{dQ} < 0$ if $\alpha+\beta > 1$.

2.8.3 Utility Maximization

Problem 2.62 *[B**] Suppose that* $U = 4Q_1^{\frac{1}{2}} + 2Q_2^{\frac{1}{2}}$ *with budget constraint* $Y = P_1Q_1 + P_2Q_2$ *where* $Y = 10, P_2 = 1$. *Show that both goods have positive and diminishing marginal utility. Find the first-order condition and solve for* Q_1^*, Q_2^*. *Show that your solution maximizes utility. Is the elasticity of demand the same for all* P_1*?*

Answer: We have positive and diminishing marginal utility as

$$\begin{aligned} MU_1(Q_1) &= U_1'(Q_1) = 2Q_1^{-\frac{1}{2}} > 0,\ MU_1'(Q_1) = U_1''(Q_1) = -Q_1^{-\frac{3}{2}} < 0 \\ MU_2(Q_2) &= U_2'(Q_2) = Q_2^{-\frac{1}{2}} > 0,\ MU_2'(Q_2) = U_2''(Q_2) = -\frac{1}{2}Q_2^{-\frac{3}{2}} < 0. \end{aligned}$$

We have

$$Y = 10, P_2 = 1, Y = P_1Q_1 + P_2Q_2 \Longrightarrow Q_2 = 10 - P_1Q_1$$

and so we can write utility as a function of Q_1 as

$$U(Q_1) = 4Q_1^{\frac{1}{2}} + 2(10 - P_1Q_1)^{\frac{1}{2}} \Longrightarrow U'(Q_1) = 2(Q_1)^{-\frac{1}{2}} - P_1(10 - P_1Q_1)^{-\frac{1}{2}}.$$

The first-order condition $U'(Q_1^*) = 0$ is

$$2(Q_1^*)^{-\frac{1}{2}} - P_1(10 - P_1Q_1^*)^{-\frac{1}{2}} = 0$$

and so yields

$$\begin{aligned} 2(Q_1^*)^{-\frac{1}{2}} &= P_1(10 - P_1Q_1^*)^{-\frac{1}{2}} \\ &\Longrightarrow \left(2(Q_1^*)^{-\frac{1}{2}}\right)^{-2} = \left(P_1(10 - P_1Q_1^*)^{-\frac{1}{2}}\right)^{-2} \\ &\Longrightarrow \frac{1}{4}Q_1^* = P_1^{-2}(10 - P_1Q_1^*) \\ &\Longrightarrow Q_1^* = \frac{40P_1^{-2}}{1 + 4P_1^{-1}} = \frac{40}{P_1(4 + P_1)}. \end{aligned}$$

Note that this doesn't have the form $y = Ax^b$ so that the elasticity of demand is not constant for all P_1.

The solution is a global maximizer since $U(Q_1)$ is globally concave as

$$U''(Q_1) = -(Q_1)^{-\frac{3}{2}} - \frac{P_1^2}{2}(10 - P_1Q_1)^{-\frac{3}{2}} < 0.$$

To find Q_2^* use

$$Q_2^* = 10 - P_1Q_1^* = 10 - P_1\frac{40P_1^{-2}}{1 + 4P_1^{-1}} = 10 - \frac{40P_1^{-1}}{1 + 4P_1^{-1}} = \frac{10}{1 + 4P_1^{-1}}.$$

Problem 2.63 *[B**] Suppose that a household has a utility function*

$$U = 2Q_1^{\frac{1}{2}} + 4Q_2^{\frac{1}{2}}$$

with budget constraint $Y = P_1Q_1 + P_2Q_2$. Find the utility maximizing Q_1^*, Q_2^*. Show that the demand curve slopes downwards and find the elasticity of demand when $Y = P_2 = 1$. Show that demand is elastic.

Answer: First from the budget constraint

$$\begin{aligned} Y &= P_1Q_1 + P_2Q_2 \\ &\implies Q_2 = \frac{Y - P_1Q_1}{P_2} = \frac{Y}{P_2} - \frac{P_1}{P_2}Q_1 \end{aligned}$$

and so we write U as a function of Q_1 as

$$U(Q_1) = 2Q_1^{\frac{1}{2}} + 4\left(\frac{Y}{P_2} - \frac{P_1}{P_2}Q_1\right)^{\frac{1}{2}}.$$

Thus

$$U'(Q_1) = Q_1^{-\frac{1}{2}} - 2\frac{P_1}{P_2}\left(\frac{Y}{P_2} - \frac{P_1}{P_2}Q_1\right)^{-\frac{1}{2}}$$

so that the first-order condition for utility maximization $U'(Q_1^*) = 0$ is

$$(Q_1^*)^{-\frac{1}{2}} - 2\frac{P_1}{P_2}\left(\frac{Y}{P_2} - \frac{P_1}{P_2}Q_1^*\right)^{-\frac{1}{2}} = 0$$

which yields

$$\begin{aligned} (Q_1^*)^{-\frac{1}{2}} &= 2\frac{P_1}{P_2}\left(\frac{Y}{P_2} - \frac{P_1}{P_2}Q_1^*\right)^{-\frac{1}{2}} \\ &\implies Q_1^* = \frac{1}{4}\left(\frac{P_1}{P_2}\right)^{-2}\left(\frac{Y}{P_2} - \frac{P_1}{P_2}Q_1^*\right) \\ &\implies Q_1^* = \frac{1}{4}\left(\frac{P_1}{P_2}\right)^{-2}\frac{Y}{P_2} - \frac{1}{4}\left(\frac{P_1}{P_2}\right)^{-1}Q_1^* \\ &\implies Q_1^* = \frac{\frac{1}{4}\left(\frac{P_1}{P_2}\right)^{-2}\frac{Y}{P_2}}{1 + \frac{1}{4}\left(\frac{P_1}{P_2}\right)^{-1}} = \frac{YP_1^{-2}}{4P_2^{-1}\left(1 + \frac{1}{4}\frac{P_2}{P_1}\right)} = \frac{YP_1^{-2}}{P_1^{-1} + 4P_2^{-1}}. \end{aligned}$$

Here Q_1^* is a global utility maximizer since for all Q_1

$$U''(Q_1) = -\frac{1}{2}Q_1^{-\frac{3}{2}} - \left(\frac{P_1}{P_2}\right)^2\left(\frac{Y}{P_2} - \frac{P_1}{P_2}Q_1\right)^{-\frac{3}{2}} < 0$$

so that $U(Q_1)$ is globally concave.

To find Q_2^* use

$$\begin{aligned} Q_2^* &= \frac{Y}{P_2} - \frac{P_1}{P_2} Q_1^* = \frac{Y}{P_2} - \frac{P_1}{P_2} \frac{YP_1^{-2}}{P_1^{-1} + 4P_2^{-1}} = \frac{Y}{P_2} - \frac{Y}{P_2} \frac{P_1^{-1}}{P_1^{-1} + 4P_2^{-1}} \\ &= \frac{Y}{P_2}\left(1 - \frac{P_1^{-1}}{P_1^{-1} + 4P_2^{-1}}\right) = \frac{Y}{P_2}\left(\frac{4P_2^{-1}}{P_1^{-1} + 4P_2^{-1}}\right) = \frac{4YP_2^{-2}}{P_1^{-1} + 4P_2^{-1}}. \end{aligned}$$

When $Y = P_2 = 1$ we have

$$Q_1^* = \frac{YP_1^{-2}}{P_1^{-1} + 4P_2^{-1}} = \frac{P_1^{-2}}{P_1^{-1} + 4} = \frac{1}{P_1^2\left(P_1^{-1} + 4\right)} = \frac{1}{P_1 + 4P_1^2}$$

so that the demand curve slopes downwards as

$$\frac{dQ_1^*}{dP_1} = -\frac{1 + 8P_1}{\left(P_1 + 4P_1^2\right)^2} < 0.$$

The elasticity of demand is

$$\eta\left(P_1\right) \equiv \frac{dQ_1^*}{dP_1} \frac{P_1}{Q_1^*} = -\frac{1 + 8P_1}{\left(P_1 + 4P_1^2\right)^2} \frac{P_1}{\frac{1}{P_1 + 4P_1^2}} = -\frac{P_1 + 8P_1^2}{P_1 + 4P_1^2}.$$

For all P_1 demand is elastic since

$$\begin{aligned} \eta\left(P_1\right) &= -\frac{P_1 + 8P_1^2}{P_1 + 4P_1^2} = -\frac{P_1 + 4P_1^2 + 4P_1^2}{P_1 + 4P_1^2} \\ &= -\frac{P_1 + 4P_1^2}{P_1 + 4P_1^2} - \frac{4P_1^2}{P_1 + 4P_1^2} \\ &= -1 - \frac{4P_1^2}{P_1 + 4P_1^2} < -1. \end{aligned}$$

Problem 2.64 *[B**] A household has a utility function given by*

$$U = -4Q_1^{-1} - Q_2^{-1}.$$

Here $U_1\left(Q_1\right) = -4Q_1^{-1}$ *is utility from* Q_1*, show that there is positive and diminishing marginal utility. Assume* $P_2 = Y = 1$ *and find the utility maximizing* Q_1^*, Q_2^**. Show that the demand curve for* Q_1 *slopes downwards.*

Answer: If $U_1\left(Q_1\right) = -4Q_1^{-1}$ then there is positive and diminishing marginal utility since

$$MU_1\left(Q_1\right) = U_1'\left(Q_1\right) = 4Q_1^{-2} > 0 \text{ and } MU_1'\left(Q_1\right) = -8Q_1^{-3} < 0.$$

From the budget constraint with $P_2 = Y = 1$ we have the budget line

$$\begin{aligned} Y &= P_1Q_1 + P_2Q_2 \\ &\implies 1 = P_1Q_1 + Q_2 \\ &\implies Q_2 = 1 - P_1Q_1. \end{aligned}$$

Substituting $Q_2 = 1 - P_1 Q_1$ for Q_2 in the utility function yields

$$U(Q_1) = -4Q_1^{-1} - (1 - P_1 Q_1)^{-1} \Longrightarrow U'(Q_1) = 4Q_1^{-2} - P_1 (1 - P_1 Q_1)^{-2}.$$

The first-order condition $U'(Q_1^*) = 0$ is

$$4(Q_1^*)^{-2} - P_1 (1 - P_1 Q_1^*)^{-2} = 0.$$

Solving for Q_1^* we have

$$\begin{aligned}
4(Q_1^*)^{-2} &= P_1 (1 - P_1 Q_1^*)^{-2} \\
&\Longrightarrow \left(4(Q_1^*)^{-2}\right)^{-\frac{1}{2}} = \left(P_1 (1 - P_1 Q_1^*)^{-2}\right)^{-\frac{1}{2}} \\
&\Longrightarrow \frac{1}{2} Q_1^* = P_1^{-\frac{1}{2}} (1 - P_1 Q_1^*) \Longrightarrow Q_1^* = 2P_1^{-\frac{1}{2}} (1 - P_1 Q_1^*) \\
&\Longrightarrow Q_1^* = 2P_1^{-\frac{1}{2}} - 2P_1^{\frac{1}{2}} Q_1^* \Longrightarrow \left(1 + 2P_1^{\frac{1}{2}}\right) Q_1^* = 2P_1^{-\frac{1}{2}} \\
&\Longrightarrow Q_1^* = \frac{2P_1^{-\frac{1}{2}}}{1 + 2P_1^{\frac{1}{2}}} = \frac{2}{P_1^{\frac{1}{2}} + 2P_1}.
\end{aligned}$$

The demand curve for fudge is downward sloping since

$$\frac{dQ_1^*}{dP_1} = -\frac{2}{\left(P_1^{\frac{1}{2}} + 2P_1\right)^2} \left(\frac{1}{2} P_1^{-\frac{1}{2}} + 2\right) < 0.$$

Problem 2.65 *[A**] Consider a household with utility*

$$U = \frac{Q_1^\beta - 1}{\beta} + \gamma \frac{Q_2^\beta - 1}{\beta}$$

where $\beta < 1$ and $\gamma > 0$. Find the demand curves Q_1^*, Q_2^*. Let $P_2 = Y = 1$ and show that the budget share of Q_1 is a decreasing function of P_1 if if $0 < \beta < 1$, and an increasing function if $\beta < 0$.

Answer: From the budget constraint $Y = P_1 Q_1 + P_2 Q_2$ we have $Q_2 = \frac{Y - P_1 Q_1}{P_2}$. Substituting $\frac{Y - P_1 Q_1}{P_2}$ for Q_2 in the utility function yields utility $U(Q_1)$ as a function of Q_1 as

$$U(Q_1) = \frac{Q_1^\beta - 1}{\beta} + \gamma \frac{\left(\frac{Y}{P_2} - \frac{P_1}{P_2} Q_1\right)^\beta - 1}{\beta}.$$

We need $U'(Q_1)$ which is

$$U'(Q_1) = Q_1^{\beta - 1} - \gamma \frac{P_1}{P_2} \left(\frac{Y}{P_2} - \frac{P_1}{P_2} Q_1\right)^{\beta - 1}$$

and so the first-order condition $U'(Q_1^*) = 0$ is

$$(Q_1^*)^{\beta-1} - \gamma \frac{P_1}{P_2}\left(\frac{Y}{P_2} - \frac{P_1}{P_2}Q_1^*\right)^{\beta-1} = 0.$$

Solving we have

$$\begin{aligned}
(Q_1^*)^{\beta-1} &= \gamma \frac{P_1}{P_2}\left(\frac{Y}{P_2} - \frac{P_1}{P_2}Q_1^*\right)^{\beta-1} \\
&\implies Q_1^* = \gamma^{\frac{1}{\beta-1}}\left(\frac{P_1}{P_2}\right)^{\frac{1}{\beta-1}}\left(\frac{Y}{P_2} - \frac{P_1}{P_2}Q_1^*\right) \\
&\implies Q_1^* = \frac{\gamma^{\frac{1}{\beta-1}}\left(\frac{P_1}{P_2}\right)^{\frac{1}{\beta-1}}\frac{Y}{P_2}}{1+\gamma^{\frac{1}{\beta-1}}\left(\frac{P_1}{P_2}\right)^{\frac{\beta}{\beta-1}}} = \frac{\gamma^{\frac{1}{\beta-1}}P_1^{\frac{1}{\beta-1}}P_2^{-\frac{1}{\beta-1}-1}Y}{1+\gamma^{\frac{1}{\beta-1}}P_1^{\frac{\beta}{\beta-1}}P_2^{-\frac{\beta}{\beta-1}}} = \frac{\gamma^{\frac{1}{\beta-1}}P_2^{-\frac{\beta}{\beta-1}}YP_1^{\frac{1}{\beta-1}}}{\gamma^{\frac{1}{\beta-1}}P_1^{\frac{\beta}{\beta-1}}P_2^{-\frac{\beta}{\beta-1}}+1} \\
&\implies Q_1^* = \frac{YP_1^{\frac{1}{\beta-1}}}{\gamma^{-\frac{1}{\beta-1}}P_2^{\frac{\beta}{\beta-1}}\left(\gamma^{\frac{1}{\beta-1}}P_1^{\frac{\beta}{\beta-1}}P_2^{-\frac{\beta}{\beta-1}}+1\right)} = \frac{YP_1^{\frac{1}{\beta-1}}}{P_1^{\frac{\beta}{\beta-1}}+\gamma^{\frac{1}{1-\beta}}P_2^{\frac{\beta}{\beta-1}}}.
\end{aligned}$$

This is a maximizer since

$$U''(Q_1) = (\beta-1)Q_1^{\beta-2} + (\beta-1)\gamma\left(\frac{P_1}{P_2}\right)^2\left(\frac{Y}{P_2} - \frac{P_1}{P_2}Q_1\right)^{\beta-2} < 0$$

since we assume that $\beta < 1$ and so $\beta - 1 < 0$.

To find Q_2^* replace Q_1^* in the budget constraint as

$$\begin{aligned}
Q_2^* &= \frac{Y}{P_2} - \frac{P_1}{P_2}Q_1^* = \frac{Y}{P_2} - \frac{P_1}{P_2}\left(\frac{YP_1^{\frac{1}{\beta-1}}}{P_1^{\frac{\beta}{\beta-1}}+\gamma^{\frac{1}{1-\beta}}P_2^{\frac{\beta}{\beta-1}}}\right) \\
&= \frac{Y}{P_2} - \frac{Y}{P_2}\left(\frac{P_1^{\frac{\beta}{\beta-1}}}{P_1^{\frac{\beta}{\beta-1}}+\gamma^{\frac{1}{1-\beta}}P_2^{\frac{\beta}{\beta-1}}}\right) = \frac{Y}{P_2}\left(1 - \frac{P_1^{\frac{\beta}{\beta-1}}}{P_1^{\frac{\beta}{\beta-1}}+\gamma^{\frac{1}{1-\beta}}P_2^{\frac{\beta}{\beta-1}}}\right) \\
&= \frac{Y}{P_2}\left(\frac{\gamma^{\frac{1}{1-\beta}}P_2^{\frac{\beta}{\beta-1}}}{P_1^{\frac{\beta}{\beta-1}}+\gamma^{\frac{1}{1-\beta}}P_2^{\frac{\beta}{\beta-1}}}\right) = \frac{Y\gamma^{\frac{1}{1-\beta}}P_2^{\frac{1}{\beta-1}}}{P_1^{\frac{\beta}{\beta-1}}+\gamma^{\frac{1}{1-\beta}}P_2^{\frac{\beta}{\beta-1}}} \\
&= \frac{YP_2^{\frac{1}{\beta-1}}}{\gamma^{-\frac{1}{1-\beta}}\left(P_1^{\frac{\beta}{\beta-1}}+\gamma^{\frac{1}{1-\beta}}P_2^{\frac{\beta}{\beta-1}}\right)} = \frac{YP_2^{\frac{1}{\beta-1}}}{\gamma^{-\frac{1}{1-\beta}}P_1^{\frac{\beta}{\beta-1}}+P_2^{\frac{\beta}{\beta-1}}}.
\end{aligned}$$

Setting $P_2 = Y = 1$ we have

$$Q_1^* = \frac{YP_1^{\frac{1}{\beta-1}}}{P_1^{\frac{\beta}{\beta-1}}+\gamma^{\frac{1}{1-\beta}}P_2^{\frac{\beta}{\beta-1}}} = \frac{P_1^{\frac{1}{\beta-1}}}{P_1^{\frac{\beta}{\beta-1}}+\gamma^{\frac{1}{1-\beta}}}$$

so that the budget share $s_1 \equiv \frac{P_1 Q_1^*}{Y}$ is

$$s_1 \equiv \frac{P_1 Q_1^*}{Y} = \frac{P_1}{1} \frac{P_1^{\frac{1}{\beta-1}}}{P_1^{\frac{\beta}{\beta-1}} + \gamma^{\frac{1}{1-\beta}}} = \frac{P_1^{\frac{\beta}{\beta-1}}}{P_1^{\frac{\beta}{\beta-1}} + \gamma^{\frac{1}{1-\beta}}} = \frac{1}{1 + \gamma^{\frac{1}{1-\beta}} P_1^{\frac{\beta}{1-\beta}}}.$$

Thus

$$\frac{ds_1}{dP_1} = \frac{d}{dP_1}\left(\frac{1}{1 + \gamma^{\frac{1}{1-\beta}} P_1^{\frac{\beta}{1-\beta}}}\right) = -\gamma^{\frac{1}{1-\beta}} \frac{\beta}{1-\beta} \frac{1}{\left(1 + \gamma^{\frac{1}{1-\beta}} P_2^{\frac{\beta}{\beta-1}} P_1^{\frac{\beta}{1-\beta}}\right)^2} P_1^{\frac{\beta}{1-\beta}-1}$$

where $1 - \beta > 0$ since $\beta < 1$. The sign of $\frac{ds_1}{dP_1}$ depends on β as

$$0 < \beta < 1 \Longrightarrow \frac{ds_1}{dP_1} < 0, \ \beta < 0 \Longrightarrow \frac{ds_1}{dP_1} > 0.$$

Problem 2.66 *[A**] Given a utility function*

$$U = U_1(Q_1) + U_2(Q_2)$$

with $U_1''(Q_1) < 0$ and $U_2''(Q_2) < 0$, find the general rule for utility maximization. If $P_1 = P_2 = 1$, then show that Q_1 and Q_2 are normal goods.

Answer: First solve for Q_2 as a function of Q_1 as

$$Q_2 = \frac{Y - P_1 Q_1}{P_2} = \frac{Y}{P_2} - \frac{P_1}{P_2} Q_1$$

and use this to write U as a function of Q_1 only as

$$U(Q_1) = U_1(Q_1) + U_2\left(\frac{Y}{P_2} - \frac{P_1}{P_2} Q_1\right).$$

Differentiating $U(Q_1)$ with respect to Q_1 and using the chain rule for the second term yields

$$U'(Q_1) = U_1'(Q_1) - \frac{P_1}{P_2} U_2'\left(\frac{Y}{P_2} - \frac{P_1}{P_2} Q_1\right)$$

so that the first-order condition $U'(Q_1^*) = 0$ yields

$$\begin{aligned} U_1'(Q_1^*) - \frac{P_1}{P_2} U_2'\left(\frac{Y}{P_2} - \frac{P_1}{P_2} Q_1^*\right) &= 0 \\ &\Longrightarrow U_1'(Q_1^*) = \frac{P_1}{P_2} U_2'\left(\frac{Y}{P_2} - \frac{P_1}{P_2} Q_1^*\right) \\ &\Longrightarrow U_1'(Q_1^*) = \frac{P_1}{P_2} U_2'(Q_2^*) \end{aligned}$$

since $Q_2^* = \frac{Y}{P_2} - \frac{P_1}{P_2}Q_1^*$. We can write this in turn as

$$\frac{U_1'(Q_1^*)}{P_1} = \frac{U_2'(Q_2^*)}{P_2}$$

or the household allocates goods so that the marginal utility of each good divided by its price is the same for both goods.

To show that both goods are normal write $Q_1^*(Y)$; that is the demand for good 1 as a function of income Y. From the first-order condition and using $P_1 = P_2 = 1$ we have

$$U_1'(Q_1^*(Y)) = \frac{P_1}{P_2}U_2'\left(\frac{Y}{P_2} - \frac{P_1}{P_2}Q_1^*(Y)\right) = U_2'(Y - Q_1^*(Y))$$

where $Q_2^*(Y) = Y - Q_1^*(Y)$. Differentiating both sides with respect to Y and using the chain rule yields

$$\begin{aligned} U_1''(Q_1^*(Y))\frac{dQ_1^*(Y)}{dY} &= U_2''(Y - Q_1^*(Y))\left(1 - \frac{dQ_1^*(Y)}{dY}\right) \\ \implies \frac{dQ_1^*(Y)}{dY} &= \frac{U_2''(Y - Q_1^*(Y))}{U_1''(Q_1^*(Y)) + U_2''(Y - Q_1^*(Y))} > 0 \end{aligned}$$

since $U_1'' < 0$ and $U_2'' < 0$ implies that both the numerator and denominator are negative. For $Q_2^*(Y) = Y - Q_1^*(Y)$ we have

$$\frac{dQ_2^*(Y)}{dY} = \frac{d}{dY}(Y - Q_1^*(Y)) = 1 - \frac{dQ_1^*(Y)}{dY} = \frac{U_1''(Y - Q_1^*(Y))}{U_1''(Q_1^*(Y)) + U_2''(Y - Q_1^*(Y))} > 0.$$

2.8.4 Mark Maximization

Problem 2.67 *[B**] Suppose that the marks on Question 1 and 2 on an exam are*

$$M_1(t_1) = 40t_1^{\frac{1}{2}},\ M_2(t_2) = 60t_2^{\frac{1}{2}}$$

where t_1 and t_2 are the time spent on each question. The total mark is $M = M_1(t_1) + M_2(t_2)$. If there are T hours to finish the exam then $T = t_1 + t_2$. Show that there is an increasing and diminishing marginal mark for both questions. Using the time constraint write your total mark as a function of t_1 as $M(t_1)$, find the first-order condition, and solve. Show that this is a global maximizer. If $T = 1$, how much time will be spent on each question and what grade will you receive? If the exam is increased in length by 1 percent, how much will the optimal t_1 increase?

Answer: We have an increasing and diminishing marginal mark as

$$\begin{aligned} M_1'(t_1) &= 20t_1^{-\frac{1}{2}} > 0,\ M_1''(t_1) = -10t_1^{-\frac{3}{2}} < 0 \\ M_2'(t_2) &= 30t_2^{-\frac{1}{2}} > 0,\ M_2''(t_2) = -15t_2^{-\frac{3}{2}} < 0. \end{aligned}$$

Now

$$\begin{aligned} M(t_1) &= M_1(t_1) + M_2(T - t_1) = 40t_1^{\frac{1}{2}} + 60(T - t_1)^{\frac{1}{2}} \\ &\implies M'(t_1) = 20t_1^{-\frac{1}{2}} - 30(T - t_1)^{-\frac{1}{2}} \end{aligned}$$

and so the first-order condition $M'(t_1^*) = 0$ yields

$$\begin{aligned} 20(t_1^*)^{-\frac{1}{2}} - 30(T - t_1^*)^{-\frac{1}{2}} &= 0 \\ &\implies (t_1^*)^{-\frac{1}{2}} = \frac{3}{2}(T - t_1^*)^{-\frac{1}{2}} \\ &\implies \left((t_1^*)^{-\frac{1}{2}}\right)^{-2} = \left(\frac{3}{2}(T - t_1^*)^{-\frac{1}{2}}\right)^{-2} \\ &\implies t_1^* = \left(\frac{3}{2}\right)^{-2}(T - t_1^*) \\ &\implies t_1^* = \frac{\left(\frac{3}{2}\right)^{-2} T}{1 + \left(\frac{3}{2}\right)^{-2}} = \frac{4}{13}T. \end{aligned}$$

This is a global maximizer since $M(t_1)$ is globally concave as

$$M''(t_1) = -10t_1^{-\frac{3}{2}} - 15(T - t_1)^{-\frac{3}{2}} < 0.$$

If $T = 1$ then $t_1^* = \frac{4}{13} \times 1 = \frac{4}{13}$ and $t_2^* = 1 - \frac{4}{13} = \frac{9}{13}$. The grade is

$$M(t_1^*) = 40\left(\frac{4}{13}\right)^{\frac{1}{2}} + 60\left(1 - \frac{4}{13}\right)^{\frac{1}{2}} \approx 72.1.$$

Since $t_1^* = \frac{4}{13}T$ has the form $y = Ax^b$ the elasticity is the exponent on T or 1. It follows that a 1% increase in T will lead to a 1% increase in t_1^*.

Problem 2.68 *[B**] Suppose the mark on Questions 1 and 2 is $M_1(t_1)$ and $M_2(t_2)$ where t_1 and t_2 are the time spent on each question. Find the rule that determines the optimal t_1^*, t_2^*. Show that the solution to the first-order condition is a global maximizer if $M_1(t_1)$ and $M_2(t_2)$ are concave. Show that if T increases (the instructor gives you more time) then t_1^*, t_2^* increase.*

Answer: We have $t_2 = T - t_1$ so that we can write the mark as a function of t_1 alone as

$$M(t_1) = M_1(t_1) + M_2(T - t_1).$$

Using the chain rule for $M_2(T - t_1)$ we have

$$M'(t_1) = M_1'(t_1) - M_2'(T - t_1)$$

and so the first-order condition $M'(t_1^*) = 0$ and $t_2^* = T - t_1^*$ yields

$$M_1'(t_1^*) - M_2'(T - t_1^*) = 0 \implies M'(t_1^*) = M_2'(t_2^*).$$

You should allocate your time so as to equate the marginal mark in each question.

Given that $M_1''(t_1) < 0$ and $M_2''(t_2) < 0$ we have

$$\begin{aligned} M'(t_1) &= M_1'(t_1) - M_2'(T - t_1) \\ &\implies M''(t_1) = M_1''(t_1) + M_2''(T - t_1) < 0 \end{aligned}$$

so that $M(t_1)$ is globally concave, and the solution to the first-order condition is a global maximizer.

Here $t_1^*(T)$, the optimal amount of time spent on Question 1, satisfies

$$M_1'(t_1^*(T)) = M_2'(T - t_1^*(T)).$$

Differentiating both sides with respect to T and using the chain rule we find that

$$\begin{aligned} M_1''(t_1^*(T)) \frac{dt_1^*(T)}{dT} &= M_2''(T - t_1^*(T)) \left(1 - \frac{dt_1^*(T)}{dT}\right) \\ &\implies \frac{dt_1^*(T)}{dT} = \frac{M_2''(T - t_1^*(T))}{M_1''(t_1^*(T)) + M_2''(T - t_1^*(T))} > 0 \end{aligned}$$

since $M_1''(t_1) < 0$, $M_2''(t_2) < 0$ and so both the numerator and denominator are negative. Furthermore since $t_2^*(T) = T - t_1^*(T)$ we have

$$\begin{aligned} \frac{dt_2^*(T)}{dT} &= 1 - \frac{dt_1^*(T)}{dT} = 1 - \frac{M_2''(T - t_1^*(T))}{M_1''(t_1^*(T)) + M_2''(T - t_1^*(T))} \\ &= \frac{M_1''(t_1^*(T))}{M_1''(t_1^*(T)) + M_2''(T - t_1^*(T))} > 0. \end{aligned}$$

2.8.5 Monopoly

Problem 2.69 *[B*] If inverse demand is $P(Q) = Q^{-\beta}$ for $0 < \beta < 1$ and the monopolist's technology is $Q = f(L) = L^{\alpha}$ for $0 < \alpha < 1$, find the revenue function $R(Q)$ from $R \equiv PQ$, the cost function $C(Q)$ from $C \equiv WL$, (assume the monopolist is a price taker in the labour market) and the profit maximizing level of output Q^*.*

Answer: Revenue $R(Q)$ as a function of Q is

$$R(Q) = P(Q)Q = Q^{-\beta}Q = Q^{1-\beta}.$$

To find $C(Q)$ use

$$Q = L^{\alpha} \implies L(Q) = Q^{\frac{1}{\alpha}}$$

so that the cost function is

$$C(Q) = WL(Q) = WQ^{\frac{1}{\alpha}}.$$

We have

$$\pi(Q) = R(Q) - C(Q) = Q^{1-\beta} - WQ^{\frac{1}{\alpha}}$$

so that

$$\pi'(Q) = (1-\beta)Q^{-\beta} - \frac{1}{\alpha}WQ^{\frac{1}{\alpha}-1}.$$

The first-order condition $\pi'(Q^*) = 0$ for profit maximization then is

$$(1-\beta)(Q^*)^{-\beta} - \frac{1}{\alpha}W(Q^*)^{\frac{1}{\alpha}-1} = 0$$

so that solving yields

$$\begin{aligned}
(1-\beta)(Q^*)^{-\beta} &= \frac{1}{\alpha}W(Q^*)^{\frac{1}{\alpha}-1} = \frac{1}{\alpha}W(Q^*)^{\frac{1-\alpha}{\alpha}} \\
&\Longrightarrow (Q^*)^{\frac{1-\alpha}{\alpha}+\beta} = (1-\beta)\alpha W^{-1} \\
&\Longrightarrow Q^* = \left((1-\beta)\alpha W^{-1}\right)^{\frac{1}{\frac{1-\alpha}{\alpha}+\beta}}.
\end{aligned}$$

To check the second-order condition we have

$$\pi''(Q) = -\beta(1-\beta)Q^{-\beta-1} - \left(\frac{1}{\alpha}-1\right)\frac{1}{\alpha}WQ^{\frac{1}{\alpha}-2} < 0$$

for all Q since

$$0 < \beta < 1 \Longrightarrow 1-\beta > 0 \text{ and } \alpha < 1 \Longrightarrow \frac{1}{\alpha} - 1 > 0.$$

Thus $\pi(Q)$ is globally concave and so Q^* is a global maximizer.

Problem 2.70 *[C**] Suppose the inverse demand curve is* $P(Q) = 2Q^{-\frac{1}{2}}$*, the wage rate is* $W = \frac{1}{16}$*, and the monopolist's technology is* $Q = L^{\frac{1}{2}}$*. Find the profit maximizing level of output.*

Answer: Revenue $R(Q)$ as a function of Q is

$$R(Q) = P(Q)Q = 2Q^{-\frac{1}{2}}Q = 2Q^{\frac{1}{2}}.$$

To find the cost function we use

$$Q = L^{\frac{1}{2}} \Longrightarrow L = Q^2$$

so that

$$C(Q) = WQ^2 = \frac{1}{16}Q^2.$$

We have

$$\pi(Q) = R(Q) - C(Q) = 2Q^{\frac{1}{2}} - \frac{1}{16}Q^2$$

so that $\pi'(Q) = Q^{-\frac{1}{2}} - \frac{1}{8}Q$ and the first-order condition $\pi'(Q^*) = 0$ yields

$$\begin{aligned}(Q^*)^{-\frac{1}{2}} - \frac{1}{8}(Q^*) &= 0 \\ &\Longrightarrow (Q^*)^{-\frac{1}{2}} = \frac{1}{8}(Q^*) \\ &\Longrightarrow (Q^*)^{\frac{3}{2}} = 8 \\ &\Longrightarrow Q^* = 8^{\frac{2}{3}} = 4.\end{aligned}$$

Since

$$\pi''(Q) = -\frac{1}{2}Q^{-\frac{3}{2}} - \frac{1}{8} < 0$$

for all Q it follows that $Q^* = 4$ is a global maximizer.

Problem 2.71 *[B**] Suppose a monopolist faces an inverse demand curve $P(Q) = 2Q^{-\frac{1}{4}}$ and cost function $C(Q) = 5Q^3 + 10$. What are the marginal revenue, marginal cost and average cost functions? Determine Q^*, the profit maximizing level of output for the monopolist and P^*, the resulting price.*

Answer: Revenue and marginal revenue are given by

$$\begin{aligned}R(Q) &= P(Q)Q = 2Q^{\frac{3}{4}} \\ &\Longrightarrow MR(Q) = R'(Q) = \frac{3}{2}Q^{-\frac{1}{4}}.\end{aligned}$$

Marginal $MC(Q)$ and average cost $AC(Q)$ are

$$MC(Q) = C'(Q) = 15Q^2 \text{ and } AC(Q) = \frac{C(Q)}{Q} = \frac{5Q^3+10}{Q} = 5Q^2 + \frac{10}{Q}.$$

From $MR(Q^*) = MC(Q^*)$ we have

$$\begin{aligned}\frac{3}{2}(Q^*)^{-\frac{1}{4}} &= 15(Q^*)^2 \\ &\Longrightarrow (Q^*)^{2+\frac{1}{4}} = \frac{3}{2\times 15} \\ &\Longrightarrow Q^* = \left(\frac{1}{10}\right)^{\frac{4}{9}} \approx 0.36.\end{aligned}$$

Therefore

$$P^* = 2(Q^*)^{-\frac{1}{4}} = 2\left(\left(\frac{1}{10}\right)^{\frac{4}{9}}\right)^{-\frac{1}{4}} = 2\times 10^{\frac{1}{9}} \approx 2.58.$$

Problem 2.72 *[A**] Consider a monopolist with revenue function $R(Q)$ and cost function $C(Q)$ which satisfy*

$$\begin{aligned}MR(Q) &\equiv R'(Q) > 0,\ MR'(Q) = R''(Q) < 0 \\ MC(Q) &\equiv C'(Q) > 0,\ MC'(Q) = C''(Q) > 0\end{aligned}$$

where $MR(Q)$ is marginal revenue and $MC(Q)$ is marginal cost. Show the monopolist equates marginal revenue with marginal cost.

Answer: Profits are given by $\pi(Q) = R(Q) - C(Q)$ so that $\pi'(Q) = R'(Q) - C'(Q)$. The first-order condition for profit maximization $\pi'(Q^*) = 0$ then yields

$$\begin{aligned} R'(Q^*) - C'(Q^*) &= 0 \\ &\Longrightarrow MR(Q^*) - MC(Q^*) = 0 \\ &\Longrightarrow MR(Q^*) = MC(Q^*). \end{aligned}$$

Since $R''(Q) < 0$ and $C''(Q) > 0$ we have

$$\pi''(Q) = R''(Q) - C''(Q) < 0$$

so that $\pi(Q)$ is globally concave and Q^* is a global maximizer.

Problem 2.73 *[B**] Suppose that the inverse demand for economy class and business class seats in an airline is given respectively by*

$$P_1 = \frac{4}{3}Q_1^{-\frac{1}{4}}, P_2 = 2Q_2^{-\frac{1}{2}}.$$

If the total number of seats is $Q = 10$ (say 10 million) then what will Q_1^*, Q_2^* be? What will prices be?

Answer: To calculate Q_1^*, Q_2^* we use

$$\begin{aligned} R_1(Q_1) &= P_1Q_1 = \frac{4}{3}Q_1^{-\frac{1}{4}}Q_1 = \frac{4}{3}Q_1^{\frac{3}{4}} \\ &\Longrightarrow MR_1(Q_1) = Q_1^{-\frac{1}{4}} \\ R_2(Q_2) &= P_2Q_2 = 2Q_2^{-\frac{1}{2}}Q_2 = 2Q_2^{\frac{1}{2}} \\ &\Longrightarrow MR_2(Q_2) = Q_2^{-\frac{1}{2}} \end{aligned}$$

and

$$\begin{aligned} Q &= 10 = Q_1 + Q_2 \\ &\Longrightarrow Q_2 = 10 - Q_1. \end{aligned}$$

Since $Q = 10$ is fixed, profit maximization requires that revenue $R \equiv R_1(Q_1) + R_2(Q_2)$ be maximized. Using $Q_2 = 10 - Q_1$ we have

$$R(Q_1) = R_1(Q_1) + R_2(10 - Q_1) = \frac{4}{3}Q_1^{\frac{3}{4}} + 2(10 - Q_1)^{\frac{1}{2}}$$

and so

$$R'(Q_1) = Q_1^{-\frac{1}{4}} - (10 - Q_1)^{-\frac{1}{2}}$$

and so the first-order condition yields

$$(Q_1^*)^{-\frac{1}{4}} - (10 - Q_1^*)^{-\frac{1}{2}} = 0.$$

From this we have

$$\begin{aligned}
(Q_1^*)^{-\frac{1}{4}} &= (10 - Q_1^*)^{-\frac{1}{2}} \\
&\implies \left((Q_1^*)^{-\frac{1}{4}}\right)^{-4} = \left((10 - Q_1^*)^{-\frac{1}{2}}\right)^{-4} \\
&\implies Q_1^* = (10 - Q_1^*)^2 = 100 - 20Q_1^* + (Q_1^*)^2 \\
&\implies (Q_1^*)^2 - 21Q_1^* + 100 = 0
\end{aligned}$$

so that

$$\begin{aligned}
Q_1^* &= \frac{-(-21) \pm \sqrt{(-21)^2 - 4 \times 100}}{2} = \frac{21 \pm \sqrt{41}}{2} \\
&\implies Q_1^* = \frac{21 - \sqrt{41}}{2} \approx 7.3
\end{aligned}$$

where we exclude the other root $Q_1^* = \frac{21+\sqrt{41}}{2} \approx 13.7$ since $10 = Q_1 + Q_2$ and so $Q_1^* > 10$ is not possible. We have

$$Q_2^* = 10 - Q_1^* = 10 - \frac{21 - \sqrt{41}}{2} = \frac{\sqrt{41} - 1}{2} \approx 2.7.$$

Prices then are

$$\begin{aligned}
P_1 &= \frac{4}{3}(Q_1^*)^{-\frac{1}{4}} = \frac{4}{3}\left(\frac{21 - \sqrt{41}}{2}\right)^{-\frac{1}{4}} \approx 0.81 \\
P_2 &= 2(Q_2^*)^{-\frac{1}{2}} = 2\left(\frac{\sqrt{41} - 1}{2}\right)^{-\frac{1}{2}} \approx 1.22.
\end{aligned}$$

Problem 2.74 *[A **] Consider a monopolist who is able to price discriminate between two markets (for example an airline which can charge one price for an economy fare to a tourist and another price for business class). Let P_1 and Q_1 be the price and quantity sold to the first market, let P_2 and Q_2 be the price and quantity sold to the second market, and let $Q = Q_1 + Q_2$ be the total quantity sold in both markets. The inverse demand curves are given by*

$$P_1 = P_1(Q_1),\ P_1'(Q_1) < 0 \text{ and } P_2 = P_2(Q_2),\ P_2'(Q_2) < 0.$$

How should the monopolist allocate any given Q between the two markets? What does this say about the prices in each market?

Answer: The monopolist needs to maximize total revenue R from the two markets where

$$R = R_1(Q_1) + R_2(Q_2)$$

and

$$\begin{aligned} R_1(Q_1) &\equiv P_1(Q_1)Q_1, MR_1(Q_1) \equiv R_1'(Q_1) \\ R_2(Q_2) &\equiv P_2(Q_2)Q_2, MR_2(Q_1) \equiv R_2'(Q_2). \end{aligned}$$

Since

$$Q = Q_1 + Q_2 \Longrightarrow Q_2 = Q - Q_1$$

with Q given, we can write R as a univariate function of Q_1 as

$$R(Q_1) = R_1(Q_1) + R_2(Q - Q_1).$$

Using the sum and chain rules we have

$$R'(Q_1) = R_1'(Q_1) - R_2'(Q - Q_1)$$

so that the first-order condition $R'(Q_1^*) = 0$ yields

$$\begin{aligned} R_1'(Q_1^*) - R_2'(Q - Q_1^*) &= 0 \\ &\Longrightarrow MR_1(Q_1^*) = MR_2(Q - Q_1^*) = MR_2(Q_2^*). \end{aligned}$$

The monopolist should allocate Q_1, Q_2 between the two markets so as to equate the marginal revenues in each market.

Now

$$\begin{aligned} MR_1(Q_1) &= \frac{d}{dQ_1}(P_1(Q_1)Q_1) = P_1(Q_1) + P_1'(Q_1)Q_1 = P_1(Q_1)(1 + \eta_1(Q_1)) \\ MR_2(Q_2) &= \frac{d}{dQ_2}(P_2(Q_2)Q_2) = P_2(Q_2) + P_2'(Q_2)Q_2 = P_2(Q_2)(1 + \eta_2(Q_2)) \end{aligned}$$

where

$$\eta_1(Q_1) \equiv \frac{P_1'(Q_1)Q_1}{P_1(Q_1)}, \eta_2(Q_2) \equiv \frac{P_2'(Q_2)Q_2}{P_2(Q_2)}$$

are the elasticities of the inverse demand functions. We have

$$\begin{aligned} MR_1(Q_1^*) &= MR_2(Q_2^*) \\ &\Longrightarrow P_1(Q_1^*)(1 + \eta_1(Q_1^*)) = P_2(Q_2^*)(1 + \eta_2(Q_2^*)) \\ &\Longrightarrow \frac{P_1(Q_1^*)}{P_2(Q_2^*)} = \frac{1 + \eta_2(Q_2^*)}{1 + \eta_1(Q_1^*)}. \end{aligned}$$

2.8.6 Econometrics

Problem 2.75 *[A**] You have consumption data for n families where Y_i is the consumption of family i. Suppose that $Y_i = \mu + e_i$ where μ is average family consumption and e_i is random noise that is on average 0. Consider the problem of finding the least squares estimator of μ, that is the value $\hat{\mu}$ that minimizes the sum of squares function*

$$S(\mu) \equiv (Y_1 - \mu)^2 + (Y_2 - \mu)^2 + \cdots + (Y_n - \mu)^2.$$

Show that the least squares estimator is $\hat{\mu} = \bar{Y}$, the sample mean. Verify the second-order condition for a global minimum.

Answer: We have using the sum and chain rules that

$$\begin{aligned}\frac{dS(\mu)}{d\mu} &= \frac{d}{d\mu}(Y_1-\mu)^2 + \frac{d}{d\mu}(Y_2-\mu)^2 + \cdots + \frac{d}{d\mu}(Y_n-\mu)^2 \\ &= -2(Y_1-\mu) + -2(Y_2-\mu) + \cdots + -2(Y_n-\mu)\end{aligned}$$

so that the first-order condition $\frac{dS(\hat{\mu})}{d\mu} = 0$ is

$$-2(Y_1-\hat{\mu}) + -2(Y_2-\hat{\mu}) + \cdots + -2(Y_n-\hat{\mu}) = 0.$$

(Instead of writing μ^* for the solution to the first-order condition it is traditional in econometrics to use the ^ symbol where $\hat{\mu}$ is pronounced 'mu hat.')

Cancelling the $-2's$ yields

$$Y_1 - \hat{\mu} + Y_2 - \hat{\mu} + \cdots + Y_n - \hat{\mu} = 0$$

or

$$\begin{aligned}Y_1 + Y_2 + \cdots + Y_n - n\hat{\mu} &= 0 \\ &\implies \hat{\mu} = \frac{1}{n}(Y_1 + Y_2 + \cdots + Y_n) = \bar{Y}.\end{aligned}$$

Now

$$\frac{d^2S(\mu)}{d\mu^2} = 2 + 2 + \cdots + 2 = 2n > 0$$

so that $S(\mu)$ is globally convex and hence $\hat{\mu}$ is a global minimizer.

Problem 2.76 *[A **] Suppose family consumption Y_i depends on family income X_i as $Y_i = \beta X_i + e_i$ where e_i is random noise that is on average 0, and β is the marginal propensity to consume. The least squares estimator $\hat{\beta}$ of the marginal propensity to consume β minimizes the sum of squares function*

$$S(\beta) = (Y_1 - \beta X_1)^2 + (Y_2 - \beta X_2)^2 + \cdots + (Y_n - \beta X_n)^2.$$

Show that

$$\hat{\beta} = \frac{X_1Y_1 + X_2Y_2 + \cdots + X_nY_n}{X_1^2 + X_2^2 + \cdots + X_n^2}.$$

Verify that $\hat{\beta}$ is a global minimizer as long as one family has a non-zero income.

Answer: We have using the sum and chain rules that

$$\begin{aligned}\frac{dS(\beta)}{d\beta} &= \frac{d}{d\beta}(Y_1-\beta X_1)^2 + \frac{d}{d\beta}(Y_2-\beta X_2)^2 + \cdots + \frac{d}{d\beta}(Y_n-\beta X_n)^2 \\ &= -2X_1(Y_1-\beta X_1) + -2X_2(Y_2-\beta X_2) + \cdots + -2X_n(Y_n-\beta X_n)\end{aligned}$$

so that the first-order condition $\frac{dS(\hat{\beta})}{d\beta} = 0$ is

$$-2X_1\left(Y_1 - \hat{\beta}X_1\right) + -2X_2\left(Y_2 - \hat{\beta}X_2\right) + \cdots + -2X_n\left(Y_n - \hat{\beta}X_n\right) = 0.$$

Cancelling the $-2's$ yields

$$X_1Y_1 - \hat{\beta}X_1^2 + X_2Y_2 - \hat{\beta}X_2^2 + \cdots + X_nY_n - \hat{\beta}X_n^2 = 0$$

or

$$\left(X_1^2 + X_2^2 + \cdots + X_n^2\right)\hat{\beta} = X_1Y_1 + X_2Y_2 + \cdots + X_nY_n$$

and so

$$\hat{\beta} = \frac{X_1Y_1 + X_2Y_2 + \cdots + X_nY_n}{X_1^2 + X_2^2 + \cdots + X_n^2}.$$

Now

$$\frac{d^2 S(\beta)}{d\beta^2} = 2\left(X_1^2 + X_2^2 + \cdots + X_n^2\right) > 0$$

since at least one family has non-zero income (i.e., $X_i > 0$)so that $S(\beta)$ is globally concave and hence $\hat{\beta}$ is a global minimizer.

2.8.7 Monotonic Transformations

Problem 2.77 *[A**] Consider the function* $f(x) = 3x^{\frac{2}{3}} - 5x$ *for* $x > 0$. *In Problem 2.40 on page 34 it was shown that* $f(x)$ *has a global maximum at* $x^* = \frac{8}{125}$. *Now consider the function*

$$h(x) = \frac{1}{1 + \frac{1}{3x^{\frac{2}{3}} - 5x}} + 3x^{\frac{2}{3}} - 5x.$$

Show, without using first or second-order conditions, that $h(x)$ *also has a global maximum at* $x^* = \frac{8}{125}$.

Answer: We have

$$\begin{aligned} h(x) &= \frac{1}{1 + \frac{1}{3x^{\frac{2}{3}} - 5x}} + 3x^{\frac{2}{3}} - 5x = \frac{1}{1 + \left(3x^{\frac{2}{3}} - 5x\right)^{-1}} + 3x^{\frac{2}{3}} - 5x \\ &= \left(1 + \left(3x^{\frac{2}{3}} - 5x\right)^{-1}\right)^{-1} + 3x^{\frac{2}{3}} - 5x = g(f(x)) \end{aligned}$$

where $g(x) = \left(1 + \frac{1}{x}\right)^{-1} + x$ is a monotonic transformation of $f(x) = 3x^{\frac{2}{3}} - 5x$ since for all x we have

$$g'(x) = \left(1 + \frac{1}{x}\right)^{-2} x^{-2} + 1 > 0.$$

Thus $f(x)$ and $h(x)$ share the same global maximum, and so $x^* = \frac{8}{125}$ is a global maximizer for $h(x)$ as well.

Problem 2.78 *[B**] Consider the function $f(x) = 2x^{\frac{1}{2}} - x$ for $x > 0$ which has a global maximum at $x^* = 1$. Now consider the function*

$$h(x) = \frac{\left(2x^{\frac{1}{2}} - x - 5\right)^3}{3} + 14x^{\frac{1}{2}} - 7x + 8.$$

Show, without using first or second-order conditions, that $h(x)$ also has a global maximum at $x^* = 1$.

Answer: We have

$$h(x) = \frac{\left(2x^{\frac{1}{2}} - x - 5\right)^3}{3} + 14x^{\frac{1}{2}} - 7x + 8 = \frac{\left(2x^{\frac{1}{2}} - x - 5\right)^3}{3} + 7\left(2x^{\frac{1}{2}} - x\right) + 8 = g(f(x))$$

where

$$g(x) = \frac{(x-5)^3}{3} + 7x + 8$$

is a monotonic transformation of $f(x) = 3x^{\frac{2}{3}} - 5x$ since for all x we have

$$g'(x) = (x-5)^2 + 7 > 0.$$

Thus $f(x)$ and $h(x)$ share the same global maximizer, and so $x^* = 1$ is a global maximizer for $h(x)$ as well.

2.9 Exponentials and Logarithms

Problem 2.79 *[D***] Calculate the derivatives of*

$$f(x) = e^{2x}, f(x) = x\ln(x), f(x) = e^{3x^2}, \text{ and } f(x) = e^{3x^2 - 2\ln(x)}.$$

Answer: We have

$$\begin{aligned} f(x) &= e^{2x} \Longrightarrow f'(x) = 2e^{2x},\ f(x) = x\ln(x) \Longrightarrow f'(x) = 1 + \ln(x) \\ f(x) &= e^{3x^2} \Longrightarrow f'(x) = 6xe^{3x^2},\ f(x) = e^{3x^2 - 2\ln(x)} \Longrightarrow f'(x) = \left(6x - \frac{2}{x}\right) e^{3x^2 - 2\ln(x)}. \end{aligned}$$

Problem 2.80 *[C***] Calculate the derivatives of*

$$f(x) = \exp(\exp(x)), f(x) = \ln(\ln(x)), f(x) = \exp(\ln(x)), f(x) = \ln(\exp(x)).$$

Answer: We have

$$\begin{aligned} f(x) &= \exp(\exp(x)) = e^{(e^x)} \Longrightarrow f'(x) = e^x e^{(e^x)} = \exp(x)\exp(\exp(x)) \\ f(x) &= \ln(\ln(x)) \Longrightarrow f'(x) = \frac{1}{x\ln(x)} \\ f(x) &= \exp(\ln(x)) = x \Longrightarrow f'(x) = 1,\ f(x) = \ln(\exp(x)) = x \Longrightarrow f'(x) = 1. \end{aligned}$$

Problem 2.81 *[D***] Calculate the derivative of* $f(x) = \ln\left(10^{10^{10^{10}}} x^5\right)$.

Answer: We have

$$f(x) = \ln\left(10^{10^{10^{10}}} x^5\right) = \ln\left(10^{10^{10^{10}}}\right) + \ln\left(x^5\right) = \ln\left(10^{10^{10^{10}}}\right) + 5\ln(x)$$

where $\ln\left(10^{10^{10^{10}}}\right)$ is a constant and hence has a derivative of 0 so that $f'(x) = \frac{5}{x}$.

Problem 2.82 *[D***] Show that* $e^{5\ln(x)} = x^5$ *two different ways, and that more generally* $e^{a\ln(x)} = x^a$. *Differentiate* $e^{5\ln(x)}$ *directly and show that* $\frac{d}{dx}e^{5\ln(x)} = 5x^4$.

Answer: We have

$$e^{5\ln(x)} = \left(e^{\ln(x)}\right)^5 = (x)^5 = x^5 \text{ and } e^{5\ln(x)} = e^{\ln\left(x^5\right)} = x^5.$$

The proof for $e^{a\ln(x)} = x^a$ is the same with a replacing 5 everywhere. For the derivative of $e^{5\ln(x)} = x^5$ we have

$$\frac{d}{dx}\left(e^{5\ln(x)}\right) = 5 \times \frac{1}{x} \times e^{5\ln(x)} = 5 \times \frac{1}{x} \times x^5 = 5x^4.$$

Problem 2.83 *[C***] For a non-integer exponent* a, x^a *is defined for* $x > 0$ *by* $x^a \equiv e^{\ln(x)a}$. *Prove that*

$$\frac{d}{dx}x^a = ax^{a-1}.$$

Answer: We have

$$\frac{d}{dx}x^a \equiv \frac{d}{dx}e^{\ln(x)a} = ax^{-1}e^{\ln(x)a} = ax^{-1}x^a = ax^{a-1}.$$

Problem 2.84 *[D***] For the functions below*

$$f(x) = \ln\left(\frac{1}{x^2}\right), f(x) = \exp\left(\frac{1}{3}\ln(x) + \ln(3)\right), f(x) = \ln\left(x^8 e^{-x^2}\right)$$

with $x > 0$, *determine if they are globally increasing or decreasing, and if they are globally concave or convex.*

Answer: For $f(x) = \ln\left(\frac{1}{x^2}\right)$ we have $f(x) = \ln\left(\frac{1}{x^2}\right) = -2\ln(x)$ and so

$$\begin{aligned} f'(x) &= -\frac{2}{x} < 0 \Longrightarrow f(x) \text{ is globally decreasing} \\ f''(x) &= \frac{2}{x^2} > 0 \Longrightarrow f(x) \text{ is globally convex.} \end{aligned}$$

For $f(x) = \exp\left(\frac{1}{3}\ln(x) + \ln(3)\right)$ we have

$$f(x) = \exp\left(\frac{1}{3}\ln(x) + \ln(3)\right) = 3x^{\frac{1}{3}}$$

so that

$$\begin{aligned} f'(x) &= x^{-\frac{2}{3}} > 0 \Longrightarrow f(x) \text{ is globally increasing} \\ f''(x) &= -\frac{2}{3}x^{-\frac{5}{3}} < 0 \Longrightarrow f(x) \text{ is globally concave.} \end{aligned}$$

For $f(x) = \ln\left(x^8 e^{-x^2}\right)$ we have

$$f(x) = \ln\left(x^8 e^{-x^2}\right) = 8\ln(x) - x^2$$

and so

$$f'(x) = \frac{8}{x} - 2x \text{ and } f''(x) = -\frac{8}{x^2} - 2 < 0 \Longrightarrow f(x) \text{ is globally concave.}$$

Since

$$f'(x) = \frac{8}{x} - 2x = \frac{2}{x}\left(4 - x^2\right) = \frac{2}{x}(2 - x)(2 + x)$$

it follows that $f'(x) > 0$ for $0 < x < 2$, $f'(x) < 0$ for $x > 2$, and $f'(x) = 0$ for $x = 2$. Thus $f(x)$ is locally increasing for $0 < x < 2$, locally decreasing for $x > 2$, and has a global maximum at $x = 2$. This is shown in the figure below.

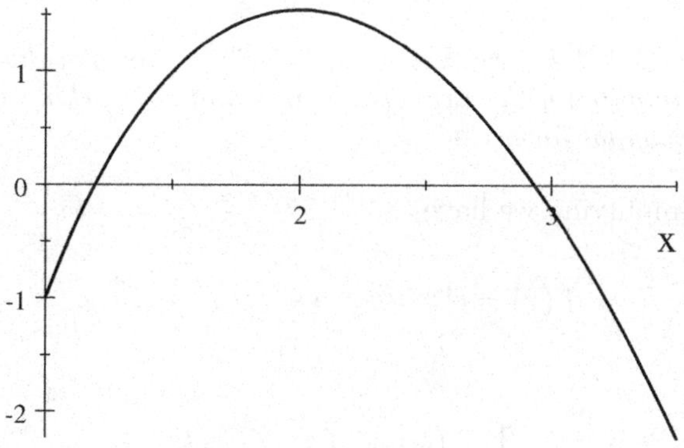

$$f(x) = \ln\left(x^8 e^{-x^2}\right)$$

Problem 2.85 *[C***] Simplify (if possible) the following functions and calculate their derivatives.*

$$f(x) = e^{\ln\left(x^5\right)} + 7 - e^0 - \ln\left(e^6\right) \text{ and } f(x) = e^{\ln(2) - 5\ln\left(\frac{1}{x}\right) + \ln\left(e^{-2x}\right)} - \frac{\ln\left(e^{x^5}\right)}{\sqrt{e^{4x}}}.$$

Answer: We have

$$f(x) = e^{\ln(x^5)} + 7 - e^0 - \ln(e^6) = x^5 + 7 - 1 - 6 = x^5 \Longrightarrow f'(x) = 5x^4.$$

For the second function we have

$$\begin{aligned} f(x) &= e^{\ln(2) - 5\ln\left(\frac{1}{x}\right) + \ln\left(e^{-2x}\right)} - \frac{\ln\left(e^{x^5}\right)}{\sqrt{e^{4x}}} = e^{\ln(2) + 5\ln(x) - 2x} - \frac{x^5}{\left(e^{4x}\right)^{\frac{1}{2}}} \\ &= e^{\ln(2)} e^{5\ln(x)} e^{-2x} - \frac{x^5}{e^{2x}} = 2e^{\ln\left(x^5\right)} e^{-2x} - x^5 e^{-2x} = 2x^5 e^{-2x} - x^5 e^{-2x} = x^5 e^{-2x} \end{aligned}$$

so that

$$f(x) = x^5 e^{-2x} \Longrightarrow f'(x) = 5x^4 e^{-2x} - 2x^5 e^{-2x} = x^4 e^{-2x}(5 - 2x).$$

Problem 2.86 *[C***] Prove that* $f(x) = \frac{(e^x)^2}{\left(e^{\frac{x}{2}}\right)^{10}}$ *is globally decreasing and convex.*

Answer: First simplifying we have

$$f(x) = \frac{(e^x)^2}{\left(e^{\frac{x}{2}}\right)^{10}} = \frac{e^{2x}}{e^{\frac{x}{2}\times 10}} = \frac{e^{2x}}{e^{5x}} = e^{2x-5x} = e^{-3x}.$$

Now $f'(x) = -3e^{-3x} < 0$ since $e^{-3x} > 0$ so that $f(x)$ is globally decreasing while $f''(x) = 9e^{-3x} > 0$ so that $f(x)$ is globally convex.

Problem 2.87 *[C***] Prove that* $f(x) = e^{\ln(x)/2}$ *is globally increasing and concave by 1) simplifying the expression first and 2) working directly with the function in its original form.*

Answer: Simplifying we have

$$f(x) = e^{\ln(x)/2} = \left(e^{\ln(x)}\right)^{\frac{1}{2}} = x^{\frac{1}{2}}$$

so that for all x

$$f'(x) = \frac{1}{2}x^{-\frac{1}{2}} > 0, f''(x) = -\frac{1}{4}x^{-\frac{3}{2}} < 0.$$

Or working directly with $f(x) = e^{\ln(x)/2}$ and using the chain rule

$$f(x) = e^{\ln(x)/2} \Longrightarrow f'(x) = \frac{1}{2x}e^{\ln(x)/2} > 0$$

and

$$f''(x) = -\frac{1}{2x^2}e^{\ln(x)/2} + \frac{1}{4x^2}e^{\ln(x)/2} = -\frac{1}{4x^2}e^{\ln(x)/2} < 0$$

Problem 2.88 *[D***] Prove that*

$$f(x) = e^{e^{-e}} + \ln(x^3) + \ln\left(\ln\left(e^{x^4}\right)\right)$$

for $x > 0$ is globally increasing and concave.

Answer: The calculations are easier if you first simplify as

$$\begin{aligned} f(x) &= e^{e^{-e}} + \ln(x^3) + \ln\left(\ln\left(e^{x^4}\right)\right) = e^{e^{-e}} + 3\ln(x) + \ln(x^4) \\ &= e^{e^{-e}} + 3\ln(x) + 4\ln(x) = e^{e^{-e}} + 7\ln(x). \end{aligned}$$

Here $e^{e^{-e}} \approx 1.07$ is just a constant with a derivative of 0, which is there just to intimidate you. Thus $f'(x) = \frac{7}{x} > 0$ since $x > 0$ and so $f(x)$ is globally increasing. Similarly $f''(x) = -\frac{7}{x^2} < 0$ so that $f(x)$ is globally concave.

Problem 2.89 *[D***] Prove that $f(x) = \ln(5x^2) + 5$ for $x > 0$ is globally increasing and concave.*

Answer: We have

$$\begin{aligned} f(x) &= \ln(5x^2) + 5 = 2\ln(x) + \ln(5) + 5 \\ &\Longrightarrow f'(x) = \frac{2}{x} > 0 \Longrightarrow f''(x) = -\frac{2}{x^2} < 0 \end{aligned}$$

so $f(x)$ is globally increasing and concave.

Problem 2.90 *[D***] Prove that*

$$f(x) = \ln(e^{-x}x^2) + x$$

for $x > 0$ is globally increasing and concave.

Answer: We have simplifying

$$f(x) = \ln(e^{-x}x^2) + x = -x + 2\ln(x) + x = 2\ln(x)$$

so that

$$f'(x) = \frac{2}{x} > 0 \text{ and } f''(x) = -\frac{2}{x^2} < 0$$

so that $f(x)$ is globally increasing and globally concave.

Problem 2.91 *[A***] For $y > 0, x > 0$ we know that that b is the elasticity of*

$$y = f(x) = Ax^b.$$

Let $\tilde{y} \equiv \ln(y), \tilde{x} \equiv \ln(x)$. Show that $\frac{d\tilde{y}}{d\tilde{x}} = b$. Now show that $\frac{d\tilde{y}}{d\tilde{x}}$ is the elasticity for any function $f(x)$. Use this result to prove that if the elasticity is b for all x then $f(x) = Ax^b$.

Answer: For $y = Ax^b$ we have for $a \equiv \ln(A)$ that $\tilde{y} = a + b\tilde{x}$ as

$$y = Ax^b \Longrightarrow \ln(y) = \ln\left(Ax^b\right) = \ln(A) + b\ln(x) \Longrightarrow \tilde{y} = a + b\tilde{x} \Longrightarrow \frac{d\tilde{y}}{d\tilde{x}} = b.$$

Now let $f(x)$ be any function with elasticity defined by $\eta(x) \equiv \frac{f'(x)x}{f(x)}$. Rewrite $y = f(x)$ as

$$y = f(x) \Longrightarrow \ln(y) = \ln\left(f\left(e^{\ln(x)}\right)\right) \Longrightarrow \tilde{y} = \ln\left(f\left(e^{\tilde{x}}\right)\right)$$

and use the chain rule to calculate $\frac{d\tilde{y}}{d\tilde{x}}$ as

$$\frac{d\tilde{y}}{d\tilde{x}} = \frac{f'\left(e^{\tilde{x}}\right)}{f\left(e^{\tilde{x}}\right)}\left(\frac{d}{d\tilde{x}}e^{\tilde{x}}\right) = \frac{f'\left(e^{\tilde{x}}\right)}{f\left(e^{\tilde{x}}\right)}e^{\tilde{x}} = \frac{f'(x)\,x}{f(x)} \equiv \eta(x).$$

Now

$$\begin{aligned}\eta(x) &= b \Longrightarrow \frac{d\tilde{y}}{d\tilde{x}} = b \Longrightarrow \tilde{y} = a + b\tilde{x} \Longrightarrow \ln(y) = a + b\ln(x)\\ &\Longrightarrow e^{\ln(y)} = e^{a+b\ln(x)} \Longrightarrow y = e^a e^{b\ln(x)} \Longrightarrow y = A\left(e^{\ln(x)}\right)^b \Longrightarrow y = Ax^b\end{aligned}$$

where $A \equiv e^a$.

Problem 2.92 *[A**] Prove that if a function $y = f(t)$ with $f(t) > 0$ has a constant growth rate ρ for all t, then $f(t) = Ae^{\rho t}$.*

Answer: If $\frac{f'(t)}{f(t)} = \rho$ then using the chain rule on $\ln(f(t))$ we have

$$\frac{d\ln(f(t))}{dt} = \frac{f'(t)}{f(t)}.$$

Thus

$$\frac{f'(t)}{f(t)} = \rho \Longrightarrow \frac{d\ln(f(t))}{dt} = \rho.$$

Since the derivative of $\ln(f(t))$ is a constant ρ, it must be that $\ln(f(t))$ is a linear function of t and so

$$\ln(f(t)) = a + \rho t \Longrightarrow f(t) = e^{a+\rho t} = e^a e^{\rho t} = Ae^{\rho t}$$

where $A = e^a$.

Problem 2.93 *[A*] Suppose $f(x)$ has the property that $f(x)^{f(x)} = x$. Show that*

$$f'(x) = \frac{1}{x\left(1 + \ln(f(x))\right)}.$$

Answer: Taking $\ln()$ of both sides of $f(x)^{f(x)} = x$ yields

$$\ln(f(x))\, f(x) = \ln(x).$$

Now differentiating both sides with respect to x yields

$$\begin{aligned} \ln(f(x))\, f'(x) + f(x)\frac{f'(x)}{f(x)} &= \frac{1}{x} \\ &\implies (1+\ln(f(x)))\, f'(x) = \frac{1}{x} \\ &\implies f'(x) = \frac{1}{x(1+\ln(f(x)))}. \end{aligned}$$

Problem 2.94 *[D***] Suppose that a firm's production is*

$$Q = f(L) = L^{\frac{1}{4}} e^{\ln\left(L^{10}\right)/80 - \ln\left(L^{-\frac{2}{16}}\right)}.$$

Calculate the marginal product of labour and show that there is a positive and diminishing marginal product of labour.

Answer: Simplifying we have

$$\begin{aligned} f(L) &= L^{\frac{1}{4}} e^{\ln\left(L^{10}\right)/80 - \ln\left(L^{-\frac{2}{16}}\right)} = L^{\frac{1}{4}} e^{10\ln(L)/80 + \frac{2}{16}\ln(L)} \\ &= L^{\frac{1}{4}} e^{\frac{1}{4}\ln(L)} = L^{\frac{1}{4}} e^{\frac{1}{4}\ln(L)} = L^{\frac{1}{4}} \left(e^{\ln(L)}\right)^{\frac{1}{4}} = L^{\frac{1}{4}} L^{\frac{1}{4}} = L^{\frac{1}{2}}. \end{aligned}$$

Thus $f(L)$ is a disguised Cobb-Douglas production function. Then

$$MP_L(L) = f'(L) = \frac{1}{2}L^{-\frac{1}{2}} > 0 \text{ and } MP_L'(L) = f''(L) = -\frac{1}{4}L^{-\frac{3}{2}} < 0$$

so there is a positive and diminishing marginal product of labour.

Problem 2.95 *Suppose that Y is the number of accidents at an intersection in a year and μ is the average number of accidents each year. The probability of there being k accidents is given by a Poisson distribution where*

$$\Pr[Y = k] = \frac{\mu^k}{k!}e^{-\mu} \text{ for } k = 0, 1, 2, \ldots \infty.$$

(The Poisson distribution is used to model events like this that occur frequently but where the probability of something occurring each time is small.) [D] If $\mu = 2$ calculate the probability that there will be $0, 1, 2, 3, 4$ accidents. [D*] Show that the sum of these probabilities is less than 1 and explain why. [A*] Show using the Taylor series for e^x that the sum of all the probabilities is 1.*

Answer: The probability of $0, 1, 2, 3, 4$ accidents is given respectively by

$$\frac{2^0}{0!}e^{-2} \approx 0.14, \frac{2^1}{1!}e^{-2} \approx 0.27, \frac{2^2}{2!}e^{-2} \approx 0.27, \frac{2^3}{3!}e^{-2} \approx 0.18, \text{ and } \frac{2^4}{4!}e^{-2} \approx 0.09.$$

These probabilities do not sum to 1 as

$$0.14 + 0.27 + 0.27 + 0.18 + 0.09 = 0.95.$$

This is because there is a probability of $1 - 0.95 = 0.05$ of having 5 or more accidents.

We can prove that the probabilities really do sum to 1 using the Taylor series for e^x which is

$$\begin{aligned} e^x &= 1 + \frac{x}{1!} + \frac{x^2}{2!} + \cdots \\ &\implies e^\mu = 1 + \frac{\mu}{1!} + \frac{\mu^2}{2!} + \cdots . \end{aligned}$$

We have

$$\begin{aligned} \Pr[Y=0] + \Pr[Y=1] + \Pr[Y=2] + \cdots &= \frac{\mu^0}{0!}e^{-\mu} + \frac{\mu^1}{1!}e^{-\mu} + \frac{\mu^2}{2!}e^{-\mu} + \cdots \\ &= e^{-\mu}\underbrace{\left(1 + \frac{\mu^1}{1!} + \frac{\mu^2}{2!} + \cdots\right)}_{e^\mu} \\ &= e^{-\mu}e^{\mu} = e^{-\mu+\mu} = e^0 = 1. \end{aligned}$$

2.9.1 Optimization with e^x and $\ln(x)$

Problem 2.96 *[D***] Consider*

$$f(x) = \ln\left(\frac{x^3}{2}\right) - 4x$$

for $x > 0$. Find the first-order condition and solve. Show that your solution corresponds to a local maximum. Is the function globally convex or concave? Does your solution yield a global maximum or minimum?

Answer: We can first simplify as

$$f(x) = \ln\left(\frac{x^3}{2}\right) - 4x = 3\ln(x) - 4x - \ln(2).$$

Therefore $f'(x) = \frac{3}{x} - 4$ and so the first-order condition $f'(x^*) = 0$ yields

$$\frac{3}{x^*} - 4 = 0 \implies x^* = \frac{3}{4} = 0.75.$$

Now $f''(x) = -\frac{3}{x^2} < 0$ so that $f(x)$ is globally concave and $x^* = \frac{3}{4}$ is a global maximizer.

Problem 2.97 *[D***] Find the global maximum of $f(x) = \ln(3x^2) - 20x$ for $x > 0$.*

Answer: First we simplify as

$$f(x) = \ln(3x^2) - 20x = \ln(3) + \ln(x^2) - 20x = \ln(3) + 2\ln(x) - 20x.$$

Thus $f'(x) = \frac{2}{x} - 20$ and so the first-order condition $f'(x^*) = 0$ yields

$$\frac{2}{x^*} - 20 = 0 \Longrightarrow \frac{2}{x^*} = 20 \Longrightarrow x^* = \frac{2}{20} = \frac{1}{10}.$$

Since for all x we have $f''(x) = -\frac{2}{x^2} < 0$ it follows that $f(x) = \ln(3x^2) - 20x$ is globally concave and $x^* = \frac{1}{10}$ is a global maximizer.

Problem 2.98 *[B***] Consider*

$$f(x) = \ln\left(\frac{x^a}{be^{cx}}\right) - dx \text{ for } x > 0.$$

How does the solution to the first-order condition change with b? What determines if the solution is a global maximizer or minimizer?

Answer: Simplifying we have

$$\begin{aligned} f(x) &= \ln\left(\frac{x^a}{be^{-cx}}\right) - dx = \ln(x^a) - \ln(be^{cx}) - dx \\ &= a\ln(x) - \ln(b) - \ln(e^{cx}) - dx = a\ln(x) - cx - dx - \ln(b) \\ &= a\ln(x) - (c+d)x - \ln(b). \end{aligned}$$

Therefore $f'(x) = \frac{a}{x} - (c+d)$ and so the first-order condition $f'(x^*) = 0$ yields

$$\frac{a}{x^*} - (c+d) = 0 \Longrightarrow x^* = \frac{a}{c+d}.$$

Here x^* does not depend on b.

If $a > 0$ then $f''(x) = -\frac{a}{x^2} < 0$ and so $f(x)$ is globally concave and $x^* = \frac{a}{c+d}$ is a global maximizer. Conversely if $a < 0$ then $f(x)$ is globally convex with $x^* = \frac{a}{c+d}$ yielding a global minimum.

Problem 2.99 *[D***] Consider*

$$f(x) = \ln\left(e^{-3x}x^5e^{-\frac{1}{2}x^2}\right) - \frac{1}{2}x^2 \text{ for } x > 0.$$

Simplify if possible, find the first-order condition and solve. Show that the solution is a global maximizer.

Answer: We have

$$f(x) = \ln\left(e^{-3x}x^5e^{-\frac{1}{2}x^2}\right) - \frac{1}{2}x^2 = -3x + 5\ln(x) - x^2.$$

We have $f'(x) = -3 + \frac{5}{x} - 2x$ and so the first-order condition $f'(x^*) = 0$ yields

$$\begin{aligned} -3 + \frac{5}{x^*} - 2x^* &= 0 \\ &\Longrightarrow 2(x^*)^2 + 3x^* - 5 = 0 \\ &\Longrightarrow x^* = \frac{-3 \pm \sqrt{3^2 - 4 \times 2 \times (-5)}}{2 \times 2} \\ &\Longrightarrow x^* = \frac{-3+7}{2 \times 2} = 1. \end{aligned}$$

(We ignore the negative root $-\frac{10}{4}$ since we require $x > 0$.) The function is globally concave since for all x we have $f''(x) = -\frac{5}{x^2} - 2 < 0$ and so $x^* = 1$ is a global maximizer.

Problem 2.100 *[D***] Consider* $f(x) = x^{10} e^{-\frac{1}{2}x^2}$ *for* $x > 0$. *Find the first-order condition and solve. Now consider* $h(x) = \ln(f(x))$. *Find the first-order condition and solve. Show that your solution is a global maximizer.*

Answer: Since

$$f'(x) = 10x^9 e^{-\frac{1}{2}x^2} - x^{11} e^{-\frac{1}{2}x^2} = x^9 e^{-\frac{1}{2}x^2}\left(10 - x^2\right)$$

the first-order condition $f'(x^*) = 0$ yields

$$(x^*)^9 e^{-\frac{1}{2}(x^*)^2}\left(10 - (x^*)^2\right) = 0 \Longrightarrow 10 - (x^*)^2 = 0 \Longrightarrow x^* = \sqrt{10} \approx 3.16$$

since $(x^*)^9 e^{-\frac{1}{2}(x^*)^2} > 0$. (We exclude the other solution $x^* = -\sqrt{10}$ since the domain is restricted to $x > 0$.)

Now

$$h(x) = \ln(f(x)) = \ln\left(x^{10} e^{-\frac{1}{2}x^2}\right) = \ln\left(x^{10}\right) + \ln\left(e^{-\frac{1}{2}x^2}\right) = 10\ln(x) - \frac{x^2}{2}.$$

Therefore $h'(x) = \frac{10}{x} - x$ and so the first-order condition $h'(x^*) = 0$ yields

$$\frac{10}{x^*} - x^* = 0 \Longrightarrow \frac{10}{x^*} = x^* \Longrightarrow 10 = (x^*)^2 \Longrightarrow x^* = \sqrt{10} \approx 3.16.$$

The function $h(x)$ is globally concave since $h''(x) = -\frac{10}{x^2} - 1 < 0$ and hence $x^* = \sqrt{10}$ is a global maximizer for $h(x)$.

Problem 2.101 *[D***] Consider the function* $f(x) = x^4 e^{-x+3\ln\left(x^2\right)}$ *for* $x > 0$. *Find the first-order condition and solve. Show that your solution is a global maximizer.*

Answer: Simplifying $f(x)$ we find that

$$f(x) = x^4 e^{-x+6\ln(x)} = x^4 e^{-x}\left(e^{\ln(x)}\right)^6 = x^4 x^6 e^{-x} = x^{10} e^{-x}.$$

We have

$$f'(x) = 10x^9e^{-x} - x^{10}e^{-x} = x^9e^{-x}(10 - x)$$

and so the first-order condition $f'(x^*) = 0$ yields

$$(x^*)^9 e^{-x^*}(10 - x^*) = 0 \Longrightarrow 10 - x^* = 0 \Longrightarrow x^* = 10$$

since $(x^*)^9 > 0$ and $e^{-x^*} > 0$. Here $x^* = 10$ is a global maximizer since $f'(x) > 0$ for $x < 10$ and $f'(x) < 0$ for $x > 10$.

Problem 2.102 *[D***] Consider $f(x) = x^3e^{-x^2}$ for $x > 0$. Find the first-order condition for $f(x)$ and solve. Show that x^* corresponds to a local maximum of $f(x)$. Calculate $h(x) = \ln(f(x))$, the first-order condition for $h(x)$, and solve and show that $h(x)$ has the same solution as $f(x)$. Show that $h(x)$ is globally concave and that x^* is a global maximizer for $h(x)$. (See Problem 2.139 on page 112 for a proof that x^* also is a global maximizer for $f(x)$.)*

Answer: We have

$$f'(x) = 3x^2e^{-x^2} - 2x^4e^{-x^2} = x^2e^{-x^2}\left(3 - 2x^2\right)$$

and so the first-order condition $f'(x^*) = 0$ yields

$$(x^*)^2 e^{-(x^*)^2}\left(3 - 2(x^*)^2\right) = 0 \Longrightarrow 3 - 2(x^*)^2 = 0 \Longrightarrow x^* = \sqrt{\frac{3}{2}}$$

since $(x^*)^2 e^{-(x^*)^2} > 0$. The other root $-\sqrt{\frac{3}{2}}$ can be excluded since $x > 0$.

Now using the product rule we have

$$f''(x) = \left(\frac{d}{dx}\left(x^2e^{-x^2}\right)\right)(3 - 2x^2) - 4x\left(x^2e^{-x^2}\right) = g(x)(3 - 2x^2) - 4x^3e^{-x^2}$$

where

$$g(x) = \frac{d}{dx}\left(x^2e^{-x^2}\right) = 2xe^{-x^2}\left(1 - x^2\right).$$

At $x^* = \sqrt{\frac{3}{2}}$ we have

$$\begin{aligned} f''\left(\sqrt{\frac{3}{2}}\right) &= g\left(\sqrt{\frac{3}{2}}\right)\left(3 - 2\left(\sqrt{\frac{3}{2}}\right)^2\right) - 4\left(\sqrt{\frac{3}{2}}\right)^3 e^{-\left(\sqrt{\frac{3}{2}}\right)^2} \\ &= -3\sqrt{6}e^{-\frac{3}{2}} < 0 \end{aligned}$$

so that $f(x) = x^3e^{-x^2}$ is locally concave at $x^* = \sqrt{\frac{3}{2}}$ and hence $x^* = \sqrt{\frac{3}{2}}$ corresponds to a local maximum.

For $h(x) = \ln(f(x))$ we have

$$h(x) = \ln(f(x)) = \ln\left(x^3e^{-x^2}\right) = \ln\left(x^3\right) + \ln\left(e^{-x^2}\right) = 3\ln(x) - x^2$$

and so $h'(x) = \frac{3}{x} - 2x$. The first-order condition $h'(x^*) = 0$ yields the same x^* as $f(x)$ as

$$\begin{aligned}\frac{3}{x^*} - 2x^* &= 0 \\ &\Longrightarrow -2x^{*2} + 3 = 0 \\ &\Longrightarrow x^* = \left(\frac{3}{2}\right)^{\frac{1}{2}}.\end{aligned}$$

The function $h(x)$ is globally concave since $h''(x) = -\frac{3}{x^2} - 2 < 0$ and so $x^* = \left(\frac{3}{2}\right)^{\frac{1}{2}}$ is a global maximizer for $h(x)$.

Problem 2.103 *[D***] Consider* $f(x) = x^6 e^{-\frac{x^2}{2}}$ *for* $x > 0$. *Find the first-order condition and solve. Show that your solution is a local maximizer.*

Answer: We have

$$f'(x) = 6x^5 e^{-\frac{x^2}{2}} - x \times x^6 e^{-\frac{x^2}{2}} = 6x^5 e^{-\frac{x^2}{2}} - x^7 e^{-\frac{x^2}{2}} = e^{-\frac{x^2}{2}} x^5 \left(6 - x^2\right)$$

and so the first-order condition yields

$$e^{-\frac{(x^*)^2}{2}} (x^*)^5 \left(6 - (x^*)^2\right) = 0 \Longrightarrow 6 - (x^*)^2 = 0 \Longrightarrow x^* = 6^{\frac{1}{2}} \approx 2.45$$

since $e^{-\frac{(x^*)^2}{2}} > 0$ and $(x^*)^5 > 0$.

Using the product rule to calculate the second derivative we have

$$f''(x) = e^{-\frac{x^2}{2}} x^5 \left(6 - x^2\right) \Longrightarrow \left(\frac{d}{dx} e^{-\frac{x^2}{2}} x^5\right) \left(6 - x^2\right) - 2x e^{-\frac{x^2}{2}} x^5 = g(x) \left(6 - x^2\right) - 2x e^{-\frac{x^2}{2}} x^5$$

where $g(x) \equiv \frac{d}{dx}\left(e^{-\frac{x^2}{2}} x^5\right)$. We do not need to evaluate $g(x)$ since the term $g(x)\left(6 - x^2\right)$ is 0 when we replace x with $x^* = 6^{\frac{1}{2}}$ as

$$f''\left(6^{\frac{1}{2}}\right) = g\left(6^{\frac{1}{2}}\right)\left(6 - \left(6^{\frac{1}{2}}\right)^2\right) - 2 \times 6^{\frac{1}{2}} e^{-\frac{\left(6^{\frac{1}{2}}\right)^2}{2}} \left(6^{\frac{1}{2}}\right)^5 = -432 e^{-3} < 0.$$

Since $f''(x^*) < 0$ it follows that $f(x)$ is locally concave at $x^* = 6^{\frac{1}{2}}$ and so $x^* = 6^{\frac{1}{2}}$ is a local maximizer.

Problem 2.104 *[B***] Consider* $f(x) = (3x - 1)e^{-2x}$ *for* $x > 0$. *Find the first-order condition and solve. Show that your solution is a local maximizer. Show that the function is neither globally convex nor globally concave. Show nevertheless that your solution is to a global maximizer.*

Answer: Using the chain and product rules we have

$$f'(x) = 3e^{-2x} - 2(3x-1)e^{-2x} = e^{-2x}(5-6x)$$

and so the first-order condition $f'(x^*) = 0$ yields

$$e^{-2x^*}(5-6x^*) = 0 \Longrightarrow 5 - 6x^* = 0 \text{ (since } e^{-2x^*} > 0) \Longrightarrow x^* = \frac{5}{6}.$$

Now

$$f''(x) = -2e^{-2x}(5-6x) - 6e^{-2x} = e^{-2x}(12x-16)$$

so that

$$f''\left(\frac{5}{6}\right) = e^{-2\times\frac{5}{6}}\left(12 \times \frac{5}{6} - 16\right) = -6e^{-\frac{5}{3}} < 0$$

so that the function is locally concave at $x^* = \frac{5}{6}$ and so $x^* = \frac{5}{6}$ is a local maximizer.

The function is neither globally concave nor convex. The function has an inflection point at $x = \frac{4}{3} \approx 1.33$ where $f''(x) = 0$ since

$$f''(x) = e^{-2x}(12x-16) = 12e^{-2x}\left(x - \frac{4}{3}\right).$$

In fact $f(x)$ is locally concave for $0 < x < \frac{4}{3}$ $(f''(x) < 0)$ and locally convex for $x > \frac{4}{3}$ $(f''(x) < 0)$. This is illustrated in the figure below.

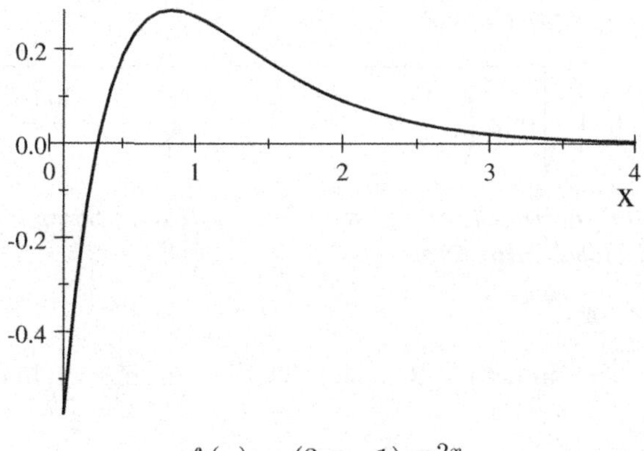

$$f(x) = (3x-1)e^{-2x}$$

From the figure it appears that $x^* = \frac{5}{6}$ is a global maximizer even though $f(x)$ is not globally concave. We can prove that $x^* = \frac{5}{6}$ is a global maximizer by noting that

$$f'(x) = e^{-2x}(5-6x) = -6e^{-2x}\left(x - \frac{5}{6}\right)$$

so that $f'(x) > 0$ for $x < \frac{5}{6}$ and $f'(x) < 0$ for $x > \frac{5}{6}$.

Problem 2.105 *[D***] Suppose that $f(x)$ is given by*

$$f(x) = \ln\left(\frac{10000x^x\left(x^3+3-e^0-\ln\left(e^2\right)\right)}{e^{3\ln(x)}}\right).$$

Calculate the first-order condition and solve. Show that your solution is a global minimizer.

Answer: We have

$$\begin{aligned} f(x) &= \ln\left(\frac{10000x^x\left(x^3+3-\overbrace{e^0}^{1}-\overbrace{\ln\left(e^2\right)}^{2}\right)}{\underbrace{e^{3\ln(x)}}_{x^3}}\right) = \ln\left(\frac{10000x^x x^3}{x^3}\right) \\ &= \ln(10000) + \ln(x^x) = \ln(10000) + x\ln(x). \end{aligned}$$

Therefore $f'(x) = 1 + \ln(x)$ and so the first-order condition $f'(x^*) = 0$ yields

$$1 + \ln(x^*) = 0 \Longrightarrow \ln(x^*) = -1 \Longrightarrow x^* = e^{-1}.$$

Furthermore $f''(x) = \frac{1}{x} > 0$ since $x > 0$ so that $f(x)$ is globally convex and $x^* = e^{-1}$ is a global minimizer.

Problem 2.106 *[D***] Consider*

$$f(x) = \ln\left(\sqrt{\exp\left(\frac{x^4\ln(x^2)\left(e^{\ln(x^5)}+7-e^0-\ln\left(e^6\right)\right)}{\ln\left(e^{x^7}\right)}\right)}\right)$$

for $x > 0$. Find the solution to the first-order condition. Show that your solution corresponds to a global minimum.

Answer: We have

$$\ln\left(e^{x^7}\right) = x^7, \ln\left(x^2\right) = 2\ln(x), e^{\ln\left(x^5\right)} = x^5, e^0 = 1, \ln\left(e^6\right) = 6$$

so that

$$\begin{aligned} f(x) &= \ln\left(\sqrt{\exp\left(\frac{x^4\ln(x^2)\left(e^{\ln(x^5)}+7-e^0-\ln\left(e^6\right)\right)}{\ln\left(e^{x^7}\right)}\right)}\right) \\ &= \ln\left(\left(\exp\left(2\frac{x^4\ln(x)\left(x^5+7-1-6\right)}{x^7}\right)\right)^{\frac{1}{2}}\right) \\ &= \frac{1}{2}\ln\left(\exp\left(2\frac{x^4\ln(x)\left(x^5+7-1-6\right)}{x^7}\right)\right) = \frac{1}{2}\left(2\frac{\ln(x)\,x^4x^5}{x^7}\right) = x^2\ln(x). \end{aligned}$$

Therefore

$$f(x) = x^2 \ln(x) \implies f'(x) = x + 2x\ln(x) = x(1 + 2\ln(x))$$

and so the first-order condition $f'(x^*) = 0$ yields

$$x^*(1 + 2\ln(x^*)) = 0 \implies 1 + 2\ln(x^*) = 0 = \ln(x^*) = -\frac{1}{2} \implies x^* = e^{-\frac{1}{2}}.$$

This is a global minimizer since

$$f'(x) = x(1 + 2\ln(x)) = x\left(1 + 2\ln\left(\frac{x}{e^{-\frac{1}{2}}}e^{-\frac{1}{2}}\right)\right) = 2x\left(\ln\left(\frac{x}{e^{-\frac{1}{2}}}\right)\right)$$

and so $f'(x) < 0$ for $x < e^{-\frac{1}{2}}$ and $f'(x) > 0$ for $x > e^{-\frac{1}{2}}$.

Problem 2.107 *[D***] Suppose that $f(x)$ is given by*

$$f(x) = \ln\left(\frac{5000(2x)^{2x}\left(x^5 + 6 - e^0 - \ln(e^5)\right)}{e^{5\ln(x)}}\right).$$

Simplify first and then calculate $f'(x)$ and the solution to the first-order condition. Show that $f(x)$ is globally convex.

Answer: We have

$$\begin{aligned} f(x) &= \ln\left(\frac{5000(2x)^{2x}\left(x^5 + 6 - \overbrace{e^0}^{1} - \overbrace{\ln(e^5)}^{5}\right)}{\underbrace{e^{5\ln(x)}}_{x^5}}\right) = \ln\left(\frac{5000(2x)^{2x}x^5}{x^5}\right) \\ &= \ln\left(5000(2x)^{2x}\right) = \ln(5000) + \ln\left((2x)^{2x}\right) = \ln(5000) + 2x\ln(2x) \\ &= \ln(5000) + 2x(\ln(2) + \ln(x)) = \ln(5000) + 2\ln(2)x + 2x\ln(x). \end{aligned}$$

Therefore

$$f'(x) = 2\ln(2) + 2 + 2\ln(x) = 2(\ln(2) + 1 + \ln(x))$$

and so the first-order condition $f'(x^*) = 0$ yields

$$2(\ln(2) + 1 + \ln(x^*)) = 0 \implies \ln(x^*) = -\ln(2) - 1 \implies x^* = e^{-\ln(2)-1} = \frac{1}{2}e^{-1} \approx 0.18.$$

Furthermore

$$f''(x) = \frac{2}{x} > 0 \text{ since } x > 0$$

so that $f(x)$ is globally convex and $x^* = \frac{1}{2}e^{-1}$ is a global minimizer.

Problem 2.108 *[D***] Consider*

$$f(x) = \sqrt{\frac{1}{x^{12}e^{-x^2}}} \text{ for } x > 0.$$

Find the first-order condition for $f(x)$ and solve. Show that you have a local minimum for $f(x)$. Calculate $h(x) = \ln(f(x))$ and show that $h(x)$ has the same solution to the first-order condition as $f(x)$. Show that $h(x)$ is globally convex and so x^* is a global minimizer for $h(x)$. In fact this also shows that x^* is a global minimizer for $f(x)$. To see why look at Problem 2.136 on page 110.

Answer: We have

$$f(x) = \sqrt{\frac{1}{x^{12}e^{-x^2}}} = \left(x^{-12}e^{x^2}\right)^{\frac{1}{2}} = x^{-6}e^{\frac{1}{2}x^2}$$

so that

$$f'(x) = -6x^{-7}e^{\frac{1}{2}x^2} + x \times x^{-6}e^{\frac{1}{2}x^2} = e^{\frac{1}{2}x^2}\left(-6x^{-7} + x^{-5}\right) = x^{-7}e^{\frac{1}{2}x^2}\left(x^2 - 6\right)$$

so that the first-order condition $f'(x^*) = 0$ as well as $(x^*)^{-7} > 0, e^{\frac{1}{2}(x^*)^2} > 0$ yields

$$e^{\frac{1}{2}(x^*)^2}(x^*)^{-7}\left((x^*)^2 - 6\right) = 0 \Longrightarrow (x^*)^2 - 6 = 0 \Longrightarrow x^* = \sqrt{6} \approx 2.45.$$

Using the product rule we have

$$\begin{aligned} f''(x) &= \frac{d}{dx}\left(x^{-7}e^{\frac{1}{2}x^2}\left(x^2 - 6\right)\right) = \left(\frac{d}{dx}\left(x^{-7}e^{\frac{1}{2}x^2}\right)\right)\left(x^2 - 6\right) + x^{-7}e^{\frac{1}{2}x^2} \times 2x \\ &= g(x)\left(x^2 - 6\right) + 2x^{-6}e^{\frac{1}{2}x^2} \end{aligned}$$

where

$$g(x) = \frac{d}{dx}\left(x^{-7}e^{\frac{1}{2}x^2}\right).$$

We will not need to evaluate $g(x)$ since $g(x)\left(x^2 - 6\right)$ is 0 when we replace x with $x^* = \sqrt{6}$ as

$$f''\left(\sqrt{6}\right) = g\left(\sqrt{6}\right)\left(\sqrt{6}^2 - 6\right) + 2\left(\sqrt{6}\right)^{-6}e^{\frac{1}{2}\left(\sqrt{6}\right)^2} = \frac{e^3}{108} > 0.$$

Since $f''(\sqrt{6}) > 0$ it follows that $f(x)$ is locally convex at $x^* = \sqrt{6}$ and hence $x^* = \sqrt{6}$ is a local minimizer.

For $h(x) = \ln(f(x))$ we have

$$h(x) = \ln(f(x)) = \ln\left(x^{-6}e^{\frac{1}{2}x^2}\right) = -6\ln(x) + \frac{1}{2}x^2$$

and $h'(x) = -\frac{6}{x} + x$ so that the first-order condition $h'(x^*) = 0$ yields

$$-\frac{6}{x^*} + x^* = 0 \Longrightarrow -6 + (x^*)^2 \Longrightarrow (x^*)^2 = 6 \Longrightarrow x^* = \sqrt{6}.$$

The function $h(x)$ is globally convex since for all x $h''(x) = \frac{6}{x^2} + 1 > 0$ and hence $x^* = \sqrt{6}$ is a global minimizer for $h(x)$.

Problem 2.109 *[D***] Consider the function*

$$f(x) = \ln\left(\frac{x^8}{(e^x)^2}\right) + \ln\left(\frac{x^7}{e^{\frac{1}{2}x^2}}\right) \quad \textit{for } x > 0.$$

Simplify if possible, find the first-order condition and solve. Determine if the solution is a global maximizer or minimizer.

Answer: We have

$$\begin{aligned}
f(x) &= \ln\left(\frac{x^8}{(e^x)^2}\right) + \ln\left(\frac{x^7}{e^{\frac{1}{2}x^2}}\right) = \ln\left(\frac{x^8}{e^{2x}}\right) + \ln\left(x^7 e^{-\frac{1}{2}x^2}\right) \\
&= \ln\left(x^8\right) - \ln\left(e^{2x}\right) + \ln\left(x^7\right) + \ln\left(e^{-\frac{1}{2}x^2}\right) = 8\ln(x) - 2x + 7\ln(x) - \frac{1}{2}x^2 \\
&= 15\ln(x) - 2x - \frac{1}{2}x^2.
\end{aligned}$$

From $f'(x) = \frac{15}{x} - 2 - x$ the first-order condition $f'(x^*) = 0$ yields

$$\begin{aligned}
\frac{15}{x^*} - 2 - x^* &= 0 \Longrightarrow (x^*)^2 + 2x^* - 15 = 0 \\
\Longrightarrow x^* &= \frac{-2 + \sqrt{(2)^2 - 4(-15)}}{2} \Longrightarrow x^* = 3
\end{aligned}$$

where we ignore the other negative root $x^* = -5$. The function is globally concave since $f''(x) = -\frac{15}{x^2} - 1 < 0$ and so $x^* = 3$ is a global maximizer.

Problem 2.110 *[D***] Consider*

$$f(x) = \ln\left(\frac{x^3 e^{\ln(x)} + 1 - e^0}{e^{2\ln(x^2)}}\right) + \ln\left(10^{e^{e^e}} x^3 e^{-\frac{x^2}{2} + \ln(x)}\right), \quad \textit{for } x > 0.$$

Find the first-order condition for $f(x)$ and solve. Determine if your solution is a global maximizer or minimizer.

Answer: We first simplify as

$$\begin{aligned}
f(x) &= \ln\left(\frac{x^3 e^{\ln(x)} + 1 - e^0}{e^{2\ln(x^2)}}\right) + \ln\left(10^{e^{e^e}} x^3 e^{-\frac{x^2}{2}+\ln(x)}\right) \\
&= \ln\left(\frac{x^3 e^{\ln(x)} + 1 - 1}{e^{4\ln(x)}}\right) + \ln\left(x^3 e^{-\frac{x^2}{2}+\ln(x)}\right) + \ln\left(10^{e^{e^e}}\right) \\
&= \ln\left(\frac{x^3 x}{\left(e^{\ln(x)}\right)^4}\right) + \ln\left(x^3\right) + \ln\left(e^{-\frac{x^2}{2}+\ln(x)}\right) + \ln\left(10^{e^{e^e}}\right) \\
&= \ln\left(\frac{x^4}{x^4}\right) + 3\ln(x) - \frac{x^2}{2} + \ln(x) + \ln\left(10^{e^{e^e}}\right) \\
&= \ln(1) + 3\ln(x) - \frac{x^2}{2} + \ln(x) + \ln\left(10^{e^{e^e}}\right) \\
&= 4\ln(x) - \frac{x^2}{2} + \ln\left(10^{e^{e^e}}\right).
\end{aligned}$$

Thus $f'(x) = \frac{4}{x} - x$ (note the seemingly terrifying $\ln\left(10^{e^{e^e}}\right)$ is an irrelevant constant and disappears when we differentiate) and so the first-order condition $f'(x^*) = 0$ yields

$$\frac{4}{x^*} - x^* = 0 \Longrightarrow (x^*)^2 = 4 \Longrightarrow x^* = 2.$$

Now $f''(x) = -\frac{4}{x^2} - 1 < 0$ for all x since $x^2 > 0$ and so $f(x)$ is globally concave and $x^* = 2$ is the global maximum of $f(x)$.

Problem 2.111 *[B***] Consider*

$$f(x) = 5 - 4e^{-5x} - 5e^{2x}.$$

Find the first-order condition and solve. Determine if the solution is a global maximizer or minimizer.

Answer: We have $f'(x) = 20e^{-5x} - 10e^{2x}$ and so the first-order condition $f'(x^*) = 0$ yields

$$20e^{-5x^*} - 10e^{2x^*} = 0 \Longrightarrow 20e^{-5x^*} = 10e^{2x^*} \Longrightarrow e^{7x^*} = 2 \Longrightarrow x^* = \frac{\ln(2)}{7} \approx 0.10.$$

Now for all x

$$f''(x) = -100e^{-5x} - 20e^{2x} < 0$$

since $e^{-5x} > 0$ and $e^{2x} > 0$, and so $f(x)$ is globally concave and $x^* = \frac{\ln(2)}{7}$ is a global maximizer.

Problem 2.112 *[C**] Consider $f(x) = (3x)^x$ for $x > 0$. Find the first-order condition and solve. Show that your solution is a global minimizer.*

Answer: Changing from base $3x$ to base e we have

$$f(x) = (3x)^x = \left(e^{\ln(3x)}\right)^x = e^{x\ln(3x)} = e^{x(\ln(3)+\ln(x))} = e^{x\ln(3)+x\ln(x)}.$$

Therefore

$$f'(x) = (\ln(3) + 1 + \ln(x))\, e^{x\ln(3)+x\ln(x)}$$

and so the first-order condition $f'(x^*) = 0$ yields

$$\begin{aligned}(\ln(3) + 1 + \ln(x^*))\, e^{x^*\ln(3)+x^*\ln(x^*)} &= 0 \\ &\Longrightarrow \ln(3) + 1 + \ln(x^*) = 0 \\ &\Longrightarrow \ln(x^*) = -1 - \ln(3) \Longrightarrow x^* = \frac{e^{-1}}{3} \approx 0.123\end{aligned}$$

since $e^{x^*\ln(3)+x^*\ln(x^*)} > 0$.

Now

$$f''(x) = \overbrace{(\ln(3) + 1 + \ln(x))^2}^{+}\overbrace{e^{x\ln(3)+x\ln(x)}}^{+} + \overbrace{\frac{1}{x}}^{+}\overbrace{e^{x\ln(3)+x\ln(x)}}^{+} > 0$$

for all x and so $x^* = \frac{e^{-1}}{3}$ is a global minimizer.

Problem 2.113 *Consider*

$$f(x) = e^{-x^2+2x-4}, \text{ for } -\infty < x < \infty.$$

*[C***] Find the first-order condition and solve. [B**] Determine whether your solution corresponds to a local maximum or minimum. [A**] Show that $f(x)$ is neither globally convex nor globally concave. [A***]Show that x^* corresponds to a global maximum.*

Answer: Using the chain rule we have

$$f'(x) = (-2x + 2)\, e^{-x^2+2x-4}$$

and so the first-order condition $f'(x^*) = 0$ yields

$$\begin{aligned}(-2x^* + 2)\, e^{-(x^*)^2+2x^*-4} &= 0 \\ &\Longrightarrow -2x^* + 2 = 0 \text{ (since } e^{-(x^*)^2+2x^*-4} > 0) \Longrightarrow x^* = 1.\end{aligned}$$

Furthermore

$$f''(x) = -2e^{-x^2+2x-4} + (-2x+2)^2\, e^{-x^2+2x-4} = e^{-x^2+2x-4}\left((-2x+2)^2 - 2\right)$$

and so

$$f''(1) = \left(e^{-x^2+2x-4}\left((-2x+2)^2 - 2\right)\right)|_{x=1} = -2e^{-3} < 0$$

so $f(x)$ is locally concave at $x^* = 1$ and hence $x^* = 1$ is a local maximizer. It turns out that $f(x)$ is not globally concave since (for example)

$$f''(3) = e^{-3^2+2\times3-4}\left((-2\times3+2)^2 - 2\right) = 14e^{-7} > 0.$$

In general $f(x)$ is locally concave for

$$1 - \frac{1}{2}\sqrt{2} < x < 1 + \frac{1}{2}\sqrt{2}$$

and locally convex outside this region since

$$\begin{aligned} f''(x) &= e^{-x^2+2x-4}\left((-2x+2)^2 - 2\right) \\ &= 4e^{-x^2+2x-4}\left(x - \left(1 - \frac{1}{2}\sqrt{2}\right)\right)\left(x - \left(1 + \frac{1}{2}\sqrt{2}\right)\right) \end{aligned}$$

and so $f(x)$ is neither globally concave nor globally convex. This is shown in the figure below.

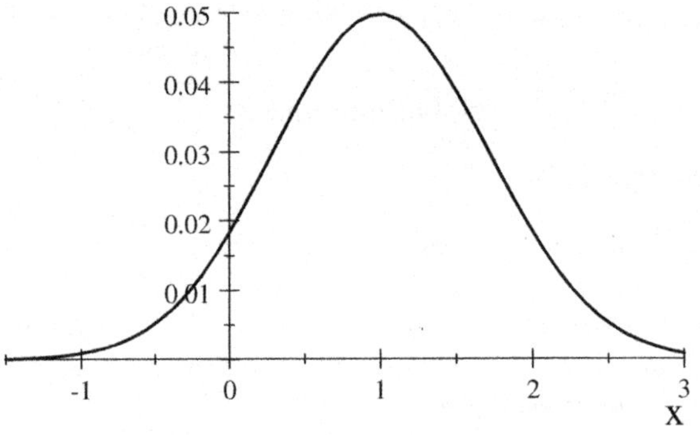

$$f(x) = e^{-x^2+2x-4}$$

We can nevertheless show that $x^* = 1$ corresponds to a global maximum in at least three different ways. First completing the square in the exponent we have

$$-x^2 + 2x - 4 = -(x-1)^2 - 3$$

so that

$$f(x) = e^{-x^2+2x-4} = e^{-(x-1)^2-3} = e^{-(x-1)^2}e^{-3}.$$

Now $e^{-(x-1)^2} \leq 1$ with equality only when $x = 1$ since for $x \neq 1$ we have $-(x-1)^2 < 0$ and $e^x < 1$ for $x < 0$. So it follows that $x^* = 1$ is a global maximizer.

Another approach is to use

$$f'(x) = (-2x+2)e^{-x^2+2x-4} = -2(x-1)\overbrace{e^{-x^2+2x-4}}^{+}$$

so that $f'(x) > 0$ for $x < 1$, $f'(x) < 0$ for $x > 1$ so that $x^* = 1$ is a global maximizer.

For yet another approach based on monotonic transformations see Problem 2.140 on page 112.

2.9.2 Profit Maximization with e^x and $\ln(x)$

Problem 2.114 *[D***] Suppose the production function is $Q = f(L) = L^{\frac{1}{3}}$. If one rewrote $f(L)$ in the form $f(L) = e^{\alpha \ln(L)}$, then what is α? [B**] Write profits as a function of $\tilde{L} \equiv \ln(L)$. Find the log of the demand for labour $\tilde{L}^* \equiv \ln(L^*)$ as a function of $\tilde{w} \equiv \ln(w)$. Find L^*. What is $\frac{\partial \tilde{L}^*}{\partial \tilde{w}}$ and what is its economic interpretation?*

Answer: We have $\alpha = \frac{1}{3}$ as

$$f(L) = e^{\alpha \ln(L)} = \left(e^{\ln(L)}\right)^{\frac{1}{3}} = e^{\frac{1}{3}\ln(L)} \Longrightarrow \alpha = \frac{1}{3}.$$

For profits as a function of $\tilde{L} \equiv \ln(L)$ we have

$$\begin{aligned} \pi\left(\tilde{L}\right) &= Pe^{\frac{1}{3}\tilde{L}} - WL = Pe^{\frac{1}{3}\tilde{L}} - We^{\ln(L)} = Pe^{\frac{1}{3}\tilde{L}} - We^{\tilde{L}} \\ &\Longrightarrow \pi'\left(\tilde{L}\right) = \frac{P}{3}e^{\frac{1}{3}\tilde{L}} - We^{\tilde{L}}. \end{aligned}$$

The first-order condition $\pi'\left(\tilde{L}^*\right) = 0$ yields

$$\begin{aligned} \frac{P}{3}e^{\frac{1}{3}\tilde{L}^*} - We^{\tilde{L}^*} &= 0 \\ &\Longrightarrow \frac{P}{3}e^{\frac{1}{3}\tilde{L}^*} = We^{\tilde{L}^*} \\ &\Longrightarrow \frac{\frac{P}{3}e^{\frac{1}{3}\tilde{L}^*}}{e^{\tilde{L}^*}} = \frac{We^{\tilde{L}^*}}{e^{\tilde{L}^*}} \\ &\Longrightarrow e^{-\frac{2}{3}\tilde{L}^*} = 3\frac{W}{P} \\ &\Longrightarrow -\frac{2}{3}\tilde{L}^* = \ln\left(3\frac{W}{P}\right) \\ &\Longrightarrow \tilde{L}^* = -\frac{3}{2}\ln\left(3\frac{W}{P}\right) \\ &\Longrightarrow \tilde{L}^* = -\frac{3}{2}\ln(3) - \frac{3}{2}\ln\left(\frac{W}{P}\right) = -\frac{3}{2}\ln(3) - \frac{3}{2}\tilde{w}. \end{aligned}$$

Here $\frac{d\tilde{L}^*}{d\tilde{w}} = -\frac{3}{2}$ is the elasticity of labour demand as

$$\tilde{L}^* = -\frac{3}{2}\ln(3) - \frac{3}{2}\tilde{w} \Longrightarrow \frac{d\tilde{L}^*}{d\tilde{w}} = \frac{d}{d\tilde{w}}\left(-\frac{3}{2}\ln(3) - \frac{3}{2}\tilde{w}\right) = -\frac{3}{2}$$

and $-\frac{3}{2}$ is the exponent on L^* as

$$\begin{aligned} \tilde{L}^* &= -\frac{3}{2}\ln(3) - \frac{3}{2}\tilde{w} \\ &\implies \ln(L^*) = -\frac{3}{2}\ln(3) - \frac{3}{2}\ln(w) \\ &\implies L^* = 3^{-\frac{3}{2}}w^{-\frac{3}{2}}. \end{aligned}$$

Problem 2.115 *[B***] Suppose for $0 < \alpha < 1$ a firm's production function is $Q = e^{\alpha\ln(L)}$. Write profits as a function of $\tilde{L} \equiv \ln(L)$. Find the log of the demand for labour $\tilde{L}^* \equiv \ln(L^*)$ as a function of $\tilde{w} \equiv \ln(w)$. What is $\frac{\partial \tilde{L}^*}{\partial \tilde{w}}$ and what is its economic interpretation?*

Answer: Here $Q = e^{\alpha\ln(L)}$ is just an ordinary Cobb-Douglas production function since

$$Q = e^{\alpha\ln(L)} = \left(e^{\ln(L)}\right)^{\alpha} = L^{\alpha}.$$

Profits are

$$\pi\left(\tilde{L}\right) = Pe^{\alpha\ln(L)} - WL = Pe^{\alpha\tilde{L}} - We^{\ln(L)} = Pe^{\alpha\tilde{L}} - We^{\tilde{L}}$$

and

$$\pi'\left(\tilde{L}\right) = \alpha Pe^{\alpha\tilde{L}} - We^{\tilde{L}}$$

so that the first-order condition $\pi'\left(\tilde{L}^*\right) = 0$ yields

$$\begin{aligned} \alpha Pe^{\alpha\tilde{L}^*} - We^{\tilde{L}^*} &= 0 \\ &\implies \alpha Pe^{\alpha\tilde{L}^*} = We^{\tilde{L}^*} \\ &\implies \frac{e^{\alpha\tilde{L}^*}}{e^{\tilde{L}^*}} = \frac{We^{\tilde{L}^*}}{\alpha Pe^{\tilde{L}^*}} \implies e^{(\alpha-1)\tilde{L}^*} = \frac{w}{\alpha} \\ &\implies (\alpha-1)\tilde{L}^* = \ln\left(\frac{w}{\alpha}\right) = -\ln(\alpha) + \ln(w) = -\ln(\alpha) + \tilde{w} \\ &\implies \tilde{L}^* = \frac{1}{1-\alpha}\ln(\alpha) - \frac{1}{1-\alpha}\tilde{w}. \end{aligned}$$

Here

$$\frac{d\tilde{L}^*}{d\tilde{w}} = \frac{d}{d\tilde{w}}\left(\frac{1}{1-\alpha}\ln(\alpha) - \frac{1}{1-\alpha}\tilde{w}\right) = -\frac{1}{1-\alpha}$$

is the elasticity of labour demand. In general $\frac{d\ln(L^*)}{d\ln(w)} = \frac{dL^*}{dw}\frac{w}{L^*}$ but a direct way to see this is to obtain the labour demand curve as

$$\begin{aligned} \tilde{L}^* &= \frac{1}{1-\alpha}\ln(\alpha) - \frac{1}{1-\alpha}\tilde{w} \\ &\implies \ln(L^*) = \frac{1}{1-\alpha}\ln(\alpha) - \frac{1}{1-\alpha}\ln(w) \\ &\implies L^* = \alpha^{\frac{1}{1-\alpha}}w^{-\frac{1}{1-\alpha}}. \end{aligned}$$

Problem 2.116 *Suppose a firm has a production function given by*

$$Q = f(L) = \ln\left(1 + 2L^{\frac{1}{2}}\right).$$

*[C***] Show that the marginal product of labour is positive and diminishing. [B***] Show that the production function is globally concave. Find the demand for labour L^* and the supply curve Q^*. Verify that the second-order condition for a global profit maximum is satisfied.*

Answer: We have

$$MP_L(L) = f'(L) = \frac{L^{-\frac{1}{2}}}{1 + 2L^{\frac{1}{2}}} = \frac{1}{L^{\frac{1}{2}}\left(1 + 2L^{\frac{1}{2}}\right)} = \frac{1}{2L + L^{\frac{1}{2}}} > 0$$

and

$$MP_L'(L) = f''(L) = -\frac{2 + \frac{1}{2}L^{-\frac{1}{2}}}{\left(2L + L^{\frac{1}{2}}\right)^2} < 0.$$

For profits as a function of L we have

$$\begin{aligned}
\pi(L) &= P\ln\left(1 + 2L^{\frac{1}{2}}\right) - WL \\
&\Longrightarrow \pi'(L) = P\left(\frac{1}{2L + L^{\frac{1}{2}}}\right) - W.
\end{aligned}$$

The first-order condition $\pi'(L^*) = 0$ yields

$$\begin{aligned}
P\left(\frac{1}{2L^* + (L^*)^{\frac{1}{2}}}\right) - W &= 0 \Longrightarrow \frac{1}{2L^* + (L^*)^{\frac{1}{2}}} = \frac{W}{P} = w \\
&\Longrightarrow 2L^* + (L^*)^{\frac{1}{2}} = \frac{1}{w} \Longrightarrow L^* + \frac{1}{2}(L^*)^{\frac{1}{2}} - \frac{1}{2w} = 0 \\
&\Longrightarrow \left((L^*)^{\frac{1}{2}}\right)^2 + \frac{1}{2}(L^*)^{\frac{1}{2}} - \frac{1}{2w} = 0 \\
&\Longrightarrow x^2 + \frac{1}{2}x - \frac{1}{2w} = 0
\end{aligned}$$

where $x = (L^*)^{\frac{1}{2}}$. This is just a quadratic so that solving we have

$$\begin{aligned}
x &= \frac{-\frac{1}{2} + \sqrt{\left(\frac{1}{2}\right)^2 - 4 \times 1 \times \left(-\frac{1}{2w}\right)}}{2} \\
&\Longrightarrow x = (L^*)^{\frac{1}{2}} = \frac{\sqrt{\frac{1}{4} + \frac{2}{w}} - \frac{1}{2}}{2} \\
&\Longrightarrow L^* = \left(\frac{\sqrt{\frac{1}{4} + \frac{2}{w}} - \frac{1}{2}}{2}\right)^2
\end{aligned}$$

where we ignore the other negative root. Now we find the supply curve using $p = w^{-1}$ as

$$\begin{aligned} Q^* &= f(L^*) = \ln\left(1 + 2(L^*)^{\frac{1}{2}}\right) \\ &= \ln\left(1 + 2\left(\left(\frac{\sqrt{\frac{1}{4}+\frac{2}{w}}-\frac{1}{2}}{2}\right)^2\right)^{\frac{1}{2}}\right) \\ &= \ln\left(1 + \sqrt{\frac{1}{4}+\frac{2}{w}}-\frac{1}{2}\right) = \ln\left(\frac{1}{2}+\sqrt{\frac{1}{4}+2p}\right). \end{aligned}$$

The second-order condition for a global profit maximum is satisfied since

$$\pi''(L) = -P\frac{2+\frac{1}{2}L^{-\frac{1}{2}}}{\left(2+L^{\frac{1}{2}}\right)^2} < 0.$$

Problem 2.117 *Suppose a firm has a production function given by*

$$Q = f(L) = \ln\left(1 + \frac{3}{2}L^{\frac{2}{3}}\right).$$

*[C***] Show that the marginal product of labour is positive and diminishing. Show that the production function is globally concave. [B**] Find the inverse demand curve for labour $w = h(L^*)$ and show that $h'(L^*) < 0$. Verify the second-order condition for a profit maximum. [A**] What mathematical trick would you need to solve for L^*?*

Answer: We have

$$\begin{aligned} MP_L(L) &= f'(L) = \frac{L^{-\frac{1}{3}}}{1+\frac{3}{2}L^{\frac{2}{3}}} \\ &= \frac{1}{L^{\frac{1}{3}}\left(1+\frac{3}{2}L^{\frac{2}{3}}\right)} = \frac{1}{L^{\frac{1}{3}}+\frac{3}{2}L} > 0 \\ MP_L'(L) &= f''(L) = -\frac{\frac{1}{3}L^{-\frac{2}{3}}+\frac{3}{2}}{\left(L^{\frac{1}{3}}+\frac{3}{2}L\right)^2} < 0. \end{aligned}$$

For profits as a function of L we have

$$\pi(L) = P\ln\left(1+\frac{3}{2}L^{\frac{2}{3}}\right) - WL \Longrightarrow \pi'(L) = P\left(\frac{1}{L^{\frac{1}{3}}+\frac{3}{2}L}\right) - W.$$

The first-order condition $\pi'(L^*) = 0$ yields

$$\begin{aligned} P\left(\frac{1}{(L^*)^{\frac{1}{3}} + \frac{3}{2}L^*}\right) - W &= 0 \\ &\Longrightarrow \frac{1}{(L^*)^{\frac{1}{3}} + \frac{3}{2}L^*} = \frac{W}{P} = w. \end{aligned}$$

Here

$$w = h(L^*) = \frac{1}{(L^*)^{\frac{1}{3}} + \frac{3}{2}L^*}$$

is the **inverse** labour demand curve. To show that it is downward sloping use

$$h'(L^*) = -\frac{\frac{1}{3}(L^*)^{-\frac{2}{3}} + \frac{3}{2}}{\left((L^*)^{\frac{1}{3}} + \frac{3}{2}L^*\right)^2} < 0.$$

The second-order condition for a global profit maximum is satisfied since

$$\pi''(L) = -P\frac{\frac{1}{3}L^{-\frac{2}{3}} + \frac{3}{2}}{\left(L^{\frac{1}{3}} + \frac{3}{2}L\right)^2} < 0.$$

Now let's try and solve for L^*. We have

$$\begin{aligned} \frac{1}{(L^*)^{\frac{1}{3}} + \frac{3}{2}L^*} &= w \Longrightarrow (L^*)^{\frac{1}{3}} + \frac{3}{2}L^* = w^{-1} \\ &\Longrightarrow \frac{3}{2}L^* + (L^*)^{\frac{1}{3}} - w^{-1} = 0 \\ &\Longrightarrow \frac{3}{2}x^3 + x - w^{-1} = 0 \end{aligned}$$

where $x = (L^*)^{\frac{1}{3}}$ and $L^* = x^3$. We require the formula for the root of a third degree polynomial, something that exists but which no normal person knows. To satisfy your curiosity the root is

$$x = \frac{1}{3w}\left(\left(9 + \sqrt{8w^2 + 81}\right)w^2\right)^{\frac{1}{3}} - \frac{2}{3}\frac{w}{\left(\left(9 + \sqrt{8w^2 + 81}\right)w^2\right)^{\frac{1}{3}}}$$

so that

$$L^* = x^3 = \left(\frac{1}{3w}\left(\left(9 + \sqrt{8w^2 + 81}\right)w^2\right)^{\frac{1}{3}} - \frac{2}{3}\frac{w}{\left(\left(9 + \sqrt{8w^2 + 81}\right)w^2\right)^{\frac{1}{3}}}\right)^3$$

as shown in the figure below.

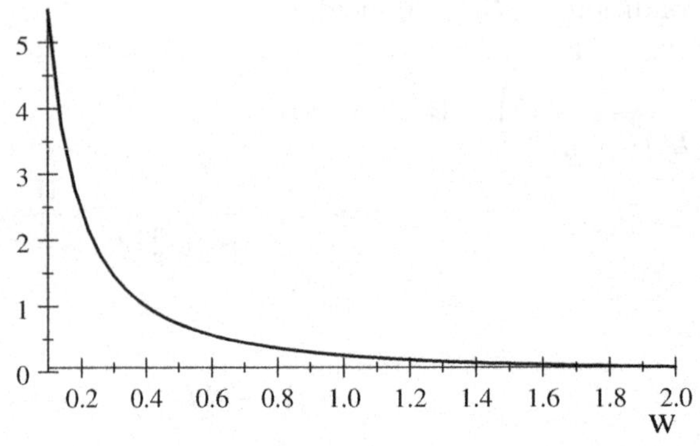

L^*

Problem 2.118 *[B**] Suppose a firm has a production function given by*

$$Q = f(L) = 1 - \frac{1}{2}e^{-2L}$$

for real wage $w \le 1$. Show that $\frac{1}{2} \le Q \le 1$, and that the marginal product of labour is positive and diminishing. Find the demand for labour L^* and the supply curve Q^*. Why do we require $w < 1$? Verify that the second-order condition for a global profit maximum is satisfied.

Answer: Since $0 < e^x \le 1$ for $x \le 0$, and since inequalities are reversed when we multiply both sides by a negative number as

$$x < a \Longrightarrow -x > -a \text{ and } x > a \Longrightarrow -x < -a$$

we have

$$\begin{aligned} 0 &< e^{-2L} \le 1 \Longrightarrow 0 < \frac{1}{2}e^{-2L} \le \frac{1}{2} \Longrightarrow -\frac{1}{2} \le -\frac{1}{2}e^{-2L} < -0 \\ &\Longrightarrow 1 - \frac{1}{2} \le \underbrace{1 - \frac{1}{2}e^{-2L}}_{Q} < 1 - 0 \Longrightarrow \frac{1}{2} \le Q < 1 \end{aligned}$$

as shown in the figure below.

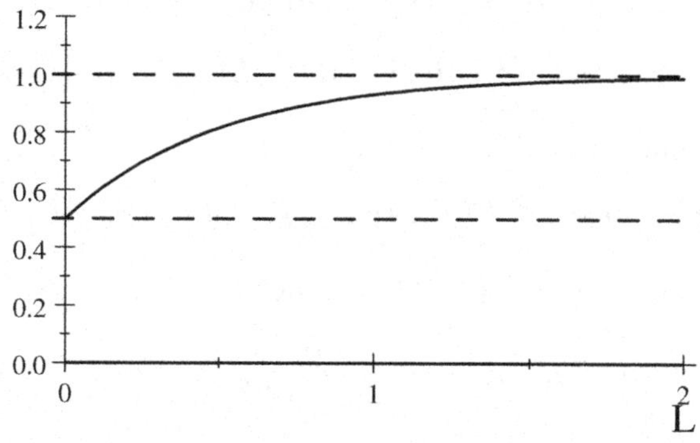

$Q = 1 - \frac{1}{2}e^{-2L}$

We have a positive and diminishing marginal product of labour as

$$MP_L(L) = f'(L) = e^{-2L} > 0 \text{ and } MP_L'(L) = f''(L) = -2e^{-2L} < 0.$$

For profits as a function of L we have

$$\pi(L) = P\left(1 - \frac{1}{2}e^{-2L}\right) - WL \Longrightarrow \pi'(L) = Pe^{-2L} - W$$

and so the first-order condition $\pi'(L^*) = 0$ yields

$$Pe^{-2L^*} - W = 0 \Longrightarrow e^{-2L^*} = \frac{W}{P} = w \Longrightarrow -2L^* = \ln(w) \Longrightarrow L^* = -\frac{1}{2}\ln(w).$$

The second-order condition is satisfied since for all L we have $\pi''(L) = -2Pe^{-2L} < 0$.

We require $w \leq 1$ since

$$w > 1 \Longrightarrow \ln(w) > 0 \Longrightarrow L^* = -\frac{1}{2}\ln(w) < 0.$$

The supply curve is found as

$$Q^* = f(L^*) = 1 - \frac{1}{2}e^{-2L^*} = 1 - \frac{1}{2}e^{-2\left(-\frac{1}{2}\ln(w)\right)} = 1 - \frac{1}{2}e^{\ln(w)} = 1 - \frac{w}{2} = 1 - \frac{1}{2p}.$$

Problem 2.119 *[A**] Suppose a firm has a production function given by*

$$Q = f(L) = 1 - \frac{e^{-\lambda L}}{\lambda}$$

where $w \leq 1$. Show that the marginal product of labour is positive and diminishing. Find the demand for labour L^* and the supply curve Q^*. Why do we require $w \leq 1$?

Answer: We have a positive and diminishing marginal product of labour as

$$MP_L(L) = f'(L) = e^{-\lambda L} > 0 \text{ and } MP_L'(L) = f''(L) = -\lambda e^{-\lambda L} < 0.$$

For profits as a function of L we have

$$\pi(L) = P\left(1 - e^{-\lambda L}\right) - WL \Longrightarrow \pi'(L) = Pe^{-\lambda L} - W$$

so that the first-order condition $\pi'(L^*) = 0$ yields

$$Pe^{-\lambda L^*} - W = 0 \Longrightarrow e^{-\lambda L^*} = \frac{W}{P} = w \Longrightarrow L^* = -\frac{1}{\lambda}\ln(w).$$

We require $w \leq 1$ since

$$w > 1 \Longrightarrow \ln(w) > 0 \Longrightarrow L^* = -\frac{1}{\lambda}\ln(w) < 0.$$

The supply curve is

$$Q^* = f(L^*) = 1 - \frac{e^{-\lambda L^*}}{\lambda} = 1 - \frac{e^{-\lambda\left(-\frac{1}{\lambda}\ln(w)\right)}}{\lambda} = 1 - \frac{e^{\ln(w)}}{\lambda} = 1 - \frac{w}{\lambda} = 1 - \frac{1}{\lambda p}.$$

2.9.3 Utility Maximization with e^x and $\ln(x)$

Problem 2.120 *[C***] Suppose that a household has utility*

$$U = \frac{2}{5}\ln(Q_1) + \frac{3}{5}\ln(Q_2)$$

with $Y = 10, P_2 = 5$ *so that the budget constraint is* $10 = P_1Q_1 + 5Q_2$*. Find the utility maximizing* Q_1^*, Q_2^* *from the first-order condition and show that* Q_1^* *is a global maximizer. What is the elasticity of demand for* Q_1^**?*

Answer: From the budget constraint we obtain the budget line as

$$10 = P_1Q_1 + 5Q_2 \Longrightarrow Q_2 = \frac{10 - P_1Q_1}{5} = 2 - \frac{P_1}{5}Q_1$$

so that

$$\begin{aligned} U &= \frac{2}{5}\ln(Q_1) + \frac{3}{5}\ln(Q_2) \\ &\Longrightarrow U(Q_1) = \frac{2}{5}\ln(Q_1) + \frac{3}{5}\ln\left(2 - \frac{P_1}{5}Q_1\right). \end{aligned}$$

Using the chain rule on the second term we have

$$U'(Q_1) = \frac{2}{5Q_1} - \frac{P_1}{5}\frac{\frac{3}{5}}{2 - \frac{P_1}{5}Q_1}.$$

The first-order condition $U'(Q_1^*) = 0$ yields

$$\begin{aligned}
\frac{2}{5Q_1^*} - \frac{P_1}{5}\frac{\frac{3}{5}}{2 - \frac{P_1}{5}Q_1^*} &= 0 \\
\implies \frac{2}{5Q_1^*} &= P_1 \frac{\frac{3}{5}}{10 - P_1 Q_1^*} \\
\implies \frac{2}{5}(10 - P_1 Q_1^*) &= \frac{3}{5}P_1 Q_1^* \\
\implies 4 &= P_1 Q_1^* \\
\implies Q_1^* &= \frac{4}{P_1} = 4P_1^{-1}.
\end{aligned}$$

Since this has the form Ax^b, the elasticity of demand is the exponent on P_1 which is -1. We show can that $Q_2^* = \frac{6}{5}$ (independent of P_1!) as

$$Q_2^* = 2 - \frac{P_1}{5}Q_1^* = 2 - \frac{P_1}{5} \times 4P_1^{-1} = 2 - \frac{4}{5} = \frac{6}{5}.$$

Taking the second derivative we find that

$$U''(Q_1) = -\frac{2}{5Q_1^2} - \left(\frac{P_1}{5}\right)^2 \frac{\frac{3}{5}}{\left(2 - \frac{P_1}{5}Q_1\right)^2} < 0$$

so that $U(Q_1)$ is globally concave and hence $Q_1^* = 4P_1^{-1}$ is a global maximizer.

Problem 2.121 *[B***] Suppose that a household has utility*

$$U = \frac{2}{5}\ln(Q_1) + \frac{3}{5}\ln(Q_2)$$

with general budget constraint is $Y = P_1Q_1 + P_2Q_2$. Find the utility maximizing Q_1^*, Q_2^* from the first-order condition and show that Q_1^* is a global maximizer. Show that the household always spends 40% of its income Y on Q_1, no matter what prices P_1, P_2 are.

Answer: From the budget constraint we obtain the budget line as

$$Y = P_1Q_1 + P_2Q_2 \implies Q_2 = \frac{Y - P_1Q_1}{P_2} = \frac{Y}{P_2} - \frac{P_1}{P_2}Q_1$$

so that

$$\begin{aligned}
U &= \frac{2}{5}\ln(Q_1) + \frac{3}{5}\ln(Q_2) \\
\implies U(Q_1) &= \frac{2}{5}\ln(Q_1) + \frac{3}{5}\ln\left(\frac{Y}{P_2} - \frac{P_1}{P_2}Q_1\right).
\end{aligned}$$

We have

$$U'(Q_1) = \frac{2}{5Q_1} - \frac{P_1}{P_2}\frac{\frac{3}{5}}{\frac{Y}{P_2} - \frac{P_1}{P_2}Q_1} = \frac{2}{5Q_1} - \frac{3}{5}\frac{P_1}{Y - P_1Q_1}.$$

The first-order condition $U'(Q_1^*) = 0$ yields

$$\begin{aligned}\frac{2}{5Q_1^*} - \frac{3}{5}\frac{P_1}{Y - P_1Q_1^*} &= 0 \\ \Longrightarrow\ 2(Y - P_1Q_1^*) &= 3P_1Q_1^* \Longrightarrow Q_1^* = \frac{2}{5}\frac{Y}{P_1}.\end{aligned}$$

To find Q_2^* use

$$Q_2^* = \frac{Y}{P_2} - \frac{P_1}{P_2}Q_1^* = \frac{Y}{P_2} - \frac{P_1}{P_2}\frac{2}{5}\frac{Y}{P_1} = \frac{3}{5}\frac{Y}{P_2}.$$

Taking the second derivative we find that

$$U''(Q_1) = -\frac{2}{5Q_1^2} - \frac{3}{5}\frac{P_1^2}{(Y - P_1Q_1)^2} < 0$$

so that $U(Q_1)$ is globally concave and hence $Q_1^* = \frac{2}{5}\frac{Y}{P_1}$ is a global maximizer.

Since

$$Q_1^* = \frac{2}{5}\frac{Y}{P_1} \Longrightarrow \frac{P_1Q_1^*}{Y} = \frac{2}{5}$$

the household always spends $\frac{2}{5}$ or 40% of its income Y on Q_1.

Problem 2.122 *[B***] Suppose that a household has utility*

$$U = \alpha\ln(Q_1) + (1-\alpha)\ln(Q_2)$$

where $0 < \alpha < 1$. Find the utility maximizing Q_1^*, Q_2^* . What does the parameter α determine?

Answer: First solve for Q_2 as a function of Q_1 as

$$Q_2 = \frac{Y - P_1Q_1}{P_2} = \frac{Y}{P_2} - \frac{P_1}{P_2}Q_1$$

and use this to write U as a function of Q_1 only as

$$U(Q_1) = \alpha\ln(Q_1) + (1-\alpha)\ln\left(\frac{Y}{P_2} - \frac{P_1}{P_2}Q_1\right).$$

Differentiating $U(Q_1)$ with respect to Q_1 and using the chain rule for the second term yields

$$U'(Q_1) = \frac{\alpha}{Q_1} - \frac{P_1}{P_2}\frac{(1-\alpha)}{\frac{Y}{P_2} - \frac{P_1}{P_2}Q_1} = \frac{\alpha}{Q_1} - \frac{(1-\alpha)P_1}{Y - P_1Q_1}.$$

Note that P_2 drops out! The first-order condition $U'(Q_1^*) = 0$ then yields

$$\begin{aligned}\frac{\alpha}{Q_1^*} - \frac{(1-\alpha)P_1}{Y - P_1Q_1^*} &= 0 \\ \Longrightarrow\ \frac{\alpha}{Q_1^*} &= \frac{(1-\alpha)P_1}{Y - P_1Q_1^*} \\ \Longrightarrow\ \alpha(Y - P_1Q_1^*) &= (1-\alpha)P_1Q_1^* \\ \Longrightarrow\ \alpha Y &= P_1Q_1^* \Longrightarrow Q_1^* = \frac{\alpha Y}{P_1}.\end{aligned}$$

To find Q_2^* use

$$Q_2^* = \frac{Y}{P_2} - \frac{P_1}{P_2}Q_1^* = \frac{Y}{P_2} - \frac{P_1}{P_2}\frac{\alpha Y}{P_1} = \frac{(1-\alpha)Y}{P_2}.$$

Since

$$Q_1^* = \frac{\alpha Y}{P_1} \Longrightarrow \frac{P_1 Q_1^*}{Y} = \alpha \text{ and } Q_2^* = \frac{(1-\alpha)Y}{P_2} \Longrightarrow \frac{P_2 Q_2^*}{Y} = 1-\alpha$$

the parameter α determines the budget shares of Q_1 and $1-\alpha$ determines the budget share of Q_2. For example if $\alpha = \frac{1}{3}$ then the household always spends one-third of its income on Q_1 and two-thirds of its income on Q_2.

Problem 2.123 *[C**] Suppose that the household has utility*

$$U = -e^{-Q_1} - e^{-Q_2}.$$

*Show that there is positive and diminishing marginal utility. [B**] Solve for Q_1^*.*

Answer: We have positive and diminishing marginal utility as

$$\begin{aligned} MU_1(Q_1) &= U_1'(Q_1) = e^{-Q_1} > 0,\ MU_1'(Q_1) = U_1''(Q_1) = -e^{-Q_1} < 0 \\ MU_2(Q_2) &= U_2'(Q_2) = e^{-Q_2} > 0,\ MU_2'(Q_2) = U_2''(Q_2) = -e^{-Q_2} < 0. \end{aligned}$$

Substituting $Q_2 = \frac{Y - P_1 Q_1}{P_2}$ in the utility function we obtain

$$\begin{aligned} U &= -e^{-Q_1} - e^{-Q_2} \\ &\Longrightarrow U(Q_1) = -e^{-Q_1} - e^{-\left(\frac{Y-P_1Q_1}{P_2}\right)} \\ &\Longrightarrow U'(Q_1) = e^{-Q_1} - \frac{P_1}{P_2}e^{-\left(\frac{Y-P_1Q_1}{P_2}\right)} \end{aligned}$$

and so the first-order condition $U'(Q_1^*) = 0$ yields

$$\begin{aligned} e^{-Q_1^*} - \frac{P_1}{P_2}e^{-\left(\frac{Y-P_1Q_1^*}{P_2}\right)} &= 0 \Longrightarrow e^{-Q_1^*} = \frac{P_1}{P_2}e^{-\left(\frac{Y-P_1Q_1^*}{P_2}\right)} \\ &\Longrightarrow -Q_1^* = \ln\left(\frac{P_1}{P_2}\right) - \left(\frac{Y-P_1Q_1^*}{P_2}\right) \\ &\Longrightarrow Q_1^* = -\ln\left(\frac{P_1}{P_2}\right) + \frac{Y}{P_2} - \frac{P_1}{P_2}Q_1^* \\ &\Longrightarrow Q_1^* = \frac{\ln\left(\frac{P_2}{P_1}\right) + \frac{Y}{P_2}}{1 + \frac{P_1}{P_2}} = \frac{P_2\ln(P_2) - P_2\ln(P_1) + Y}{P_1 + P_2}. \end{aligned}$$

The solution Q_1^* is a global maximizer since

$$U''(Q_1) = -e^{-Q_1} - \left(\frac{P_1}{P_2}\right)^2 e^{-\left(\frac{Y-P_1Q_1}{P_2}\right)} < 0$$

and hence $U(Q_1)$ is globally concave.

Problem 2.124 *[A**] Suppose a household has utility*

$$U = -Q_1^{-1} + \ln(Q_2).$$

Find the utility maximizing Q_1^* from the first-order condition and show that Q_1^* is a global maximizer.

Answer: From the budget constraint we obtain the budget line as

$$Y = P_1 Q_1 + P_2 Q_2 \Longrightarrow Q_2 = \frac{Y - P_1 Q_1}{P_2} = \frac{Y}{P_2} - \frac{P_1}{P_2} Q_1$$

and so we have

$$U(Q_1) = -Q_1^{-1} + \ln\left(\frac{Y}{P_2} - \frac{P_1}{P_2} Q_1\right)$$

so that

$$U'(Q_1) = Q_1^{-2} - \frac{P_1}{P_2} \frac{1}{\frac{Y}{P_2} - \frac{P_1}{P_2} Q_1} = Q_1^{-2} - \frac{P_1}{Y - P_1 Q_1}.$$

The first-order condition $U'(Q_1^*) = 0$ then yields

$$\begin{aligned} (Q_1^*)^{-2} - \frac{P_1}{Y - P_1 Q_1^*} &= 0 \\ &\Longrightarrow (Q_1^*)^{-2} = \frac{P_1}{Y - P_1 Q_1^*} \\ &\Longrightarrow Y - P_1 Q_1^* = P_1 (Q_1^*)^2 \\ &\Longrightarrow P_1 (Q_1^*)^2 + P_1 Q_1^* - Y = 0 \\ &\Longrightarrow (Q_1^*)^2 + Q_1^* - \frac{Y}{P_1} = 0. \end{aligned}$$

This is a quadratic of the form $x^2 + x - \frac{Y}{P_1} = 0$ where $x = Q_1^*$. Using $x = \frac{-b \pm \sqrt{b^2 - 4ac}}{2a}$ we have choosing the positive root

$$Q_1^* = \frac{-1 + \sqrt{1^2 - 4\left(-\frac{Y}{P_1}\right)}}{2} = \frac{\sqrt{1 + 4\frac{Y}{P_1}} - 1}{2}.$$

Notice that $Q_1^* > 0$ since

$$1 + 4\frac{Y}{P_2} > 1 \Longrightarrow \sqrt{1 + 4\frac{Y}{P_1}} - 1 > 0.$$

Taking the second derivative we find that

$$U''(Q_1) = -2Q_1^{-3} - \frac{P_1^2}{(Y - P_1 Q_1)^2} < 0$$

so that $U(Q_1)$ is globally concave and hence Q_1^* is a global maximizer.

Problem 2.125 *[A **] Suppose a household has utility*

$$U = 2Q_1^{\frac{1}{2}} + \ln(Q_2).$$

Find the utility maximizing Q_1^* from the first-order condition and show that Q_1^* is a global maximizer. Show for $P_1 = 1$ that Q_1 is a normal good.

Answer: From the budget constraint we obtain the budget line as

$$Y = P_1Q_1 + P_2Q_2 \Longrightarrow Q_2 = \frac{Y - P_1Q_1}{P_2} = \frac{Y}{P_2} - \frac{P_1}{P_2}Q_1$$

and so we have

$$U(Q_1) = 2Q_1^{\frac{1}{2}} + \ln\left(\frac{Y}{P_2} - \frac{P_1}{P_2}Q_1\right)$$

so that

$$U'(Q_1) = Q_1^{-\frac{1}{2}} - \frac{P_1}{P_2}\frac{1}{\frac{Y}{P_2} - \frac{P_1}{P_2}Q_1} = Q_1^{-\frac{1}{2}} - \frac{P_1}{Y - P_1Q_1}.$$

The first-order condition $U'(Q_1^*) = 0$ then is

$$\begin{aligned}
(Q_1^*)^{-\frac{1}{2}} - \frac{P_1}{Y - P_1Q_1^*} &= 0 \\
&\Longrightarrow (Q_1^*)^{-\frac{1}{2}} = \frac{P_1}{Y - P_1Q_1^*} \\
&\Longrightarrow Y - P_1Q_1^* = P_1(Q_1^*)^{\frac{1}{2}} \\
&\Longrightarrow P_1Q_1^* + P_1(Q_1^*)^{\frac{1}{2}} - Y = 0 \\
&\Longrightarrow Q_1^* + (Q_1^*)^{\frac{1}{2}} - \frac{Y}{P_1} = 0.
\end{aligned}$$

This is a quadratic of the form

$$x^2 + x - \frac{Y}{P_1} = 0$$

where $x = (Q_1^*)^{\frac{1}{2}}$ so that using $x = \frac{-b \pm \sqrt{b^2 - 4ac}}{2a}$ we have choosing the positive root

$$\begin{aligned}
(Q_1^*)^{\frac{1}{2}} &= \frac{-1 + \sqrt{1^2 - 4\left(-\frac{Y}{P_1}\right)}}{2} = \frac{\sqrt{1 + 4\frac{Y}{P_1}} - 1}{2} \\
&\Longrightarrow Q_1^* = \left(\frac{\sqrt{1 + 4\frac{Y}{P_1}} - 1}{2}\right)^2.
\end{aligned}$$

Taking the second derivative we find that

$$U''(Q_1) = -\frac{1}{2}Q_1^{-\frac{3}{2}} - \frac{P_1^2}{(Y - P_1Q_1)^2} < 0$$

so that $U(Q_1)$ is globally concave and hence Q_1^* is a global maximizer.

For $P_1 = 1$ we can show that Q_1^* is a normal good as

$$\begin{aligned} P_1 &= 1 \\ &\Longrightarrow Q_1^* = \frac{d}{dY}\left(\frac{\sqrt{1+4Y}-1}{2}\right)^2 \\ &\Longrightarrow \frac{dQ_1^*}{dY} = \frac{\sqrt{1+4Y}-1}{\sqrt{1+4Y}} > 0 \end{aligned}$$

since $Y > 0$ insures that the numerator of $\frac{dQ_1^*}{dY}$ is positive.

2.9.4 Mark Maximization with e^x and $\ln(x)$

Problem 2.126 *[B**] An exam has two questions. Let t_1 and t_2 be the time spent on Questions 1 and 2 respectively. The mark on Question 1 and Question 2 is*

$$M_1(t_1) = 60\left(1 - e^{-t_1}\right), \; M_2(t_2) = 40\left(1 - e^{-t_2}\right).$$

Show that the marginal mark is positive and diminishing for each question. If the exam is 1 hour long so that $t_1 + t_2 = 1$, find the optimal amount of time t_1^* and t_2^* spent on each question. What grade would the student receive?

Answer: We have a positive and diminishing marginal mark for each question as

$$\begin{aligned} M_1'(t_1) &= \overbrace{60e^{-t_1}}^{+} > 0, \; M_1''(t_1) = -\overbrace{60e^{-t_1}}^{+} < 0 \\ M_2'(t_2) &= \overbrace{40e^{-t_2}}^{+} > 0, \; M_2''(t_2) = -\overbrace{40e^{-t_2}}^{+} < 0. \end{aligned}$$

Since $t_2 = 1 - t_1$ we can write the mark as a function of t_1 alone as

$$M(t_1) = M_1(t_1) + M_2(1 - t_1) = 60\left(1 - e^{-t_1}\right) + 40\left(1 - e^{-(1-t_1)}\right)$$

so that $M'(t_1) = 60e^{-t_1} - 40e^{-(1-t_1)}$ and the first-order condition $M'(t_1^*) = 0$ yields

$$\begin{aligned} 60e^{-t_1^*} - 40e^{-(1-t_1^*)} &= 0 \Longrightarrow 60e^{-t_1^*} = 40e^{-(1-t_1^*)} \\ &\Longrightarrow \frac{60}{40} = \frac{3}{2} = e^{-(1-t_1^*)}e^{-(-t_1^*)} = e^{2t_1^*-1} \\ &\Longrightarrow 2t_1^* - 1 = \ln\left(\frac{3}{2}\right) \\ &\Longrightarrow t_1^* = \frac{1}{2} + \frac{1}{2}\ln\left(\frac{3}{2}\right) \approx 0.7. \end{aligned}$$

We have

$$t_2^* = 1 - t_1^* = 1 - \left(\frac{1}{2} + \frac{1}{2}\ln\left(\frac{3}{2}\right)\right) = \frac{1}{2} - \frac{1}{2}\ln\left(\frac{3}{2}\right) \approx 0.3.$$

The student's grade will be

$$M(t_1^*) = 60\left(1 - e^{-0.7}\right) + 40\left(1 - e^{-0.3}\right) \approx 41.$$

Problem 2.127 *[B**] Suppose the marks on Questions 1 and 2 are*

$$M_1(t_1) = 70\left(1 - e^{-t_1}\right), M_2(t_2) = 30\left(1 - e^{-t_2}\right).$$

Show for Question 1 that the marginal mark is positive and diminishing. If the exam is 1 hour long so that $t_1 + t_2 = 1$, find the optimal amount of time t_1^*, t_2^*. Show this is a global maximizer. What would the grade be?

Answer: We have a positive and diminishing marginal mark for Question 1 as

$$M_1'(t_1) = \overbrace{70e^{-t_1}}^{+} > 0 \text{ and } M_1''(t_1) = -\overbrace{70e^{-t_1}}^{+} < 0.$$

Since $t_2 = 1 - t_1$ we can write the mark as a function of t_1 alone as

$$M(t_1) = M_1(t_1) + M_2(1 - t_1) = 70\left(1 - e^{-t_1}\right) + 30\left(1 - e^{-(1-t_1)}\right)$$

so that $M'(t_1) = 70e^{-t_1} - 30e^{-(1-t_1)}$ and the first-order condition $M'(t_1^*) = 0$ yields

$$\begin{aligned}
70e^{-t_1^*} - 30e^{-(1-t_1^*)} &= 0 \\
&\implies 70e^{-t_1^*} = 30e^{-(1-t_1^*)} \\
&\implies \frac{70}{30} = \frac{7}{3} = e^{-(1-t_1^*)}e^{-(-t_1^*)} = e^{2t_1^*-1} \\
&\implies 2t_1^* - 1 = \ln\left(\frac{7}{3}\right) \\
&\implies t_1^* = \frac{1}{2} + \frac{1}{2}\ln\left(\frac{7}{3}\right) \approx 0.92.
\end{aligned}$$

We have

$$t_2^* = 1 - t_1^* = 1 - \left(\frac{1}{2} + \frac{1}{2}\ln\left(\frac{7}{3}\right)\right) \approx 0.08.$$

The grade is

$$M(t_1^*) = 70\left(1 - e^{-0.92}\right) + 30\left(1 - e^{-0.08}\right) \approx 44.$$

This is a global maximizer since $M(t_1)$ is globally concave as

$$M''(t_1) = -70e^{-t_1} - 30e^{-(1-t_1)} < 0.$$

Problem 2.128 *[B**] An exam has two questions. Let t_1 and t_2 be the time spent on Questions 1 and 2 respectively. The mark on Question 1 and Question 2 is*

$$M_1(t_1) = 60\ln(1+t_1), M_2(t_2) = 40\ln(1+t_2).$$

Show that the marginal mark is positive and diminishing for each question. If the exam has a length of one hour so that $t_1 + t_2 = 1$, find the optimal amount of time t_1^*, t_2^* spent on each question. What would the student's grade be? Now generalize this problem so that the exam has a length of T so that $t_1 + t_2 = T$. Find the optimal amount of time t_1^* and t_2^* spent on each question.

Answer: The marginal mark is positive and diminishing for each question as

$$\begin{aligned} M_1'(t_1) &= \frac{60}{1+t_1} > 0, M_1''(t_1) = -\frac{60}{(1+t_1)^2} < 0 \\ M_2'(t_2) &= \frac{40}{1+t_2} > 0, M_2''(t_2) = -\frac{40}{(1+t_2)^2} < 0. \end{aligned}$$

Replacing t_2 with $1 - t_1$ we have

$$M(t_1) = 60\ln(1+t_1) + 40\ln(1+1-t_1) = 60\ln(1+t_1) + 40\ln(2-t_1)$$

so that $M'(t_1) = \frac{60}{1+t_1} - \frac{40}{2-t_1}$. The first-order condition $M'(t_1^*) = 0$ yields

$$\begin{aligned} \frac{60}{1+t_1^*} - \frac{40}{2-t_1^*} &= 0 \\ &\implies \frac{60}{1+t_1^*} = \frac{40}{2-t_1^*} \\ &\implies 60(2-t_1^*) = 40(1+t_1^*) \\ &\implies 100t_1^* = 80 \implies t_1^* = \frac{80}{100} = 0.8. \end{aligned}$$

Since $t_2 = 1 - t_1$ we have $t_2^* = 1 - 0.8 = 0.2$. The student would receive a grade of

$$M(t_1^*) = 60\ln(1+0.8) + 40\ln(2-0.8) \approx 43.$$

Here $t_1^* = 0.8$ is a global maximizer since $M(t_1)$ is globally concave as

$$M''(t_1) = -\frac{60}{(1+t_1)^2} - \frac{40}{(2-t_1)^2} < 0.$$

Now if $t_1 + t_2 = T$ we have $t_2 = 1 - t_1$ so that

$$\begin{aligned} M(t_1) &= 60\ln(1+t_1) + 40\ln(T+1-t_1) \\ &\implies M'(t_1) = \frac{60}{1+t_1} - \frac{40}{T+1-t_1}. \end{aligned}$$

The first-order condition $M'(t_1^*) = 0$ yields

$$\begin{aligned}
\frac{60}{1+t_1^*} - \frac{40}{T+1-t_1^*} &= 0 \\
&\Longrightarrow \frac{60}{1+t_1^*} = \frac{40}{T+1-t_1^*} \\
&\Longrightarrow 60\,(T+1-t_1^*) = 40\,(1+t_1^*) \\
&\Longrightarrow 100t_1^* = 60T + 20 \\
&\Longrightarrow t_1^* = 0.6T + 0.2.
\end{aligned}$$

Since $t_2 = T - t_1$ we have

$$t_2^* = T - (0.6T + 0.2) = 0.4T - 0.2.$$

Here t_1^* is a global maximizer since

$$M''(t_1) = -\frac{60}{(1+t_1)^2} - \frac{40}{(T+1-t_1)^2} < 0$$

and $M(t_1)$ is globally concave.

Problem 2.129 *[B**] Suppose marks on Questions 1 and 2 are*

$$M_1(t_1) = 50\ln(1+2t_1)\,, M_2(t_2) = 50\ln(1+2t_2)\,.$$

If the exam has a length of T so that $t_1 + t_2 = T$, show that it is optimal to spend an equal amount of time on each question. How long must the exam be for the grade to be an 80?

Answer: Replacing t_2 with $1 - t_1$ we have

$$M(t_1) = 50\ln(1+2t_1) + 50\ln(1+2(T-t_1)) = 50\ln(1+2t_1) + 50\ln(2T+1-2t_1)$$

so that

$$M'(t_1) = \frac{50}{1+2t_1} \times 2 + \frac{50}{2T+1-2t_1} \times (-2) = \frac{100}{1+2t_1} - \frac{100}{2T+1-2t_1}$$

and so the first-order condition $M'(t_1^*) = 0$ yields

$$\begin{aligned}
\frac{100}{1+2t_1^*} - \frac{100}{2T+1-2t_1^*} &= 0 \\
&\Longrightarrow \frac{100}{1+2t_1^*} = \frac{100}{2T+1-2t_1^*} \\
&\Longrightarrow 2T+1-2t_1^* = 1+2t_1^* \\
&\Longrightarrow 4t_1^* = 2T \Longrightarrow t_1^* = \frac{2}{4}T = \frac{T}{2}.
\end{aligned}$$

Since $t_2 = T - t_1$ we have $t_2^* = T - \frac{T}{2} = \frac{T}{2}$. Here t_1^* is a global maximizer since

$$M''(t_1) = -\frac{200}{(1+2t_1)^2} - \frac{200}{(2T+1-2t_1)^2} < 0$$

so that $M(t_1)$ is globally concave.

The grade is

$$\begin{aligned} M(t_1^*) &= 50\ln(1+2t_1^*) + 50\ln(2T+1-2t_1^*) \\ &= 50\ln(1+2t_1^*) + 50\ln(2T+1-2t_1^*) \\ &= 50\ln\left(1+2\frac{T}{2}\right) + 50\ln\left(2T+1-2\frac{T}{2}\right) \\ &= 50\ln(1+T) + 50\ln(T+1) = 100\ln(1+T). \end{aligned}$$

For the grade to be 80 the exam has to be a length T such that

$$\begin{aligned} 100\ln(1+T) &= 80 \\ &\implies \ln(1+T) = \frac{80}{100} = 0.8 \\ &\implies 1+T = e^{0.8} \implies T = e^{0.8} - 1 \approx 1.2. \end{aligned}$$

2.9.5 Exponential Growth

Problem 2.130 *[B**] If t is time measured in years and the economy grows as $y = f(t) = 10e^{0.04t}$, show that the economy grows at four percent per year. Find the exact time it will take the economy to double in size, and compare this with the value predicted by the rule of 72. If instead $y = t^{10}e^{0.04t}$, find the growth rate $\rho(t)$ and show that the growth rate will converge to four percent a year.*

Answer: The growth rate is

$$\rho(t) \equiv \frac{f'(t)}{f(t)} = \frac{0.04 \times 10e^{0.04t}}{10e^{0.04t}} = 0.04$$

or four percent a year. The exact time Δt it takes the economy to double in size satisfies

$$\begin{aligned} f(t+\Delta t) &= 2f(t) \\ &\implies 10e^{0.04(t+\Delta t)} = 2\times 10e^{0.04t} \\ &\implies e^{0.04(t+\Delta t)} = 2e^{0.04t} \\ &\implies \frac{e^{0.04(t+\Delta t)}}{e^{0.04t}} = 2 \\ &\implies e^{0.04\Delta t} = 2 \\ &\implies 0.04\Delta t = \ln(2) \implies \Delta t = \frac{\ln(2)}{0.04} \approx 17.3 \text{ years.} \end{aligned}$$

The rule of 72 predicts instead that $\Delta t = \frac{72}{4} = 18$ years.

If $f(t) = t^{10}e^{0.04t}$ then the growth rate is

$$\rho(t) = \frac{d}{dt}\ln\left(t^{10}e^{0.04t}\right) = \frac{d}{dt}\left(10\ln(t) + 0.04t\right) = \frac{10}{t} + 0.04$$

so that

$$\lim_{t\to\infty}\rho(t) = \lim_{t\to\infty}\left(\frac{10}{t} + 0.04\right) = 0.04.$$

Problem 2.131 *[B**] Suppose $y = f(t)$ is the population of the world at time t, where t is measured in years. Let's define today as $t = 0$ and suppose that the population of the world today is 6 (billion). If population grows at a constant 1.8% per year, then what is $f(t)$? How long will it take for the world's population to double? How long with it take for the world's population to reach twenty billion? How long ago was it that the first person appeared on the earth? The second person? Suppose using family planning you wanted to reduce the population growth rate so that in 100 years there were no more than 10 (billion) people. What population growth rates would be consistent with this objective?*

Answer: For constant growth of 1.8% with $f(0) = 6$ we require that $f(t) = 6e^{0.018t}$. If the population doubles at time t from 6 to 12 billion then

$$\begin{aligned} 12 &= 6e^{0.018t} \\ &\Longrightarrow e^{0.018t} = \frac{12}{6} = 2 \\ &\Longrightarrow 0.018t = \ln(2) \\ &\Longrightarrow t = \frac{\ln(2)}{0.018} \approx 39. \end{aligned}$$

World population will double in about 39 years. If at time t the population reaches 20 billion then

$$\begin{aligned} 20 &= 6e^{0.018t} \Longrightarrow e^{0.018t} = \frac{20}{6} \\ &\Longrightarrow 0.018t = \ln\left(\frac{20}{6}\right) \\ &\Longrightarrow t = \frac{\ln\left(\frac{20}{6}\right)}{0.018} \approx 67. \end{aligned}$$

World population will double will reach 20 billion in about 67 years.

The first person existed at t when

$$\begin{aligned} f(t) &= \frac{1}{1000000000} = 6e^{0.018t} \\ &\Longrightarrow e^{0.018t} = \frac{1}{6000000000} \\ &\Longrightarrow 0.018t = \ln\left(\frac{1}{6000000000}\right) \\ &\Longrightarrow t = \frac{\ln\left(\frac{1}{6000000000}\right)}{0.018} \approx -1251 \end{aligned}$$

or about 1251 years ago. Similarly there were 2 people when

$$\begin{aligned} f(t) &= \frac{2}{1000000000} = 6e^{0.018t} \\ \Longrightarrow e^{0.018t} &= \frac{2}{6 \times 1000000000} \\ \Longrightarrow 0.018t &= \ln\left(\frac{2}{6000000000}\right) \\ \Longrightarrow t &= \frac{\ln\left(\frac{2}{6000000000}\right)}{0.018} \approx -1212 \end{aligned}$$

or about 1212 years ago. Historically we know people have been around much longer than 1300 years, which illustrates the difficulty of projecting exponential growth too far into either the future or past.

If in 100 years there are to be 10 billion people we require a growth rate $\mu \leq \mu_{\max}$ where

$$\begin{aligned} 10 &= 6e^{\mu_{\max} \times 100} \Longrightarrow e^{\mu_{\max} \times 100} = \frac{5}{3} \\ \Longrightarrow \mu_{\max} \times 100 &= \ln\left(\frac{5}{3}\right) \\ \Longrightarrow \mu_{\max} &= \frac{\ln\left(\frac{5}{3}\right)}{100} \approx 0.005. \end{aligned}$$

So we require $\mu \leq 0.005$ or that growth must be less than 0.5% per year.

2.9.6 Monotonic Transformations with e^x and $\ln(x)$

Problem 2.132 *[A***] In Problem 2.46 on page 38 it is shown that the global maximum of* $f(x) = 3x^{\frac{1}{3}} - 2x^{\frac{3}{2}}$ *for* $x > 0$ *is at* $x^* = 3^{-\frac{6}{7}}$. *If*

$$h(x) = \frac{e^{3x^{\frac{1}{3}}}}{\left(e^{x^{\frac{3}{2}}}\right)^2}$$

find the global maximum of $h(x)$ *without calculating* $h'(x)$.

Answer: We have

$$h(x) = \frac{e^{3x^{\frac{1}{3}}}}{\left(e^{x^{\frac{3}{2}}}\right)^2} = \frac{e^{3x^{\frac{1}{3}}}}{e^{2x^{\frac{3}{2}}}} = e^{3x^{\frac{1}{3}} - 2x^{\frac{3}{2}}} = e^{f(x)} = g(f(x)).$$

Since $g(x) = e^x$ is globally increasing it follows that $h(x)$ is a monotonic transformation of $f(x)$. The global maximum is an ordinal property, and so it follows that the global maximum of $h(x)$ is also $x^* = 3^{-\frac{6}{7}}$.

Problem 2.133 *[A ***] In Problem 2.47 on page 38 it is shown that the maximum of*

$$f(x) = -\frac{1}{5}x^5 - \frac{2}{x^2} \text{ for } x > 0$$

occurs at $x^* = 4^{\frac{1}{7}}$. *If*

$$h(x) = \frac{1}{\left(e^{x^5}\right)^{\frac{1}{5}}\left(e^{x^{-2}}\right)^2}$$

find the global maximum of $h(x)$ *without using the first-order condition.*

Answer: We have

$$h(x) = \frac{1}{\left(e^{x^5}\right)^{\frac{1}{5}}\left(e^{x^{-2}}\right)^2} = e^{-\frac{1}{5}x^5}e^{-2x^{-2}} = \exp\left(-\frac{1}{5}x^5 - \frac{2}{x^2}\right) = e^{f(x)}.$$

Since e^x is globally increasing it follows that $h(x)$ is a monotonic transformation of $f(x)$. It follows that the global maximum of $h(x)$ is $x^* = 4^{\frac{1}{7}}$.

Problem 2.134 *[A ***] In Problem 2.41 on page 34 it is shown that the global maximum of*

$$f(x) = x^{\frac{1}{2}} - x$$

for $x > 0$ *occurs at* $x^* = \frac{1}{4}$. *If*

$$h(x) = \exp\left(\frac{\exp(\sqrt{x})}{e^x}\right)$$

find the value x^* *that maximizes* $h(x)$ *without using the first-order condition for* $h(x)$.

Answer: We have

$$h(x) = \exp\left(\frac{\exp(\sqrt{x})}{e^x}\right) = \exp\left(\frac{e^{x^{\frac{1}{2}}}}{e^x}\right) = \exp\left(e^{x^{\frac{1}{2}}-x}\right) = g(f(x))$$

where $g(x) = e^{e^x}$ is a monotonic transformation as $g'(x) = e^x e^{e^x} > 0$ and so $x^* = \frac{1}{4}$ also maximizes $h(x)$.

Problem 2.135 *[A ***] The function*

$$f(x) = x^{\frac{2}{3}} - 2x \text{ for } x > 0$$

is globally concave and has a global maximum at $x^* = \frac{1}{27}$. *Show that*

$$h(x) = \exp\left(\left(e^{\ln(x)}\right)^{\frac{2}{3}} - \frac{e^{\ln(2)}}{e^{-\ln(x)}}\right)$$

is quasi-concave and find its global maximum.

Answer: We have

$$\begin{aligned} h(x) &= \exp\left(\left(e^{\ln(x)}\right)^{\frac{2}{3}} - \frac{e^{\ln(2)}}{e^{-\ln(x)}}\right) = \exp\left((x)^{\frac{2}{3}} - \frac{2}{\frac{1}{x}}\right) \\ &= \exp\left(x^{\frac{2}{3}} - 2x\right) = \exp(f(x)). \end{aligned}$$

Since $\exp(x)$ is globally increasing, it follows that $h(x)$ is a monotonic transformation of $f(x)$, and so $h(x)$ and $f(x)$ have the same global maximum at $x^* = \frac{1}{27}$. Since $f(x)$ is globally concave it follows that $h(x)$ is quasi-concave.

Problem 2.136 *[A***] Consider*

$$f(x) = \sqrt{\frac{1}{x^{12}e^{-x^2}}} \text{ for } x > 0.$$

In Problem 2.108 on page 84 we showed that $x^* = \sqrt{6}$ is the solution to the first-order condition for $f(x)$. Show that $x^* = \sqrt{6}$ is a global minimizer for $f(x)$.

Answer: Consider the monotonic transformation

$$h(x) = \ln(f(x)) = \ln\left(\sqrt{\frac{1}{x^{12}e^{-x^2}}}\right) = -6\ln(x) + \frac{x^2}{2}.$$

We have $h'(x) = -6x^{-1} + x$ and so the first-order condition yields

$$-6(x^*)^{-1} + x^* = 0 \Longrightarrow x^* = \sqrt{6}.$$

Now $h(x)$ is globally convex since for all x we have

$$h''(x) = 6x^{-2} + 1 > 0$$

and hence $x^* = \sqrt{6}$ is a global minimizer of $h(x)$. Since $f(x) = \exp(h(x))$ is a monotonic transformation of $h(x)$, it follows that $f(x) = \sqrt{\frac{1}{x^{12}e^{-x^2}}}$ has a global minimum at $x^* = \sqrt{6}$.

Problem 2.137 *[B***] Consider the functions*

$$f(x) = x^3 e^{-\frac{1}{2}x^2} \text{ and } h(x) = 3\ln(x) - \frac{x^2}{2} \text{ for } x > 0.$$

Find the x^* that is a global maximizer for $h(x)$ and show that x^* also is a global maximizer for $f(x)$.

Answer: We have $h'(x) = \frac{3}{x} - x$ and so the first-order condition $h'(x^*) = 0$ yields

$$\begin{aligned} \frac{3}{x^*} - x^* &= 0 \\ &\Longrightarrow \frac{3}{x^*} = x^* \Longrightarrow 3 = (x^*)^2 \\ &\Longrightarrow x^* = \sqrt{3}. \end{aligned}$$

The function $h(x)$ is globally concave since $h''(x) = -3x^{-2} - 1 < 0$ so $x^* = \sqrt{3}$ is a global maximizer for $h(x)$. Since

$$\ln(f(x)) = \ln\left(x^3 e^{-\frac{1}{2}x^2}\right) = \ln\left(x^3\right) + \ln\left(e^{-\frac{1}{2}x^2}\right) = 3\ln(x) - \frac{x^2}{2} = h(x)$$

and since $\ln(x)$ is globally increasing ($\frac{d\ln(x)}{dx} = \frac{1}{x} > 0$ for $x > 0$) it follows that $x^* = \sqrt{3}$ also is a global maximizer for $f(x)$. This is illustrated in the figure below.

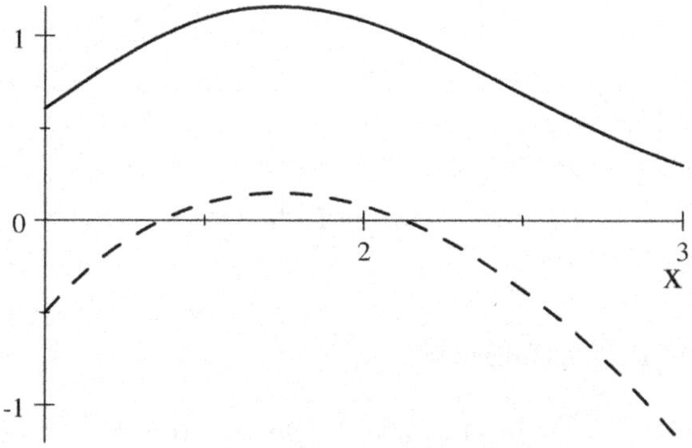

$x^3 e^{-\frac{x^2}{2}}$ — and $3\ln(x) - \frac{x^2}{2}$ - - -

Problem 2.138 *[B***] Consider $f(x) = x^4 e^{-\frac{1}{2}x^2}$ for $x > 0$. Find the first-order condition and solve. Show that your solution is a global maximizer.*

Answer: Using the product and chain rules we have

$$f'(x) = 4x^3 e^{-\frac{1}{2}x^2} - x^5 e^{-\frac{1}{2}x^2} = x^3 e^{-\frac{1}{2}x^2}\left(4 - x^2\right).$$

Therefore the first-order condition $f'(x^*) = 0$ yields

$$\begin{aligned} \underbrace{(x^*)^3}_{+}\underbrace{e^{-\frac{1}{2}(x^*)^2}}_{+}\left(4 - (x^*)^2\right) &= 0 \\ &\Longrightarrow 4 - (x^*)^2 = 0 \Longrightarrow (x^*)^2 = 4 \\ &\Longrightarrow x^* = 2. \end{aligned}$$

Consider the monotonic transformation $h(x) = \ln(f(x))$ as

$$h(x) = \ln\left(x^4 e^{-\frac{1}{2}x^2}\right) = \ln\left(x^4\right) + \ln\left(e^{-\frac{1}{2}x^2}\right) = 4\ln(x) - \frac{1}{2}x^2.$$

Then $h''(x) = -\frac{4}{x^2} - 1 < 0$ so that $h(x)$ is globally concave. It follows that $x^* = 2$ is a global maximizer for $h(x)$. Since $h(x)$ is a monotonic transformation of $f(x)$, it follows that $x^* = 2$ is a global maximizer for a $f(x)$ as well. This is illustrated in the figure below.

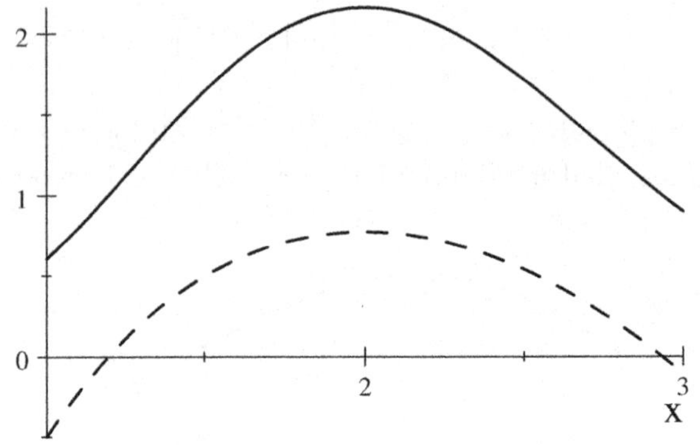

$x^4e^{-\frac{1}{2}x^2}$— and $\ln\left(x^4e^{-\frac{1}{2}x^2}\right)$ - - -

Problem 2.139 *[B***] Consider*

$$f(x) = x^3e^{-x^2} \text{ for } x > 0.$$

In Problem 2.102 on page 79 we showed that $x^* = \sqrt{\frac{3}{2}}$ *corresponds to a local maximum for* $f(x)$*. Use* $h(x) = \ln(f(x))$ *to show that* $x^* = \sqrt{\frac{3}{2}}$ *corresponds to a global maximum of* $f(x)$*.*

Answer: For $h(x) = \ln(f(x))$ we have

$$h(x) = \ln(f(x)) = \ln\left(x^3e^{-x^2}\right) = \ln\left(x^3\right) + \ln\left(e^{-x^2}\right) = 3\ln(x) - x^2$$

and so $h'(x) = \frac{3}{x} - 2x$ so that $h(x)$ is globally concave as $h''(x) = -\frac{3}{x^2} - 2 < 0.$ Thus $x^* = \sqrt{\frac{3}{2}}$ is a global maximizer for both $h(x)$ and $f(x)$.

Problem 2.140 *[B***] Consider*

$$f(x) = e^{-x^2+2x-4} \text{ for } -\infty < x < \infty.$$

In Problem 2.113 on page 87 we showed that $f(x)$ *is neither globally convex nor globally concave. Show that* $f(x)$ *is quasi-concave so that the solution to the first-order condition* $x^* = 1$ *corresponds to a global maximum.*

Answer: We first show that $h(x) = \ln(f(x))$ is globally concave as

$$\begin{aligned} h(x) &= \ln(f(x)) = -x^2 + 2x - 4 \\ &\Longrightarrow h''(x) = -2 < 0. \end{aligned}$$

Thus $f(x)$ is a monotonic transformation of a globally concave function as $f(x) = e^{h(x)}$ (with $\frac{d}{dx}e^x = e^x > 0$) and so $f(x)$ is quasi-concave. It follows that $x^* = 1$ is a global maximizer for both $h(x)$ and $f(x)$ as illustrated in the figures below.

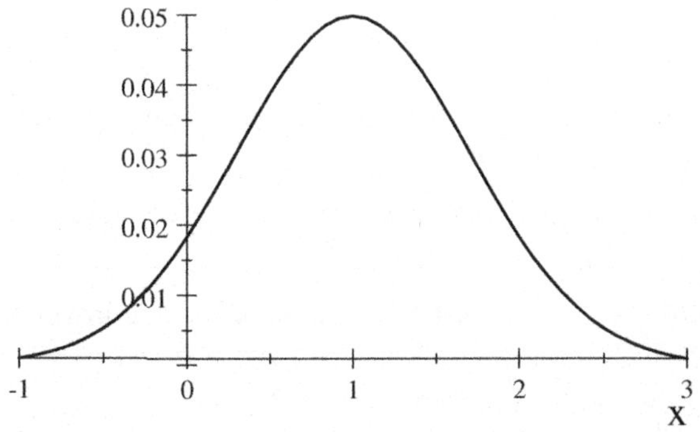

$$f(x) = e^{-x^2+2x-4}$$

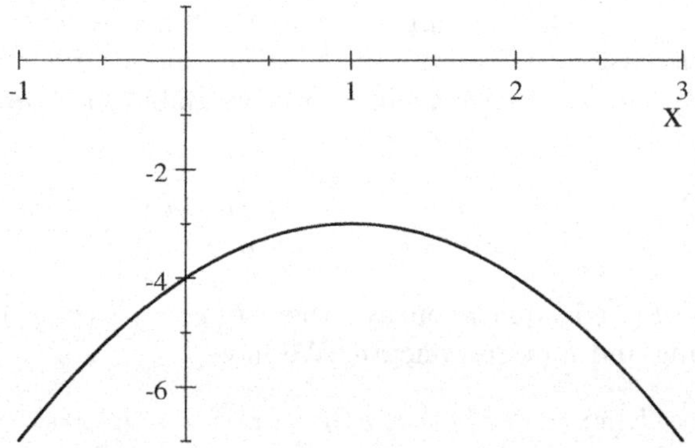

$$h(x) = -x^2 + 2x - 4$$

Problem 2.141 *[B***] Consider* $f(x) = x^3 e^{-x^2+2x+1}$ *for* $x > 0$. *Find the first-order condition and solve. Calculate* $h(x) = \ln(f(x))$, *the first-order condition for* $h(x)$, *and solve. Show that* $h(x)$ *is globally concave and that* $f(x)$ *is quasi-concave. Show that the solution to the first-order condition is a global maximizer for both* $f(x)$ *and* $h(x)$.

Answer: We have

$$\begin{aligned} f'(x) &= 3x^2 e^{-x^2+2x+1} + x^3(-2x+2)e^{-x^2+2x+1} \\ &= x^2 e^{-x^2+2x+1}\left(3 - 2x^2 + 2x\right) = \left(-2x^2 + 2x + 3\right) x^2 e^{-x^2+2x+1} \end{aligned}$$

and so the first-order condition $f'(x^*) = 0$ yields

$$\begin{aligned} \left(-2(x^*)^2 + 2x^* + 3\right)(x^*)^2 e^{-(x^*)^2+2x^*+1} &= 0 \\ &\Longrightarrow -2(x^*)^2 + 2x^* + 3 = 0 \\ &\Longrightarrow x^* = \frac{1}{2} + \frac{1}{2}\sqrt{7} \end{aligned}$$

Note the other root $x^* = \frac{1}{2} - \frac{1}{2}\sqrt{7} = -0.822$ can be excluded since $x > 0$.

For $h(x) = \ln(f(x))$ we have

$$h(x) = \ln(f(x)) = \ln\left(x^3 e^{-x^2+2x+1}\right) = \ln(x^3) + \ln\left(e^{-x^2+2x+1}\right) = 3\ln(x) - x^2 + 2x + 1.$$

Therefore $h'(x) = \frac{3}{x} - 2x + 2$ and so the first-order condition $h'(x^*) = 0$ yields

$$\begin{aligned} \frac{3}{x^*} - 2x^* + 2 &= 0 \\ &\Longrightarrow -2x^{*2} + 2x^* + 3 = 0 \\ &\Longrightarrow x^* = \frac{1}{2} + \frac{1}{2}\sqrt{7} \approx 1.82. \end{aligned}$$

The function $h(x)$ is globally concave since $h''(x) = -\frac{3}{x^2} - 2 < 0$ and hence $x^* = \frac{1}{2} + \frac{1}{2}\sqrt{7}$ is a global maximizer for $h(x)$. Since $h(x) = \ln(f(x))$ with $\ln(x)$ globally increasing, it follows that $h(x)$ is a monotonic transformation of $f(x)$. Thus $f(x)$ and $h(x)$ have their respective global maximums at the same $x^* = \frac{1}{2} + \frac{1}{2}\sqrt{7}$.

Problem 2.142 *[B***] Prove that if $f(x)$ is quasi-concave and $r(x)$ is globally increasing then $h(x) = r(f(x))$ is quasi-concave.*

Answer: If $f(x)$ is quasi-concave then $f(x) = g(h(x))$ where $g(x)$ is globally increasing and $h(x)$ is concave. We have

$$h(x) = r(f(x)) = r(g(h(x))) = s(h(x))$$

where $s(x) = r(g(x))$ is globally increasing since $s'(x) = r'(g(x))\, g'(x) > 0$. Since $h(x)$ is concave it follows by definition that $h(x)$ is quasi-concave.

Problem 2.143 *Prove that if $f(x)$ is quasi-concave then $-f(x)$ is quasi-convex.*

Answer: If $f(x)$ is quasi-concave then by definition it can be written as

$$f(x) = g(h(x))$$

where $g(x)$ is globally increasing and $h(x)$ is concave or $h''(x) < 0$. We have

$$-f(x) = -g(h(x)) = -g\left(-\underbrace{(-h(x))}_{s(x)}\right) = r(s(x))$$

where $r(x) = -g(-x)$ and $s(x) = -h(x)$. Now $r(x)$ is globally increasing since $r'(x) = g'(-x) > 0$ and $s(x)$ is convex since $s''(x) = -h''(x) > 0$ as $h''(x) < 0$. It follows then that $-f(x)$ is quasi-convex.

Problem 2.144 *[B***] Prove that if $f(x)$ is globally increasing then it is both quasi-concave and quasi-convex.*

Answer: If $f(x)$ is globally increasing then $f'(x) > 0$ for all x and $f(x)$ has a globally increasing inverse function $g(x)$ with $g'(x) > 0$ for all x that satisfies $g(f(x)) = x$. From this we obtain

$$g(f(x)) = x \Longrightarrow e^{g(f(x))} = e^x \Longrightarrow h(f(x)) = e^x$$

where $h(x) = e^{g(x)}$. Now e^x is globally convex (since $\frac{d^2}{dx^2}e^x = e^x > 0$) and $h'(x) = g'(x)e^{g(x)} > 0$ since $g'(x) > 0$. Hence $h(x)$ is a monotonic transformation. Thus a monotonic transformation of $f(x)$ is convex, and hence $f(x)$ is quasi-convex.

To prove that $f(x)$ is quasi-concave use

$$\begin{aligned} g(f(x)) &= x \\ &\Longrightarrow -e^{-g(f(x))} = -e^{-x} \\ &\Longrightarrow h(f(x)) = -e^{-x} \end{aligned}$$

where $h(x) = -e^{-g(x)}$. Now $-e^{-x}$ is globally concave (since $\frac{d^2}{dx^2}(-e^{-x}) = -e^{-x} < 0$) and $h'(x) = g'(x)e^{-g(x)} > 0$ since $g'(x) > 0$ and hence $h(x)$ is a monotonic transformation. Thus a monotonic transformation of $f(x)$ is concave and hence $f(x)$ is quasi-concave.

2.10 Taylor Series

Problem 2.145 *[B**] Calculate a first-order Taylor series of $f(x) = \frac{3}{2}x^{\frac{2}{3}}$ around $x_0 = 1$. Now calculate a second-order Taylor series around $x_0 = 1$.*

Answer: Since

$$\begin{aligned} f(x) &= \frac{3}{2}x^{\frac{2}{3}} \Longrightarrow f(x_0) = f(1) = \frac{3}{2} \times 1^{\frac{2}{3}} = \frac{3}{2} \\ f'(x) &= x^{-\frac{1}{3}} \Longrightarrow f'(x_0) = f'(1) = (1)^{-\frac{1}{3}} = 1 \end{aligned}$$

the first-order Taylor series is

$$\bar{f}(x) = f(x_0) + f'(x_0)(x - x_0) = \frac{3}{2} + 1 \times (x - 1) = \frac{1}{2} + x.$$

Since

$$f''(x) = -\frac{1}{3}x^{-\frac{4}{3}} \Longrightarrow f''(x_0) = f''(1) = -\frac{1}{3}(1)^{-\frac{4}{3}} = -\frac{1}{3}$$

the second-order Taylor series is

$$\begin{aligned}\bar{f}(x) &= f(x_0) + f'(x_0)(x - x_0) + \frac{f''(x_0)}{2}(x - x_0)^2 = \frac{1}{2} + x + \frac{f''(x_0)}{2}(x - x_0)^2 \\ &= \frac{1}{2} + x + \frac{-\frac{1}{3}}{2}(x-1)^2 = \frac{1}{3} + \frac{4}{3}x - \frac{1}{6}x^2.\end{aligned}$$

Problem 2.146 *[B**] Calculate a first-order Taylor series of* $f(x) = \frac{4}{3}x^{\frac{3}{4}}$ *around* $x_0 = 1$. *Now calculate a second-order Taylor series around* $x_0 = 1$.

Answer: Since

$$\begin{aligned}f(x) &= \frac{4}{3}x^{\frac{3}{4}} \Longrightarrow f(x_0) = f(1) = \frac{4}{3} \times 1^{\frac{3}{4}} = \frac{4}{3} \\ f'(x) &= x^{-\frac{1}{4}} \Longrightarrow f'(x_0) = f'(1) = (1)^{-\frac{1}{4}} = 1\end{aligned}$$

the first-order Taylor series is

$$\bar{f}(x) = f(x_0) + f'(x_0)(x - x_0) = \frac{4}{3} + 1 \times (x - 1) = \frac{1}{3} + x.$$

Since

$$\begin{aligned}f''(x) &= -\frac{1}{4}x^{-\frac{5}{4}} \\ &\Longrightarrow f''(x) = -\frac{1}{4}x^{-\frac{5}{4}} \\ &\Longrightarrow f''(x_0) = f''(1) = -\frac{1}{4}(1)^{-\frac{5}{4}} = -\frac{1}{4}\end{aligned}$$

the second-order Taylor series is

$$\begin{aligned}\bar{f}(x) &= f(x_0) + f'(x_0)(x - x_0) + \frac{f''(x_0)}{2}(x - x_0)^2 \\ &= \frac{1}{2} + x + \frac{f''(x_0)}{2}(x - x_0)^2 \\ &= \frac{1}{3} + x + \frac{-\frac{1}{4}}{2}(x-1)^2 = \frac{5}{24} + \frac{5}{4}x - \frac{1}{8}x^2.\end{aligned}$$

Problem 2.147 *[B**] Find the second-order Taylor series of* $f(x) = x^2e^{-x}$ *around* $x_o = 1$.

Answer: We have

$$\begin{aligned}f(1) &= 1^2e^{-1} = e^{-1},\ f'(x) = 2xe^{-x} - x^2e^{-x} \\ &\Longrightarrow f'(1) = 2 \times 1 \times e^{-1} - 1^2e^{-1} = e^{-1} \\ f''(x) &= 2e^{-x} - 4xe^{-x} + x^2e^{-x} \\ &\Longrightarrow f''(1) = 2e^{-1} - 4e^{-1} + 1^2 \times e^{-1} = -e^{-1}\end{aligned}$$

so that the Taylor series is

$$\begin{aligned}\bar{f}(x) &= f(x_0) + f'(x_0)(x - x_0) + \frac{f''(x_0)}{2}(x - x_0)^2 \\ &= e^{-1} + e^{-1}(x-1) - \frac{e^{-1}}{2}(x-1)^2 = e^{-1}\left(x - \frac{(x-1)^2}{2}\right).\end{aligned}$$

The actual function $f(x) = x^2 e^{-x}$ and the Taylor series around $x_0 = 1$ are shown below.

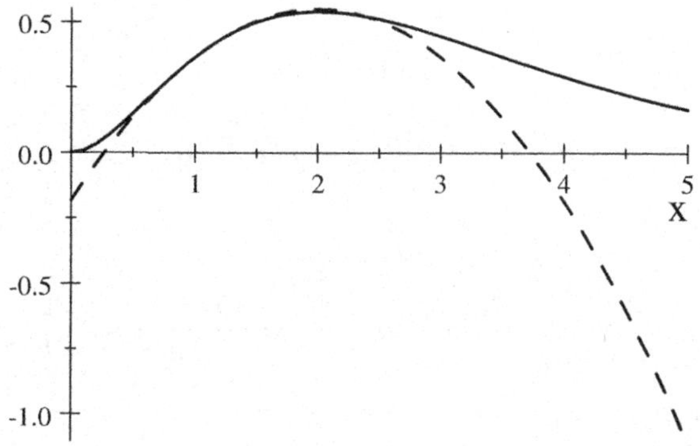

$x^2 e^{-x}$ and Taylor series - - -

Problem 2.148 *[B**] Construct a first-order Taylor series of* $f(x) = \ln(1+x)$ *around* $x_o = 0$. *Use this result to show that* $\ln(x_2) - \ln(x_1)$ *can be interpreted as the percentage difference between* x_2 *and* x_1 *normally given by* $\frac{x_2 - x_1}{x_1}$. *Compare the two definitions of percentage when* $x_2 = 110$ *and* $x_1 = 100$.

Answer: We have

$$f(0) = \ln(1+0) = 0, f'(x) = \frac{1}{1+x} \Longrightarrow f'(0) = \frac{1}{1+0} = 1$$

so that the Taylor series is

$$\bar{f}(x) = f(x_0) + f'(x_0)(x - x_0) = 0 + 1 \times (x - 0) = x.$$

Now using $\ln(1+x) \approx x$ we have

$$\ln(x_2) - \ln(x_1) = \ln\left(\frac{x_2}{x_1}\right) = \ln\left(1 + \frac{x_2 - x_1}{x_1}\right) \approx \frac{x_2 - x_1}{x_1}.$$

When $x_2 = 110$ and $x_1 = 100$ the conventional definition of percentage is

$$\frac{x_2 - x_1}{x_1} = \frac{110 - 100}{100} = 0.1$$

or 10%, while

$$\ln(x_2) - \ln(x_1) = \ln(110) - \ln(100) = \ln\left(\frac{110}{100}\right) = \ln(1.1) \approx 0.095$$

or 9.5%.

Problem 2.149 *[B**] For the function $f(x) = \ln\left(1 - \frac{x}{2}\right)$ calculate the second-order Taylor series around $x_o = 0$.*

Answer: We have $f(0) = \ln\left(1 - \frac{0}{2}\right) = 0$ and

$$\begin{aligned} f'(x) &= -\frac{1}{2\left(1 - \frac{1}{2}x\right)} \\ &\implies f'(0) = -\frac{1}{2\left(1 - \frac{1}{2}0\right)} = -\frac{1}{2} \\ f''(x) &= -\frac{1}{4\left(1 - \frac{1}{2}x\right)^2} \\ &\implies f''(0) = -\frac{1}{4\left(1 - \frac{1}{2}0\right)^2} = -\frac{1}{4} \end{aligned}$$

so that the Taylor series is

$$\begin{aligned} \bar{f}(x) &= f(x_0) + f'(x_0)(x - x_0) + \frac{f''(x_0)}{2}(x - x_0)^2 \\ &= 0 + \left(-\frac{1}{2}\right)(x - 0) + \frac{\left(-\frac{1}{4}\right)(x - 0)^2}{2} \\ &= -\frac{1}{2}x - \frac{1}{8}x^2. \end{aligned}$$

The function $f(x) = \ln\left(1 - \frac{x}{2}\right)$ and the Taylor series approximation around $x_0 = 0$ are shown below.

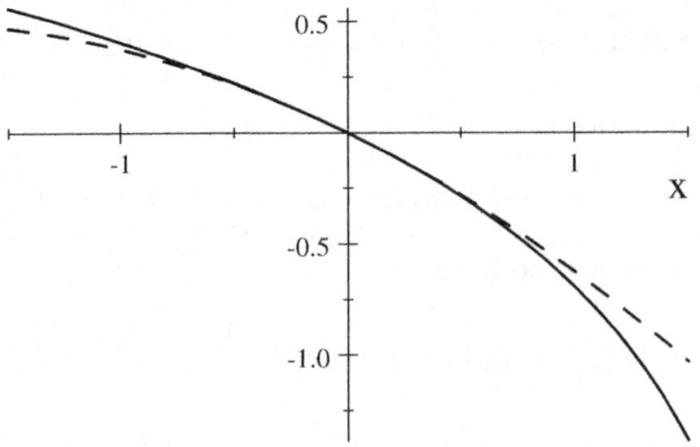

$\ln\left(1 - \frac{x}{2}\right)$ — and Taylor series - - -

Problem 2.150 *[A***] Use the first-order Taylor series with error*

$$f(x) = f(x_o) + f'(x_o)(x - x_o) + \frac{f''(\tilde{x})}{2!}(x - x_o)^2$$

to prove that a globally concave function has a unique global maximum where $f'(x^*) = 0$.

Answer: Calculating the Taylor series for $x_0 = x^*$ and using $f'(x^*) = 0$ we have

$$f(x) = f(x^*) + \overbrace{f'(x^*)}^{0}(x - x^*) + \frac{f''(\tilde{x})}{2!}(x - x^*)^2 = f(x^*) + \frac{f''(\tilde{x})}{2!}(x - x^*)^2 .$$

Now since $f(x)$ is globally concave $f''(\tilde{x}) < 0$. If $x \neq x^*$ then $(x - x^*)^2 > 0$ and

$$f(x) = f(x^*) + \overbrace{\frac{f''(\tilde{x})}{2!}(x - x^*)^2}^{-} < f(x^*) .$$

Since $f(x) < f(x^*)$ for $x \neq x^*$, it follows that x^* is a unique global maximizer. You can follow exactly the same steps to show that a solution to the first-order condition is a global minimizer when $f(x)$ is globally convex.

Problem 2.151 *[B**] Prove from the Taylor series*

$$e^x \equiv 1 + x + \frac{x^2}{2!} + \frac{x^3}{3!} + \frac{x^4}{4!} + \cdots$$

that $\frac{d}{dx}e^x = e^x$.

Answer: Appealing to the sum rule and differentiating each term (since there are an infinite number of terms, this is not generally correct to do without additional arguments, but here it is correct) we have

$$\begin{aligned} \frac{d}{dx}e^x &= \frac{d}{dx}\left(1 + x + \frac{x^2}{2!} + \frac{x^3}{3!} + \cdots\right) = \frac{d}{dx}(1) + \frac{d}{dx}(x) + \frac{1}{2!}\frac{d}{dx}(x^2) + \frac{1}{3!}\frac{d}{dx}(x^3) + \cdots \\ &= 0 + 1 + \frac{2}{2!}x + \frac{3}{3!}x^2 + \frac{4}{4!}x^3 + \cdots = 1 + x + \frac{x^2}{2!} + \frac{x^3}{3!} + \frac{x^4}{4!} + \cdots = e^x \end{aligned}$$

since $\frac{n}{n!} = \frac{1}{(n-1)!}$.

2.10.1 L'Hôpital's Rule

Problem 2.152 *[B**] Show that* $\lim_{x\to\infty} xe^{-x} = 0$.

Answer: Since

$$\lim_{x\to\infty} xe^{-x} = \lim_{x\to\infty}\frac{x}{e^x} \text{ and } \lim_{x\to\infty} x = \infty, \lim_{x\to\infty} e^x = \infty$$

we can use L'Hôpital's rule as

$$\lim_{x\to\infty} xe^{-x} = \lim_{x\to\infty} \frac{\frac{d}{dx}x}{\frac{d}{dx}e^x} = \lim_{x\to\infty} \frac{1}{e^x} = 0.$$

Problem 2.153 *[A**] Show for integer $n > 0$ that $\lim_{x\to\infty} x^n e^{-x} = 0$.*

Answer: Since

$$\lim_{x\to\infty} x^n e^{-x} = \lim_{x\to\infty} \frac{x^n}{e^x}, \lim_{x\to\infty} x = \infty, \lim_{x\to\infty} e^x = \infty$$

we can use L'Hôpital's rule as

$$\begin{aligned}\lim_{x\to\infty} xe^{-x} &= \lim_{x\to\infty} \frac{x^n}{e^x} = \lim_{x\to\infty} \frac{\frac{d}{dx}(x^n)}{\frac{d}{dx}(e^x)} = \lim_{x\to\infty} \frac{nx^{n-1}}{e^x} \\ &= \lim_{x\to\infty} \frac{n(n-1)x^{n-2}}{e^x} = \lim_{x\to\infty} \frac{n!}{e^x} = 0.\end{aligned}$$

Problem 2.154 *[A**] Show for $n > 0$ that $\lim_{x\to 0} x^n \ln(x) = 0$.*

Answer: Since

$$\lim_{x\to 0} x^n \ln(x) = \lim_{x\to 0} \frac{\ln(x)}{x^{-n}} \text{ and } \lim_{x\to 0} \ln(x) = -\infty, \lim_{x\to 0} x^{-n} = \infty$$

we can use L'Hôpital's rule as

$$\lim_{x\to 0} \frac{\ln(x)}{x^{-n}} = \lim_{x\to 0} \frac{\frac{d}{dx}(\ln(x))}{\frac{d}{dx}(x^{-n})} = \lim_{x\to 0} \frac{\frac{1}{x}}{-nx^{-n-1}} = \lim_{x\to 0} \frac{1}{-nx^{-n}} = -\frac{1}{n} \lim_{x\to 0} \frac{1}{x^{-n}} = 0.$$

Problem 2.155 *[A**] Show that for all $x > 0$ that $\lim_{\rho\to 0} \frac{x^\rho - 1}{\rho} = \ln(x)$. This is the basis for the Box-Cox transformation often used in econometrics.*

Answer: Note that it is ρ that approaches 0 here and not x. We have

$$\lim_{\rho\to 0} \frac{x^\rho - 1}{\rho} = \lim_{\rho\to 0} \frac{e^{\ln(x)\rho} - 1}{\rho} = \lim_{\rho\to 0} \frac{\frac{d}{d\rho}\left(e^{\ln(x)\rho} - 1\right)}{\frac{d}{d\rho}(\rho)} = \lim_{\rho\to 0} \frac{\ln(x)e^{\ln(x)\rho}}{1} = \ln(x).$$

Problem 2.156 *[B**] Using L'Hôpital's rule find $\lim_{x\to 0} x^x$.*

Answer: This is not directly in the form with which we can use L'Hôpital's rule. Consider though rewriting it x^x as

$$e^{\ln(x)x} = \exp\left(\frac{\ln(x)}{\frac{1}{x}}\right)$$

so that

$$\lim_{x\to 0} x^x = \exp\left(\lim_{x\to 0} \frac{\ln(x)}{\frac{1}{x}}\right).$$

Since

$$\lim_{x\to 0} \ln(x) = -\infty, \lim_{x\to 0} \frac{1}{x} = \infty$$

we can apply L'Hôpital's rule as

$$\lim_{x\to 0} \frac{\ln(x)}{\frac{1}{x}} = \lim_{x\to 0} \frac{\frac{d}{dx}(\ln(x))}{\frac{d}{dx}\left(\frac{1}{x}\right)} = \lim_{x\to 0} \frac{\frac{1}{x}}{-\frac{1}{x^2}} = -\lim_{x\to 0} x = 0$$

so that

$$\lim_{x\to 0} x^x = \exp\left(\lim_{x\to 0} \frac{\ln(x)}{\frac{1}{x}}\right) = e^0 = 1.$$

This is illustrated in the figure below where $x^x \to 1$ as $x \to 0$.

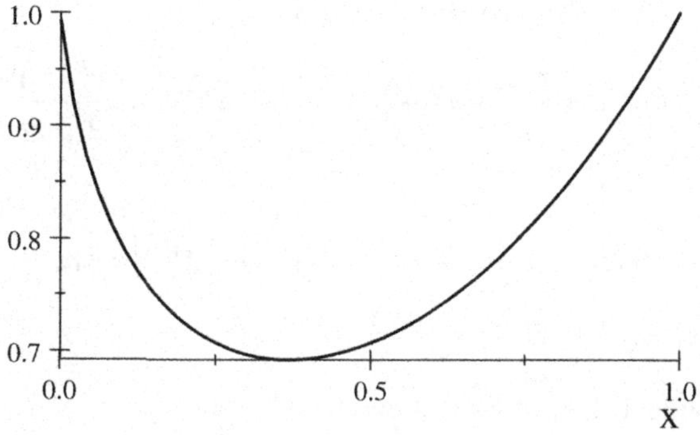

$f(x) = x^x$ as $x \to 0$

Problem 2.157 *[A **] Use L'Hôpital's rule to prove that*

$$\lim_{n\to\infty} \left(1 + \frac{x}{n}\right)^n = e^x.$$

Relate this result to the Taylor series $\ln(1+x) \approx x$ *for* x *small.*

Answer: We have

$$\left(1 + \frac{x}{n}\right)^n = e^{n\ln\left(1+\frac{x}{n}\right)} = \exp\left(\frac{\ln\left(1+\frac{x}{n}\right)}{\frac{1}{n}}\right)$$

where

$$\lim_{n\to\infty} \ln\left(1 + \frac{x}{n}\right) = \ln(1) = 0, \lim_{n\to\infty} \frac{1}{n} = 0$$

and so we can use L'Hôpital's rule as

$$\begin{aligned}\lim_{n\to\infty}\left(1+\frac{x}{n}\right)^n &= \exp\left(\frac{\ln\left(1+\frac{x}{n}\right)}{\frac{1}{n}}\right)=\exp\left(\lim_{n\to\infty}\frac{\ln\left(1+\frac{x}{n}\right)}{\frac{1}{n}}\right)\\ &= \exp\left(\lim_{n\to\infty}\frac{\frac{d}{dn}\ln\left(1+\frac{x}{n}\right)}{\frac{d}{dn}\frac{1}{n}}\right)=\exp\left(\lim_{n\to\infty}\frac{\frac{1}{1+\frac{x}{n}}\frac{-x}{n^2}}{-\frac{1}{n^2}}\right)\\ &= \exp\left(\lim_{n\to\infty}\frac{x}{1+\frac{x}{n}}\right)=\exp(x)=e^x.\end{aligned}$$

Since for n large $\ln\left(1+\frac{x}{n}\right)\approx\frac{x}{n}$ and so we have

$$\left(1+\frac{x}{n}\right)^n=e^{n\ln\left(1+\frac{x}{n}\right)}\approx e^{n\frac{x}{n}}=e^x.$$

Problem 2.158 *[A*] The geometric series*

$$S(x)\equiv 1+x+x^2+x^3+\cdots+x^{n-1}=\frac{x^n-1}{x-1}$$

is valid for $x\neq 1$. Now

$$S(1)=1+1+1^2+1^3+\cdots+1^{n-1}=n.$$

Show that $\lim_{x\to 1}\frac{x^n-1}{x-1}=S(1)=n$.

Answer: Using L'Hôpital's rule we have

$$\lim_{x\to 1}\frac{x^n-1}{x-1}=\lim_{x\to 1}\frac{\frac{d}{dx}(x^n-1)}{\frac{d}{dx}(x-1)}=\lim_{x\to 1}\frac{nx^{n-1}}{1}=n\lim_{x\to 1}x^{n-1}=n.$$

2.10.2 Newton's Method

Problem 2.159 *[C**] Consider the cubic*

$$f(x)=x^3-3x^2-16x-12.$$

Verify that this can be factored as

$$f(x)=x^3-3x^2-16x-12=(x-6)(x+2)(x+1).$$

*Use this to find the roots of $f(x)$. [A**] Now suppose you did not know the factorization or the roots of $f(x)$. Use three iterations of Newton's method four digit accuracy and with $x_0=10$ to calculate one root.*

Answer: Expanding we have

$$(x-6)(x+2)(x+1)=(x-6)\left(x^2+3x+2\right)=x^3-3x^2-16x-12=f(x).$$

We have

$$\begin{aligned} f(x) &= 0 \\ &\Longrightarrow (x-6)(x+2)(x+1)=0 \\ &\Longrightarrow x=6, \text{or } x=-2, \text{or } x=-1. \end{aligned}$$

To use Newton's method we need the derivative $f'(x) = 3x^2 - 6x - 16$. Let $x_0 = 10$ be our initial estimate of the root. Then since

$$\begin{aligned} f(x_0) &= f(10) = 10^3 - 3 \times 10^2 - 16 \times 10 - 12 = 528 \\ f'(x_0) &= f'(10) = 3 \times 10^2 - 6 \times 10 - 16 = 224 \end{aligned}$$

we have the Taylor series around $x_0 = 10$ as

$$\bar{f}(x) = 528 + 224(x-10) = -1712 + 224x.$$

Solving $\bar{f}(x_1) = 0$ then yields

$$\begin{aligned} \bar{f}(x_1) &= 0 \Longrightarrow -1712 + 224x_1 = 0 \\ &\Longrightarrow x_1 = \frac{1712}{224} = \frac{107}{14} = 7.6429. \end{aligned}$$

Now $x_1 = 7.6429$ is our new estimate of the root, which is closer to the actual root $x = 6$ than the initial estimate $x_0 = 10$. To get a better estimate of the root we construct a new Taylor series at $x_1 = 7.6429$. For this we need

$$\begin{aligned} f(x_1) &= f(7.6429) = (7.6429)^3 - 3(7.6429)^2 - 16(7.6429) - 12 = 136.9236 \\ f'(x_1) &= f'(7.6429) = 3(7.6429)^2 - 6(7.6429) - 16 = 113.3844 \end{aligned}$$

so that

$$\bar{f}(x) = 136.9236 + 113.3844(x - 7.6429) = -729.6620 + 113.3844x.$$

Now solving for $\bar{f}(x_2) = 0$ yields

$$\begin{aligned} \bar{f}(x_2) &= -729.6620 + 113.3844x_2 = 0 \\ &\Longrightarrow x_2 = \frac{729.6620}{113.3844} = 6.4353. \end{aligned}$$

Now $x_2 = 6.4353$ is our new estimate of the root, which is closer to $x = 6$ than the previous estimate $x_0 = 7.6429$. To get a better estimate of the root we construct a new Taylor series at $x_2 = 6.4353$. For this we need

$$\begin{aligned} f(x_2) &= f(6.4353) = (6.4353)^3 - 3(6.4353)^2 - 16 \times 6.4353 - 12 = 27.3016 \\ f'(x_1) &= f'(6.4353) = 3(6.4353)^2 - 6 \times 6.4353 - 16 = 69.6275 \end{aligned}$$

so that

$$\bar{f}(x) = 27.3016 + 69.6275 \times (x - 6.4353) = -420.7723 + 69.6275x.$$

Now solving for $\bar{f}(x_3) = 0$ yields

$$\begin{aligned} \bar{f}(x_3) &= -420.7723 + 69.6275x_3 = 0 \\ &\implies x_3 = \frac{420.7723}{69.6275} = 6.043. \end{aligned}$$

Here $x_3 = 6.043$ is quite close to the actual root of $x = 6$. Better estimates can be found by repeated iterations, something that is boring for humans but easy for computers.

2.11 When the First-Order Condition Fails

Problem 2.160 *[A*] Prove that*

$$f(x) = e^{-3x} - 2x \text{ for } x \geq 0$$

is globally decreasing and convex. Does the first-order condition yield a global minimum? Is there a global maximum?

Answer: We have

$$f'(x) = -3\underbrace{e^{-3x}}_{+} - 2 < 0 \text{ and } f''(x) = 9\underbrace{e^{-3x}}_{+} > 0$$

so that $f(x)$ is globally decreasing and globally convex. This function does not have a global minimum since $f'(x) < 0$ for all x. The function has a global maximum at $x^* = 0$, a corner solution. This is illustrated in the figure below.

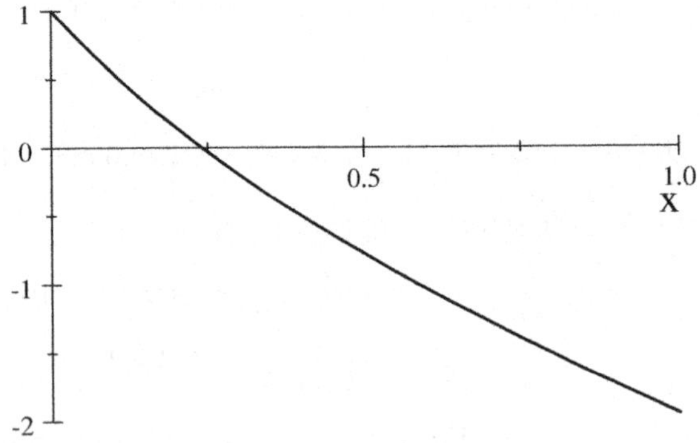

$f(x) = e^{-3x} - 2x$

Problem 2.161 *[A*] Consider a firm with production function*

$$Q = f(L) = 12L - 3L^2 \text{ for } 0 \leq L < 2.$$

What do profits as a function of L look like when $W = 13$ and $P = 1$? What must the real wage w be in order for production to take place?

Answer: Profits are

$$\begin{aligned}\pi(L) &= P\left(12L - 3L^2\right) - WL \\ &\Longrightarrow \pi'(L) = P(12 - 6L) - W.\end{aligned}$$

When $W = 13$ and $P = 1$ we have for $L > 0$ that

$$\pi(L) = -L - 3L^2 \text{ and } \pi'(L) = -1 - 6L < 0.$$

Since $\pi(0) = 0$ and $\pi'(L) < 0$ for all $L > 0$, the firm makes a loss for any $L > 0$, but makes zero profits for $L = 0$. So $L^* = 0$ is the profit maximizing L. But at $L^* = 0$ the first-order condition $\pi'(L^*) = 0$ fails since

$$\pi'(0) = -1 - 6 \times 0 = -1 < 0.$$

Here $\pi'(L^*) < 0$ because L^* occurs on the boundary of the domain of $\pi(L)$. This is shown in the figure below.

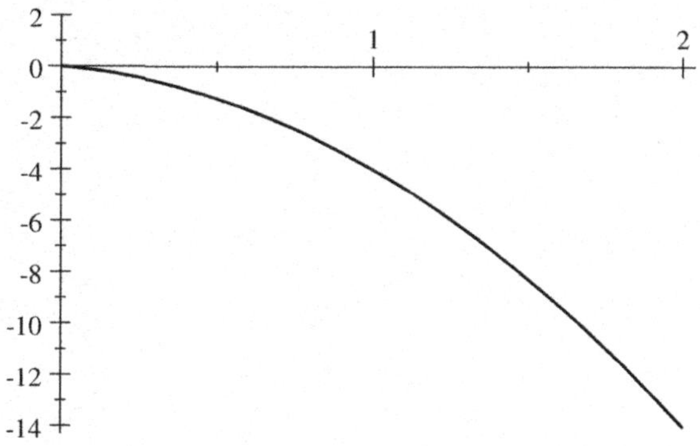

$$\pi(L) = -L - 3L^2$$

For general P, W we have

$$\pi'(L) = P(12 - 6L) - W$$

assuming $L^* > 0$ the first-order condition $\pi'(L^*) = 0$ is

$$P(12 - 6L^*) - W = 0 \Longrightarrow (12 - 6L^*) = \frac{W}{P} = w \Longrightarrow L^* = 2 - \frac{1}{6}w.$$

For production to take place $L^* > 0$ we require $w < 12$ as

$$L^* = 2 - \frac{1}{6}w > 0 \Longrightarrow w < 12.$$

Chapter 3

Matrix Algebra

3.1 Matrix Calculations

Problem 3.1 *[D***] Consider the matrices*

$$A = \begin{bmatrix} 3 & 2 \\ 1 & 3 \\ 1 & 2 \end{bmatrix}, B = \begin{bmatrix} 6 & 4 \\ 2 & 6 \\ 2 & 4 \end{bmatrix}.$$

Calculate $A + B, A - B, AB, A^T B,$ and AB^T.

Answer: We have

$$\begin{aligned} A + B &= \begin{bmatrix} 3 & 2 \\ 1 & 3 \\ 1 & 2 \end{bmatrix} + \begin{bmatrix} 6 & 4 \\ 2 & 6 \\ 2 & 4 \end{bmatrix} = \begin{bmatrix} 9 & 6 \\ 3 & 9 \\ 3 & 6 \end{bmatrix} \\ A - B &= \begin{bmatrix} 3 & 2 \\ 1 & 3 \\ 1 & 2 \end{bmatrix} - \begin{bmatrix} 6 & 4 \\ 2 & 6 \\ 2 & 4 \end{bmatrix} = \begin{bmatrix} -3 & -2 \\ -1 & -3 \\ -1 & -2 \end{bmatrix}. \end{aligned}$$

The product AB is not defined since both A and B are 3×2, so the number of columns of A is 2 which does not match the number of rows of B which is 3. Now

$$\begin{aligned} A^T B &= \begin{bmatrix} 3 & 1 & 1 \\ 2 & 3 & 2 \end{bmatrix} \begin{bmatrix} 6 & 4 \\ 2 & 6 \\ 2 & 4 \end{bmatrix} = \begin{bmatrix} 22 & 22 \\ 22 & 34 \end{bmatrix} \\ AB^T &= \begin{bmatrix} 3 & 2 \\ 1 & 3 \\ 1 & 2 \end{bmatrix} \begin{bmatrix} 6 & 2 & 2 \\ 4 & 6 & 4 \end{bmatrix} = \begin{bmatrix} 26 & 18 & 14 \\ 18 & 20 & 14 \\ 14 & 14 & 10 \end{bmatrix}. \end{aligned}$$

Problem 3.2 *[D***]Given the matrix*

$$A = \begin{bmatrix} 1 & 1 \\ 1 & 2 \\ 1 & 3 \end{bmatrix}$$

calculate $A + 3A, A - A, A^TA, AA^T$, *and* A^TA^T.

Answer: We have

$$\begin{aligned}
A+3A &= \begin{bmatrix} 1 & 1 \\ 1 & 2 \\ 1 & 3 \end{bmatrix} + 3\begin{bmatrix} 1 & 1 \\ 1 & 2 \\ 1 & 3 \end{bmatrix} = \begin{bmatrix} 1 & 1 \\ 1 & 2 \\ 1 & 3 \end{bmatrix} + \begin{bmatrix} 3 & 3 \\ 3 & 6 \\ 3 & 9 \end{bmatrix} = \begin{bmatrix} 4 & 4 \\ 4 & 8 \\ 4 & 12 \end{bmatrix} \\
A-A &= \begin{bmatrix} 1 & 1 \\ 1 & 2 \\ 1 & 3 \end{bmatrix} - \begin{bmatrix} 1 & 1 \\ 1 & 2 \\ 1 & 3 \end{bmatrix} = \begin{bmatrix} 0 & 0 \\ 0 & 0 \\ 0 & 0 \end{bmatrix} \\
A^TA &= \begin{bmatrix} 1 & 1 & 1 \\ 1 & 2 & 3 \end{bmatrix}\begin{bmatrix} 1 & 1 \\ 1 & 2 \\ 1 & 3 \end{bmatrix} = \begin{bmatrix} 3 & 6 \\ 6 & 14 \end{bmatrix} \\
AA^T &= \begin{bmatrix} 1 & 1 \\ 1 & 2 \\ 1 & 3 \end{bmatrix}\begin{bmatrix} 1 & 1 & 1 \\ 1 & 2 & 3 \end{bmatrix} = \begin{bmatrix} 2 & 3 & 4 \\ 3 & 5 & 7 \\ 4 & 7 & 10 \end{bmatrix}.
\end{aligned}$$

The product A^TA^T is not defined.

Problem 3.3 *[D***] Given the matrices*

$$A = \begin{bmatrix} 1 & 1 \\ 2 & 0 \\ 3 & 1 \end{bmatrix}, \; B = \begin{bmatrix} 1 & -1 & 1 \\ 1 & 2 & 3 \end{bmatrix}$$

calculate AB, B^TA^T, *and* A^TA.

Answer: We have

$$\begin{aligned}
AB &= \begin{bmatrix} 1 & 1 \\ 2 & 0 \\ 3 & 1 \end{bmatrix}\begin{bmatrix} 1 & -1 & 1 \\ 1 & 2 & 3 \end{bmatrix} = \begin{bmatrix} 2 & 1 & 4 \\ 2 & -2 & 2 \\ 4 & -1 & 6 \end{bmatrix} \\
B^TA^T &= (AB)^T = \begin{bmatrix} 2 & 1 & 4 \\ 2 & -2 & 2 \\ 4 & -1 & 6 \end{bmatrix}^T = \begin{bmatrix} 2 & 2 & 4 \\ 1 & -2 & -1 \\ 4 & 2 & 6 \end{bmatrix} \\
A^TA &= \begin{bmatrix} 1 & 2 & 3 \\ 1 & 0 & 1 \end{bmatrix}\begin{bmatrix} 1 & 1 \\ 2 & 0 \\ 3 & 1 \end{bmatrix} = \begin{bmatrix} 14 & 4 \\ 4 & 2 \end{bmatrix}.
\end{aligned}$$

Problem 3.4 *[D***] Given*

$$A = \begin{bmatrix} 2 & 2 & 2 \\ 2 & 4 & 6 \end{bmatrix}, \; B = \begin{bmatrix} 1 & 1 \\ 1 & 2 \\ 1 & 3 \end{bmatrix}$$

calculate AB, A^TB, BA, *and* B^TA^T.

Answer: We have

$$\begin{aligned} AB &= \begin{bmatrix} 2 & 2 & 2 \\ 2 & 4 & 6 \end{bmatrix} \begin{bmatrix} 1 & 1 \\ 1 & 2 \\ 1 & 3 \end{bmatrix} = \begin{bmatrix} 6 & 12 \\ 12 & 28 \end{bmatrix} \\ B^T A^T &= (AB)^T = \begin{bmatrix} 6 & 12 \\ 12 & 28 \end{bmatrix}^T = \begin{bmatrix} 6 & 12 \\ 12 & 28 \end{bmatrix} \\ BA &= \begin{bmatrix} 1 & 1 \\ 1 & 2 \\ 1 & 3 \end{bmatrix} \begin{bmatrix} 2 & 2 & 2 \\ 2 & 4 & 6 \end{bmatrix} = \begin{bmatrix} 4 & 6 & 8 \\ 6 & 10 & 14 \\ 8 & 14 & 20 \end{bmatrix}. \end{aligned}$$

The product $A^T B$ is not defined.

Problem 3.5 *[D***] Given*

$$A = \begin{bmatrix} 3 & 3 & 3 \\ 12 & 4 & -4 \end{bmatrix}, \; B = \begin{bmatrix} 9 & 18 \\ 3 & 6 \\ 4 & 8 \end{bmatrix}$$

calculate $AB, B^T A^T, BA$, *and* $A^T B$.

Answer: We have

$$\begin{aligned} AB &= \begin{bmatrix} 3 & 3 & 3 \\ 12 & 4 & -4 \end{bmatrix} \begin{bmatrix} 9 & 18 \\ 3 & 6 \\ 4 & 8 \end{bmatrix} = \begin{bmatrix} 48 & 96 \\ 104 & 208 \end{bmatrix} \\ B^T A^T &= (AB)^T = \begin{bmatrix} 48 & 96 \\ 104 & 208 \end{bmatrix}^T = \begin{bmatrix} 48 & 104 \\ 96 & 208 \end{bmatrix} \\ BA &= \begin{bmatrix} 9 & 18 \\ 3 & 6 \\ 4 & 8 \end{bmatrix} \begin{bmatrix} 3 & 3 & 3 \\ 12 & 4 & -4 \end{bmatrix} = \begin{bmatrix} 243 & 99 & -45 \\ 81 & 33 & -15 \\ 108 & 44 & -20 \end{bmatrix}. \end{aligned}$$

The product $A^T B$ is not defined.

Problem 3.6 *[D***] Given the two matrices A and B and the vector x as*

$$A = \begin{bmatrix} 1 & 1 \\ 2 & 1 \\ 3 & 1 \end{bmatrix}, \; B = \begin{bmatrix} 3 & 3 & 3 \\ 1 & 2 & 3 \end{bmatrix}, \; x = \begin{bmatrix} 1 \\ -2 \\ 1 \end{bmatrix}$$

calculate Bx *and* ABx.

Answer: We have $Bx = 0$ as

$$Bx = \begin{bmatrix} 3 & 3 & 3 \\ 1 & 2 & 3 \end{bmatrix} \begin{bmatrix} 1 \\ -2 \\ 1 \end{bmatrix} = \begin{bmatrix} 0 \\ 0 \end{bmatrix}$$

and $ABx = 0$ as

$$ABx = \begin{bmatrix} 1 & 1 \\ 2 & 1 \\ 3 & 1 \end{bmatrix} \times Bx = \begin{bmatrix} 1 & 1 \\ 2 & 1 \\ 3 & 1 \end{bmatrix} \begin{bmatrix} 0 \\ 0 \end{bmatrix} = \begin{bmatrix} 0 \\ 0 \\ 0 \end{bmatrix}.$$

Problem 3.7 *[D***] Consider*

$$A = \begin{bmatrix} 3 & 5 & -2 \\ 1 & -4 & 3 \\ 1 & -1 & 1 \end{bmatrix}, b = \begin{bmatrix} 7 \\ 4 \\ 10 \end{bmatrix}.$$

Is A or b symmetric?

Answer: Neither A nor b is symmetric since

$$A^T = \begin{bmatrix} 3 & 1 & 1 \\ 5 & -4 & -1 \\ -2 & 3 & 1 \end{bmatrix} \neq \begin{bmatrix} 3 & 5 & -2 \\ 1 & -4 & 3 \\ 1 & -1 & 1 \end{bmatrix} = A,\ b^T = \begin{bmatrix} 7 & 4 & 10 \end{bmatrix} \neq \begin{bmatrix} 7 \\ 4 \\ 10 \end{bmatrix} = b.$$

Problem 3.8 *[B***] Suppose A and B are $n \times n$ symmetric matrices. Is AB necessarily symmetric? [B***] Show that ABA is symmetric.*

Answer: Since A and B are symmetric by definition $A^T = A$ and $B^T = B$. To show that AB is symmetric we would have to show that $(AB)^T = AB$. But

$$(AB)^T = B^T A^T = BA.$$

Since for matrices $AB \neq BA$ we cannot conclude that AB is symmetric. For example

$$\begin{aligned} A &= \begin{bmatrix} 4 & 1 \\ 1 & 5 \end{bmatrix}, B = \begin{bmatrix} 7 & 2 \\ 2 & 5 \end{bmatrix} \\ &\implies AB = \begin{bmatrix} 4 & 1 \\ 1 & 5 \end{bmatrix} \begin{bmatrix} 7 & 2 \\ 2 & 5 \end{bmatrix} = \begin{bmatrix} 30 & 13 \\ 17 & 27 \end{bmatrix} \end{aligned}$$

is **not** symmetric.

For ABA on the other hand

$$(ABA)^T = A^T B^T A^T = ABA.$$

In the above example

$$ABA = \begin{bmatrix} 4 & 1 \\ 1 & 5 \end{bmatrix} \begin{bmatrix} 7 & 2 \\ 2 & 5 \end{bmatrix} \begin{bmatrix} 4 & 1 \\ 1 & 5 \end{bmatrix} = \begin{bmatrix} 133 & 95 \\ 95 & 152 \end{bmatrix}$$

is symmetric.

Problem 3.9 *[C**] For scalars a and b we have $(a-b)(a+b) = a^2 - b^2$. Given two square $n \times n$ matrices A and B, does it follow that*

$$(A - B)(A + B) = A^2 - B^2?$$

Answer: No since

$$(A-B)(A+B) = AA - BA + AB - BB = A^2 - B^2 + (AB - BA) \neq A^2 - B^2$$

since $AB \neq BA$.

Problem 3.10 *[A ***] Prove that if Q is an $m \times m$, non-singular, symmetric matrix, and X is any $m \times n$ matrix, then $C = X^T Q X$ is an $n \times n$ symmetric matrix.*

Answer: The matrix C is symmetric since

$$\begin{aligned} C^T &= \left(X^T Q X\right)^T \text{ (by definition)} = X^T \left(X^T Q\right)^T \text{ (since } (AB)^T = B^T A^T) \\ &= X^T Q^T \left(X^T\right)^T \text{ (since } (AB)^T = B^T A^T) = X^T Q^T X \text{ (since } \left(A^T\right)^T = A) \\ &= X^T Q X \text{ (since } Q = Q^T \text{ is symmetric)} = C \text{ (by definition).} \end{aligned}$$

The matrix C is $n \times n$ since X^T is $n \times m$, Q is $m \times m$, and X is $m \times n$.

3.2 Vectors

Problem 3.11 *[D***] Given the two vectors*

$$x = \begin{bmatrix} 2 \\ 2 \\ 2 \end{bmatrix}, \; y = \begin{bmatrix} 1 & 2 & 3 \end{bmatrix}$$

calculate xy, yx, and yx^T.

Answer: We have

$$xy = \begin{bmatrix} 2 \\ 2 \\ 2 \end{bmatrix} \begin{bmatrix} 1 & 2 & 3 \end{bmatrix} = \begin{bmatrix} 2 & 4 & 6 \\ 2 & 4 & 6 \\ 2 & 4 & 6 \end{bmatrix} \text{ and } yx = \begin{bmatrix} 1 & 2 & 3 \end{bmatrix} \begin{bmatrix} 2 \\ 2 \\ 2 \end{bmatrix} = 12.$$

The product yx^T is **not** defined.

Problem 3.12 *[D***] Given the two vectors*

$$x = \begin{bmatrix} 2 \\ 3 \\ 4 \end{bmatrix}, \; y = \begin{bmatrix} 2 & 1 & 2 \end{bmatrix}$$

what is x^{-1}? Calculate $\det[xy]$, $\det[yx]$, $(xy)^{-1}$, and $(yx)^{-1}$.

Answer: Here x^{-1} does not exist since x is not a square matrix. We have

$$\begin{aligned} xy &= \begin{bmatrix} 2 \\ 3 \\ 4 \end{bmatrix} \begin{bmatrix} 2 & 1 & 2 \end{bmatrix} = \begin{bmatrix} 4 & 2 & 4 \\ 6 & 3 & 6 \\ 8 & 4 & 8 \end{bmatrix} \\ \det[xy] &= \det \begin{bmatrix} 4 & 2 & 4 \\ 6 & 3 & 6 \\ 8 & 4 & 8 \end{bmatrix} = 0, \; yx = \begin{bmatrix} 2 & 1 & 2 \end{bmatrix} \begin{bmatrix} 2 \\ 3 \\ 4 \end{bmatrix} = 15 \end{aligned}$$

and where yx^T is **not** defined. Since yx is a scalar $\det[yx] = \det[15] = 15$ and $(yx)^{-1} = 15^{-1} = \frac{1}{15}$.Since $\det[xy] = 0$ it follows that $(xy)^{-1}$ does not exist.

Problem 3.13 *[B***] Consider the two vectors*

$$x = \begin{bmatrix} 3 \\ -2 \\ 7 \end{bmatrix}, \; y = \begin{bmatrix} 2 \\ 1 \\ 4 \end{bmatrix}.$$

Are x and y orthogonal? If they are not write down two other non-zero 3×1 column vectors which are orthogonal. Find $x^T x$, $\|x\|$, $y^T y$, and $\|y\|$.

Answer: We have

$$x^T y = \begin{bmatrix} 3 & -2 & 7 \end{bmatrix} \begin{bmatrix} 2 \\ 1 \\ 4 \end{bmatrix} = 32 \neq 0$$

so that x and y are not orthogonal to each other. An example of 2 orthogonal 3×1 vectors would be

$$x = \begin{bmatrix} 0 \\ 1 \\ 0 \end{bmatrix}, \; y = \begin{bmatrix} 1 \\ 0 \\ 0 \end{bmatrix}$$

as

$$x^T y = \begin{bmatrix} 0 & 1 & 0 \end{bmatrix} \begin{bmatrix} 1 \\ 0 \\ 0 \end{bmatrix} = 0.$$

We have

$$\begin{aligned} x^T x &= \begin{bmatrix} 3 & -2 & 7 \end{bmatrix} \begin{bmatrix} 3 \\ -2 \\ 7 \end{bmatrix} = 62 \Longrightarrow \|x\| = \sqrt{x^T x} = \sqrt{62} \approx 7.9 \\ y^T y &= \begin{bmatrix} 2 & 1 & 4 \end{bmatrix} \begin{bmatrix} 2 \\ 1 \\ 4 \end{bmatrix} = 21 \Longrightarrow \|y\| = \sqrt{y^T y} = \sqrt{21} \approx 4.6. \end{aligned}$$

Problem 3.14 *[B***] An orthogonal matrix C has the property that $C^{-1} = C^T$. [A***] Show that $\det[C] = \pm 1$ and that for any $n \times 1$ vector w, if $y = Cw$ then $\|y\| = \|w\|$.*

Answer: Since $C^T = C^{-1}$ we have

$$\det[C]^2 = \det[C]\det\left[C^T\right] = \det\left[CC^T\right] = \det\left[CC^{-1}\right] = \det[I] = 1 \Longrightarrow\Longrightarrow \det[C] = \pm 1.$$

Now if $y = Cw$ then

$$\begin{aligned} \|y\|^2 &= \|Cw\|^2 = (Cw)^T(Cw) = w^TC^TCw = w^TC^{-1}Cw = w^TIw = w^Tw = \|w\|^2 \\ &\Longrightarrow \|y\|^2 = \|w\|^2 \Longrightarrow \|y\| = \|w\|. \end{aligned}$$

Problem 3.15 *[A**] Prove the Cauchy-Schwarz Inequality: For any two $n \times 1$ vectors x and y*

$$\left|x^Ty\right| \leq \|x\|\,\|y\|$$

where $\|x\| = \sqrt{x^Tx}$. Show that equality holds if and only if y is a scalar multiple of x.

Answer: Let α be a scalar and define $f(\alpha)$ by

$$f(\alpha) \equiv (\alpha x - y)^T(\alpha x - y) \equiv \|\alpha x - y\|^2 \geq 0.$$

Now

$$f(\alpha) = (\alpha x - y)^T(\alpha x - y) = \alpha^2 x^Tx - 2\alpha x^Ty + y^Ty = \alpha^2\|x\|^2 - 2\alpha x^Ty + \|y\|^2.$$

Let's find the α that minimizes $f(\alpha)$. The first-order condition $f'(\alpha^*) = 0$ yields

$$2\alpha^*\|x\|^2 - 2x^Ty = 0 \Longrightarrow \alpha^* = \frac{x^Ty}{\|x\|^2}.$$

Since $f''(\alpha) = 2\|x\|^2 > 0$ this is a global minimizer. Now replacing α with $\alpha^* = \frac{x^Ty}{\|x\|^2}$ and using $f(\alpha^*) \geq 0$ yields

$$\begin{aligned} \left(\frac{x^Ty}{\|x\|^2}\right)^2\|x\|^2 - 2\left(\frac{x^Ty}{\|x\|^2}\right)x^Ty + \|y\|^2 &= \frac{(x^Ty)^2}{\|x\|^2} - 2\frac{(x^Ty)^2}{\|x\|^2} + \|y\|^2 \\ &= \|y\|^2 - \frac{(x^Ty)^2}{\|x\|^2} \geq 0 \\ &\Longrightarrow \|y\|^2 \geq \frac{(x^Ty)^2}{\|x\|^2} \Longrightarrow (x^Ty)^2 \leq \|x\|^2\|y\|^2 \\ &\Longrightarrow \left|x^Ty\right| \leq \|x\|\,\|y\|. \end{aligned}$$

Now suppose that $\left|x^Ty\right| = \|x\|\,\|y\|$. Then

$$f(\alpha^*) = \|\alpha^*x - y\| = \|y\|^2 - \frac{(x^Ty)^2}{\|x\|^2} = \|y\|^2 - \frac{(\|x\|\,\|y\|)^2}{\|x\|^2} = 0.$$

But

$$\|\alpha^* x - y\| = 0 \Longrightarrow \alpha^* x - y = 0 \Longrightarrow y = \alpha^* x$$

and so y is a scalar multiple of x. If on the other hand $y = \alpha x$ for some scalar α then

$$\|x\| \|y\| = \|x\| \|\alpha x\| = |\alpha| \|x\|^2 \text{ and } |x^T y| = |x^T \alpha y| = |\alpha| |x^T x| = |\alpha| \|x\|^2$$

so $|x^T y| = \|x\| \|y\|$.

3.3 Inverse and Determinants

Problem 3.16 *[D***] Calculate* $\det[A]$ *and* A^{-1} *where*

$$A = \begin{bmatrix} 3 & 2 \\ 1 & 5 \end{bmatrix}$$

and calculate $(A^T)^{-1}$.

Answer: We have

$$\begin{aligned} \det[A] &= \det \begin{bmatrix} 3 & 2 \\ 1 & 5 \end{bmatrix} = 3 \times 5 - 2 \times 1 = 13 \\ A^{-1} &= \begin{bmatrix} 3 & 2 \\ 1 & 5 \end{bmatrix}^{-1} = \frac{1}{13} \begin{bmatrix} 5 & -2 \\ -1 & 3 \end{bmatrix} = \begin{bmatrix} \frac{5}{13} & -\frac{2}{13} \\ -\frac{1}{13} & \frac{3}{13} \end{bmatrix} \\ (A^T)^{-1} &= (A^{-1})^T = \begin{bmatrix} \frac{5}{13} & -\frac{2}{13} \\ -\frac{1}{13} & \frac{3}{13} \end{bmatrix}^T = \begin{bmatrix} \frac{5}{13} & -\frac{1}{13} \\ -\frac{2}{13} & \frac{3}{13} \end{bmatrix}. \end{aligned}$$

Problem 3.17 *[B**] Prove that if* A *is a* 2×2 *matrix then* A *is invertible (*A^{-1}*exists) if and only if* $\det[A] = a_{11}a_{22} - a_{12}a_{21} \neq 0$ *and*

$$A^{-1} = \begin{bmatrix} a_{11} & a_{12} \\ a_{21} & a_{22} \end{bmatrix}^{-1} = \frac{1}{\det[A]} \begin{bmatrix} a_{22} & -a_{12} \\ -a_{21} & a_{11} \end{bmatrix}.$$

Answer: Let B be given by

$$B = \begin{bmatrix} a_{22} & -a_{12} \\ -a_{21} & a_{11} \end{bmatrix}.$$

We have

$$BA = AB = (a_{11}a_{22} - a_{12}a_{21}) I = \det[A] I$$

as

$$\begin{aligned} \begin{bmatrix} a_{22} & -a_{12} \\ -a_{21} & a_{11} \end{bmatrix} \times \begin{bmatrix} a_{11} & a_{12} \\ a_{21} & a_{22} \end{bmatrix} &= \begin{bmatrix} a_{11} & a_{12} \\ a_{21} & a_{22} \end{bmatrix} \times \begin{bmatrix} a_{22} & -a_{12} \\ -a_{21} & a_{11} \end{bmatrix} \\ &= (a_{11}a_{22} - a_{12}a_{21}) \begin{bmatrix} 1 & 0 \\ 0 & 1 \end{bmatrix}. \end{aligned}$$

If $\det[A] \neq 0$ then we can divide both sides $\det[A]$ as

$$BA = AB = \det[A]\, I \Longrightarrow \left(\frac{1}{\det[A]}B\right) A = A\left(\frac{1}{\det[A]}B\right) = I$$

so that

$$A^{-1} = \frac{1}{\det[A]}B = \frac{1}{a_{11}a_{22} - a_{12}a_{21}}\begin{bmatrix} a_{22} & -a_{12} \\ -a_{21} & a_{11} \end{bmatrix}$$

satisfies the definition of an inverse: that $A^{-1}A = AA^{-1} = I$.

We now use proof by contradiction to show that the existence of A^{-1} implies that $a_{11}a_{22} - a_{12}a_{21} \neq 0$. Suppose that A^{-1} exists but $a_{11}a_{22} - a_{12}a_{21} = 0$. Then

$$AB = (a_{11}a_{22} - a_{12}a_{21})\, I = 0 \times I = 0 \Longrightarrow B = A^{-1} \times 0 = 0.$$

But from the definition of B above we have

$$B = 0 \Longrightarrow A = 0.$$

But $A = 0$ does not have an inverse, which contradicts the existence of A^{-1}. Thus $a_{11}a_{22} - a_{12}a_{21} \neq 0$.

Problem 3.18 *[D***] Given the matrix*

$$A = \begin{bmatrix} 3 & 1 & 0 \\ 0 & 4 & 0 \\ 0 & 0 & 5 \end{bmatrix}$$

calculate $\det[A]$, $\det\left[A^T\right]$, $\det\left[A^{-1}\right]$, $tr\,[A]$, *and* A^{-1}.

Answer: Since A is upper triangular its determinant is the product of the diagonal elements and so

$$\det\begin{bmatrix} 3 & 1 & 0 \\ 0 & 4 & 0 \\ 0 & 0 & 5 \end{bmatrix} = 3 \times 4 \times 5 = 60 = \det\left[A^T\right], \det\left[A^{-1}\right] = \frac{1}{\det[A]} = \frac{1}{60}$$

$$A^{-1} = \frac{1}{60}\begin{bmatrix} 20 & -5 & 0 \\ 0 & 15 & 0 \\ 0 & 0 & 12 \end{bmatrix}, \quad tr\,[A] = 3 + 4 + 5 = 12.$$

Problem 3.19 *[C***] Given the matrix*

$$A = \begin{bmatrix} 1 & 2 & 4 \\ -2 & 3 & 2 \\ 3 & 1 & 3 \end{bmatrix}$$

find $\det[A]$ *using the Laplace expansion by: 1) going across the first row, 2) going down the second column. [B*] Calculate the cofactor matrix C and the adjoint matrix $adj\,[A]$. Calculate $adj\,[A] \times A$ and $A \times adj\,[A]$. Calculate A^{-1} and $\left(A^T\right)^{-1}$.*

Answer: Going across the first row

$$\det[A] = 1\det\begin{bmatrix} 3 & 2 \\ 1 & 3 \end{bmatrix} - 2\det\begin{bmatrix} -2 & 2 \\ 3 & 3 \end{bmatrix} + 4\det\begin{bmatrix} -2 & 3 \\ 3 & 1 \end{bmatrix} = -13.$$

Going down the second column

$$\det[A] = -2\det\begin{bmatrix} -2 & 2 \\ 3 & 3 \end{bmatrix} + 3\det\begin{bmatrix} 1 & 4 \\ 3 & 3 \end{bmatrix} - 1\det\begin{bmatrix} 1 & 4 \\ -2 & 2 \end{bmatrix} = -13.$$

The cofactor matrix is

$$C = \begin{bmatrix} \det\begin{bmatrix} 3 & 2 \\ 1 & 3 \end{bmatrix} & -\det\begin{bmatrix} -2 & 2 \\ 3 & 3 \end{bmatrix} & \det\begin{bmatrix} -2 & 3 \\ 3 & 1 \end{bmatrix} \\ -\det\begin{bmatrix} 2 & 4 \\ 1 & 3 \end{bmatrix} & \det\begin{bmatrix} 1 & 4 \\ 3 & 3 \end{bmatrix} & -\det\begin{bmatrix} 1 & 2 \\ 3 & 1 \end{bmatrix} \\ \det\begin{bmatrix} 2 & 4 \\ 3 & 2 \end{bmatrix} & -\det\begin{bmatrix} 1 & 4 \\ -2 & 2 \end{bmatrix} & \det\begin{bmatrix} 1 & 2 \\ -2 & 3 \end{bmatrix} \end{bmatrix} = \begin{bmatrix} 7 & 12 & -11 \\ -2 & -9 & 5 \\ -8 & -10 & 7 \end{bmatrix}$$

so that

$$adj\,[A] = C^T = \begin{bmatrix} 7 & 12 & -11 \\ -2 & -9 & 5 \\ -8 & -10 & 7 \end{bmatrix}^T = \begin{bmatrix} 7 & -2 & -8 \\ 12 & -9 & -10 \\ -11 & 5 & 7 \end{bmatrix}.$$

Here $adj\,[A] \times A$ and $A \times adj\,[A]$ are given by direct calculation as

$$\begin{aligned} \begin{bmatrix} 7 & -2 & -8 \\ 12 & -9 & -10 \\ -11 & 5 & 7 \end{bmatrix}\begin{bmatrix} 1 & 2 & 4 \\ -2 & 3 & 2 \\ 3 & 1 & 3 \end{bmatrix} &= \begin{bmatrix} -13 & 0 & 0 \\ 0 & -13 & 0 \\ 0 & 0 & -13 \end{bmatrix} \\ &= -13\begin{bmatrix} 1 & 0 & 0 \\ 0 & 1 & 0 \\ 0 & 0 & 1 \end{bmatrix} \\ \begin{bmatrix} 1 & 2 & 4 \\ -2 & 3 & 2 \\ 3 & 1 & 3 \end{bmatrix}\begin{bmatrix} 7 & -2 & -8 \\ 12 & -9 & -10 \\ -11 & 5 & 7 \end{bmatrix} &= \begin{bmatrix} -13 & 0 & 0 \\ 0 & -13 & 0 \\ 0 & 0 & -13 \end{bmatrix} \\ &= -13\begin{bmatrix} 1 & 0 & 0 \\ 0 & 1 & 0 \\ 0 & 0 & 1 \end{bmatrix} \end{aligned}$$

which illustrates

$$adj\,[A] \times A = A \times adj\,[A] = \det[A] \times I.$$

Finally

$$\begin{aligned} A^{-1} &= \frac{1}{\det[A]} adj\,[A] = \frac{1}{-13}\begin{bmatrix} 7 & -2 & -8 \\ 12 & -9 & -10 \\ -11 & 5 & 7 \end{bmatrix} \\ &= \begin{bmatrix} -\frac{7}{13} & \frac{2}{13} & \frac{8}{13} \\ -\frac{12}{13} & \frac{9}{13} & \frac{10}{13} \\ \frac{11}{13} & -\frac{5}{13} & -\frac{7}{13} \end{bmatrix}. \end{aligned}$$

For $\left(A^T\right)^{-1}$ we have

$$\begin{aligned}\left(A^T\right)^{-1} &= \left(A^{-1}\right)^T = \begin{bmatrix} -\frac{7}{13} & \frac{2}{13} & \frac{8}{13} \\ -\frac{12}{13} & \frac{9}{13} & \frac{10}{13} \\ \frac{11}{13} & -\frac{5}{13} & -\frac{7}{13} \end{bmatrix}^T \\ &= \begin{bmatrix} -\frac{7}{13} & -\frac{12}{13} & \frac{11}{13} \\ \frac{2}{13} & \frac{9}{13} & -\frac{5}{13} \\ \frac{8}{13} & \frac{10}{13} & -\frac{7}{13} \end{bmatrix}.\end{aligned}$$

Problem 3.20 *[D***] Consider the matrices*

$$A = \begin{bmatrix} 3 & 2 \\ 1 & 3 \\ 1 & 2 \end{bmatrix}, B = \begin{bmatrix} 6 & 4 \\ 2 & 6 \\ 2 & 4 \end{bmatrix}.$$

What is $\det[A]$*? What are* $\det\left[A^TB\right]$ *and* $\det\left[AB^T\right]$*? What are* A^{-1}, $\left(A^TB\right)^{-1}$, *and* $\left(AB^T\right)^{-1}$*?*

Answer: Here $\det[A]$ is not defined since A is not a square matrix. We have

$$\begin{aligned} A^TB &= \begin{bmatrix} 3 & 1 & 1 \\ 2 & 3 & 2 \end{bmatrix}\begin{bmatrix} 6 & 4 \\ 2 & 6 \\ 2 & 4 \end{bmatrix} = \begin{bmatrix} 22 & 22 \\ 22 & 34 \end{bmatrix} \\ &\Longrightarrow \det\left[A^TB\right] = 264 \\ AB^T &= \begin{bmatrix} 3 & 2 \\ 1 & 3 \\ 1 & 2 \end{bmatrix}\begin{bmatrix} 6 & 2 & 2 \\ 4 & 6 & 4 \end{bmatrix} = \begin{bmatrix} 26 & 18 & 14 \\ 18 & 20 & 14 \\ 14 & 14 & 10 \end{bmatrix} \\ \det\left[AB^T\right] &= 26\times\det\begin{bmatrix} 20 & 14 \\ 14 & 10 \end{bmatrix} - 18\times\det\begin{bmatrix} 18 & 14 \\ 14 & 10 \end{bmatrix} + 14\times\det\begin{bmatrix} 18 & 20 \\ 14 & 14 \end{bmatrix} = 0. \end{aligned}$$

Since A is not square A^{-1} does **not** exist. The matrix $\left(AB^T\right)^{-1}$ does not exist since $\det\left[AB^T\right] = 0$. We have

$$\left(A^TB\right)^{-1} = \frac{1}{264}\begin{bmatrix} 34 & -22 \\ -22 & 22 \end{bmatrix}.$$

Problem 3.21 *[D***] Given*

$$A = \begin{bmatrix} 1 & 1 \\ 1 & 2 \\ 1 & 3 \end{bmatrix}$$

find $\left(A^TA\right)^{-1}$ *and* $\left(AA^T\right)^{-1}$ *if they exist.*

Answer: We have

$$
\begin{aligned}
A^T A &= \begin{bmatrix} 1 & 1 & 1 \\ 1 & 2 & 3 \end{bmatrix} \begin{bmatrix} 1 & 1 \\ 1 & 2 \\ 1 & 3 \end{bmatrix} = \begin{bmatrix} 3 & 6 \\ 6 & 14 \end{bmatrix} \\
AA^T &= \begin{bmatrix} 1 & 1 \\ 1 & 2 \\ 1 & 3 \end{bmatrix} \begin{bmatrix} 1 & 1 & 1 \\ 1 & 2 & 3 \end{bmatrix} = \begin{bmatrix} 2 & 3 & 4 \\ 3 & 5 & 7 \\ 4 & 7 & 10 \end{bmatrix} \\
\left(A^T A\right)^{-1} &= \begin{bmatrix} 3 & 6 \\ 6 & 14 \end{bmatrix}^{-1} = \frac{1}{3 \times 14 - 6 \times 6} \begin{bmatrix} 14 & -6 \\ -6 & 3 \end{bmatrix} \\
&= \begin{bmatrix} \frac{7}{3} & -1 \\ -1 & \frac{1}{2} \end{bmatrix}.
\end{aligned}
$$

The matrix AA^T does not have an inverse since

$$
\det\left[AA^T\right] = \det \begin{bmatrix} 2 & 3 & 4 \\ 3 & 5 & 7 \\ 4 & 7 & 10 \end{bmatrix} = 0.
$$

Problem 3.22 *[D***] Given the matrices*

$$
A = \begin{bmatrix} 1 & 1 \\ 2 & 0 \\ 3 & 1 \end{bmatrix}, \; B = \begin{bmatrix} 1 & -1 & 1 \\ 1 & 2 & 3 \end{bmatrix}
$$

find $\left(A^T A\right)^{-1}$ *and* $(AB)^{-1}$ *if they exist.*

Answer: We have

$$
\begin{aligned}
A^T A &= \begin{bmatrix} 1 & 2 & 3 \\ 1 & 0 & 1 \end{bmatrix} \begin{bmatrix} 1 & 1 \\ 2 & 0 \\ 3 & 1 \end{bmatrix} = \begin{bmatrix} 14 & 4 \\ 4 & 2 \end{bmatrix} \\
\det\left[A^T A\right] &= \det \begin{bmatrix} 14 & 4 \\ 4 & 2 \end{bmatrix} = 12 \\
AB &= \begin{bmatrix} 1 & 1 \\ 2 & 0 \\ 3 & 1 \end{bmatrix} \begin{bmatrix} 1 & -1 & 1 \\ 1 & 2 & 3 \end{bmatrix} = \begin{bmatrix} 2 & 1 & 4 \\ 2 & -2 & 2 \\ 4 & -1 & 6 \end{bmatrix} \\
\det[AB] &= \det \begin{bmatrix} 2 & 1 & 4 \\ 2 & -2 & 2 \\ 4 & -1 & 6 \end{bmatrix} = 0
\end{aligned}
$$

so that

$$
\left(A^T A\right)^{-1} = \begin{bmatrix} 14 & 4 \\ 4 & 2 \end{bmatrix}^{-1} = \frac{1}{12} \begin{bmatrix} 2 & -4 \\ -4 & 14 \end{bmatrix} = \begin{bmatrix} \frac{1}{6} & -\frac{1}{3} \\ -\frac{1}{3} & \frac{7}{6} \end{bmatrix}.
$$

The inverse of $(AB)^{-1}$ does not exist since $\det[AB] = 0$.

Problem 3.23 *[D***] Consider*

$$A = \begin{bmatrix} 2 & 2 & 2 \\ 2 & 4 & 6 \end{bmatrix}, \; B = \begin{bmatrix} 1 & 1 \\ 1 & 2 \\ 1 & 3 \end{bmatrix}, \; x = \begin{bmatrix} 1 \\ -2 \\ 1 \end{bmatrix}.$$

What is $(AB)^{-1}$. *Calculate* $\det[AB]$, $\det\left[(AB)^{-1}\right]$, *and* $\det[A]$. *Calculate* Ax *and use this result to calculate* $\det[BA]$.

Answer: We have

$$AB = \begin{bmatrix} 2 & 2 & 2 \\ 2 & 4 & 6 \end{bmatrix} \begin{bmatrix} 1 & 1 \\ 1 & 2 \\ 1 & 3 \end{bmatrix} = \begin{bmatrix} 6 & 12 \\ 12 & 28 \end{bmatrix}$$

so that

$$\det[AB] = \det \begin{bmatrix} 6 & 12 \\ 12 & 28 \end{bmatrix} = 24$$

and

$$\begin{aligned} (AB)^{-1} &= \begin{bmatrix} 6 & 12 \\ 12 & 28 \end{bmatrix}^{-1} = \frac{1}{24} \begin{bmatrix} 28 & -12 \\ -12 & 6 \end{bmatrix} = \begin{bmatrix} \frac{7}{6} & -\frac{1}{2} \\ -\frac{1}{2} & \frac{1}{4} \end{bmatrix} \\ \det\left[(AB)^{-1}\right] &= \frac{1}{\det[AB]} = \frac{1}{24}. \end{aligned}$$

Here $\det[A]$ is not defined since A is not square. We have

$$\begin{aligned} Ax &= \begin{bmatrix} 2 & 2 & 2 \\ 2 & 4 & 6 \end{bmatrix} \begin{bmatrix} 1 \\ -2 \\ 1 \end{bmatrix} = \begin{bmatrix} 0 \\ 0 \end{bmatrix} \\ &\Longrightarrow BAx = B0 = 0 \Longrightarrow \det[BA] = 0 \end{aligned}$$

since for any square matrix C, if $Cx = 0$ for $x \neq 0$ then C does not have an inverse and so $\det[C] = 0$.

Problem 3.24 *[D***] Given*

$$A = \begin{bmatrix} 3 & 3 & 3 \\ 12 & 4 & -4 \end{bmatrix}, \; B = \begin{bmatrix} 9 & 18 \\ 3 & 6 \\ 4 & 8 \end{bmatrix}$$

calculate $\det[AB]$ *and* $(AB)^{-1}$.

Answer: We have

$$AB = \begin{bmatrix} 3 & 3 & 3 \\ 12 & 4 & -4 \end{bmatrix} \begin{bmatrix} 9 & 18 \\ 3 & 6 \\ 4 & 8 \end{bmatrix} = \begin{bmatrix} 48 & 96 \\ 104 & 208 \end{bmatrix}, \det[AB] = \det \begin{bmatrix} 48 & 96 \\ 104 & 208 \end{bmatrix} = 0.$$

Since $\det[AB] = 0$ it follows that $(AB)^{-1}$ does not exist.

Problem 3.25 *[A**] Prove that if x is a $n \times 1$ vector with $n > 1$ and $x \neq 0$, then $\left(x^T x\right)^{-1}$ exists but $\left(x x^T\right)^{-1}$ does not exist. What is* $\det\left[\left(x^T x\right)^{-1}\right]$ *and* $\det\left[x x^T\right]$?

Answer: Since $x \neq 0$ it follows that

$$x^T x = x_1^2 + x_2^2 + \cdots + x_n^2 > 0$$

since at least one of the components of $x = [x_i]$ is not equal to zero. Since $x^T x > 0$ is a positive scalar we have

$$\left(x^T x\right)^{-1} = \det\left[\left(x^T x\right)^{-1}\right] = \frac{1}{x^T x} = \frac{1}{x_1^2 + x_2^2 + \cdots + x_n^2}.$$

Suppose for the purpose of a proof by contradiction that $x x^T$ did have an inverse, say B. Then $x x^T B = I$. Let $y^T = x^T B$, which is a $1 \times n$ row vector. Then

$$\begin{aligned} x y^T &= \begin{bmatrix} x_1 \\ x_2 \\ \vdots \\ x_n \end{bmatrix} \begin{bmatrix} y_1 & y_2 & \cdots & y_n \end{bmatrix} = \begin{bmatrix} x_1 y_1 & x_1 y_2 & \cdots & x_1 y_n \\ x_2 y_1 & x_2 y_2 & \cdots & x_2 y_n \\ \vdots & \vdots & \ddots & \vdots \\ x_n y_1 & \cdots & x_n y_{n-1} & x_n y_n \end{bmatrix} \\ &= \begin{bmatrix} 1 & 0 & \cdots & 0 \\ 0 & 1 & \cdots & 0 \\ \vdots & \vdots & \ddots & \vdots \\ 0 & \cdots & 0 & 1 \end{bmatrix}. \end{aligned}$$

For the ith diagonal element we have $x_i y_i = 1$ so that $x_i \neq 0$ and $y_i = \frac{1}{x_i} \neq 0$ for all i. Since $n > 1$ there is an off-diagonal i, j^{th} element with $j \neq i$ where $x_i y_j = 0$. Since $x_i \neq 0$ this implies that $y_j = 0$. But this contradicts $y_j = \frac{1}{x_j} \neq 0$. Consequently $x x^T$ does not have an inverse and so $\det\left[x x^T\right] = 0$.

Problem 3.26 *[A**] Suppose A is an $m \times n$ matrix, B is an $n \times m$ matrix, and x is an $m \times 1$ vector such that $Bx = 0$. Suppose further that the 5th element of x is equal to the GNP of Canada. Calculate* $\det[AB]$.

Answer: Since the GNP of Canada is not zero it follows that $x \neq 0$. Now from $Bx = 0$ we have

$$ABx = A0 = 0 \text{ for } x \neq 0.$$

Since AB *is* an $m \times m$ square matrix, and since for any square matrix

$$Cx = 0 \text{ for } x \neq 0 \Longrightarrow C^{-1} \text{ does not exist } \Longrightarrow \det[C] = 0$$

it follows that $\det[AB] = 0$.

Problem 3.27 *Prove that if Q is any $m \times m$, non-singular and symmetric matrix and X is any $m \times n$ matrix then $C = X^T Q^{-1} X$ is an $n \times n$ symmetric matrix.*

Answer: The matrix C is symmetric since

$$\begin{aligned} C^T &= \left(X^T Q^{-1} X\right)^T \text{ (by definition)} = X^T \left(X^T Q^{-1}\right)^T \text{ (since } (AB)^T = B^T A^T) \\ &= X^T \left(Q^{-1}\right)^T \left(X^T\right)^T \text{ (since } (AB)^T = B^T A^T) \\ &= X^T \left(Q^T\right)^{-1} X \text{ (since } \left(A^T\right)^T = A \text{ and } \left(A^{-1}\right)^T = \left(A^T\right)^{-1}) \\ &= X^T Q^{-1} X \text{ (since } Q = Q^T \text{ is symmetric)} = C \text{ (by definition)}. \end{aligned}$$

The matrix C is $n \times n$ since X^T is $n \times m$, Q^{-1} is $m \times m$ and X is $m \times n$.

Problem 3.28 *[B**] Prove for any $n \times p$ matrix X and $n \times n$ symmetric matrix B that $C = \left(X^T B X\right)^{-1}$ is a $p \times p$ symmetric matrix.*

Answer: We have

$$\begin{aligned} C^T &= \left(\left(X^T B X\right)^{-1}\right)^T = \left(\left(X^T B X\right)^T\right)^{-1} = \left((BX)^T \left(X^T\right)^T\right)^{-1} = \left(X^T B^T X\right)^{-1} \\ &= \left(X^T B X\right)^{-1} = C \end{aligned}$$

The matrix C is $p \times p$ since X^T is $p \times n$, B is $n \times n$, and X is $n \times p$ so that $X^T B X$ is $p \times p$ and hence so too is $C = \left(X^T B X\right)^{-1}$.

Problem 3.29 *[B**] Given the matrix*

$$A = \begin{bmatrix} 1 & 2 \\ 1 & 4 \\ 1 & 6 \end{bmatrix}$$

calculate $A\left(A^T A\right)^{-1} A^T \left(I - A\left(A^T A\right)^{-1} A^T\right)$.

Answer: Rather than slogging through a direct calculation we obtain

$$A\left(A^T A\right)^{-1} A^T \left(I - A\left(A^T A\right)^{-1} A^T\right) = \begin{bmatrix} 0 & 0 & 0 \\ 0 & 0 & 0 \\ 0 & 0 & 0 \end{bmatrix}$$

as follows

$$\begin{aligned} A\left(A^T A\right)^{-1} A^T \left(I - A\left(A^T A\right)^{-1} A^T\right) &= A\left(A^T A\right)^{-1} A^T - A\left(A^T A\right)^{-1} A^T A \left(A^T A\right)^{-1} A^T \\ &= A\left(A^T A\right)^{-1} A^T - A\left(A^T A\right)^{-1} A^T = 0. \end{aligned}$$

Problem 3.30 *[C***] Given the two matrices A and B and the vector x as*

$$A = \begin{bmatrix} 1 & 1 \\ 2 & 1 \\ 3 & 1 \end{bmatrix}, \quad B = \begin{bmatrix} 3 & 3 & 3 \\ 1 & 2 & 3 \end{bmatrix}, \quad x = \begin{bmatrix} 1 \\ -2 \\ 1 \end{bmatrix}$$

what is $(BA)^{-1}$? Is $\det[AB] = \det[A]\det[B]$? Is $(AB)^{-1} = B^{-1}A^{-1}$? Calculate Bx, ABx and $\det[AB]$.

Answer: We have

$$\begin{aligned} BA &= \begin{bmatrix} 3 & 3 & 3 \\ 1 & 2 & 3 \end{bmatrix} \begin{bmatrix} 1 & 1 \\ 2 & 1 \\ 3 & 1 \end{bmatrix} = \begin{bmatrix} 18 & 9 \\ 14 & 6 \end{bmatrix} \\ (BA)^{-1} &= \begin{bmatrix} 18 & 9 \\ 14 & 6 \end{bmatrix}^{-1} = \frac{1}{-18} \begin{bmatrix} 6 & -9 \\ -14 & 18 \end{bmatrix} = \begin{bmatrix} -\frac{1}{3} & \frac{1}{2} \\ \frac{7}{9} & -1 \end{bmatrix}. \end{aligned}$$

Here $\det[AB] \neq \det[A]\det[B]$ since A and B are not square matrices so $\det[A]$ and $\det[B]$ are not defined. Similarly $(AB)^{-1} \neq B^{-1}A^{-1}$ since A and B are not square, and hence B^{-1} and A^{-1} are not defined.

We have

$$\begin{aligned} Bx &= \begin{bmatrix} 3 & 3 & 3 \\ 1 & 2 & 3 \end{bmatrix} \begin{bmatrix} 1 \\ -2 \\ 1 \end{bmatrix} = \begin{bmatrix} 0 \\ 0 \end{bmatrix} \\ &\implies ABx = A0 = 0 \\ &\implies \det[AB] = 0 \end{aligned}$$

since $x \neq 0$ and

$$ABx = 0 \implies \det[AB] = 0.$$

Problem 3.31 *[A**] Use proof by contradiction to show that if $Ax = 0$ for $x \neq 0$ then A^{-1} does not exist.*

Answer: Suppose $Ax = 0$ for $x \neq 0$ and A^{-1} does exist. Then

$$Ax = 0 \implies A^{-1}Ax = A^{-1}0 \implies x = 0.$$

But $x = 0$ contradicts $x \neq 0$. So A^{-1} does not exist.

Problem 3.32 *[A**] If A has an inverse prove that A^{-1} is unique.*

Answer: Suppose B and C are both inverses of A. We show that A^{-1} is unique by showing that $B = C$. Since B and C are both inverses of A we have by definition

$$BA = AB = I \text{ and } CA = AC = I.$$

Multiplying both sides of $BA = I$ by C yields

$$BA = I \implies B\underbrace{AC}_{I} = I \times C \implies B = C.$$

Problem 3.33 *[A**] If A^{-1} exists, then prove that $\left(A^{-1}\right)^{-1} = A$.*

Answer: Since $AA^{-1} = A^{-1}A = I$, A satisfies the definition of the inverse of A^{-1} and so $\left(A^{-1}\right)^{-1} = A$.

Problem 3.34 *Prove that if A^{-1} exists then $\left(A^T\right)^{-1} = \left(A^{-1}\right)^T$.*

Answer: Using $(AB)^T = B^T A^T$ and $I = I^T$ we have

$$\begin{aligned} AA^{-1} &= A^{-1}A = I \\ &\Longrightarrow \left(AA^{-1}\right)^T = \left(A^{-1}A\right)^T = I^T = I \\ &\Longrightarrow \left(A^{-1}\right)^T A^T = A^T\left(A^{-1}\right)^T = I. \end{aligned}$$

The matrix $\left(A^{-1}\right)^T$ satisfies the definition of the inverse of A^T and so $\left(A^T\right)^{-1} = \left(A^{-1}\right)^T$.

Problem 3.35 *[B**] Prove that if A^{-1} and B^{-1} exist, and the product AB is defined, then $(AB)^{-1} = B^{-1}A^{-1}$.*

Answer: Let $C = AB$ and $D = B^{-1}A^{-1}$. By definition D is the inverse of C if $DC = CD = I$. We can verify that this is true as

$$\begin{aligned} DC &= B^{-1}A^{-1}AB = B^{-1}\overbrace{A^{-1}A}^{I}B = B^{-1}IB = B^{-1}B = I \\ CD &= A\overbrace{BB^{-1}}^{I}A^{-1} = AIA^{-1} = AA^{-1} = I. \end{aligned}$$

Problem 3.36 *[B**] Use $\det[AB] = \det[A]\det[B]$ to prove that if $\det[A] = 0$ then A does not have an inverse.*

Answer: Here we use proof by contradiction. Suppose $\det[A] = 0$ but A^{-1} did exist. Then $AA^{-1} = I$ and so

$$\det\left[AA^{-1}\right] = \det[I] = 1.$$

Now since $\det[A] = 0$

$$\det\left[AA^{-1}\right] = \det[A]\det\left[A^{-1}\right] = 0 \times \det\left[A^{-1}\right] = 0.$$

So we have shown $\det\left[AA^{-1}\right] = 1$ and $\det\left[AA^{-1}\right] = 0$, a contradiction. Thus A^{-1} does not exist if $\det[A] = 0$.

Problem 3.37 *[C***] Prove that $P = X(X^TX)^{-1}X^T$ and $I - P$ are symmetric where X is an $n \times p$ matrix.*

Answer: First $X^T X$ is symmetric as

$$\left(X^T X\right)^T = X^T \left(X^T\right)^T = X^T X$$

since $(AB)^T = B^T A^T$ and $\left(X^T\right)^T = X$. From $\left(X^T X\right)^T = X^T X$ it follows that $(X^T X)^{-1}$ is symmetric as

$$\left(\left(X^T X\right)^{-1}\right)^T = \left(\left(X^T X\right)^T\right)^{-1} = \left(X^T X\right)^{-1}.$$

Now

$$\begin{aligned} P^T &= \Big(\underbrace{X(X^T X)^{-1}}_{A}\underbrace{X^T}_{B}\Big)^T = \left(X^T\right)^T \Big(\underbrace{X}_{A}\underbrace{(X^T X)^{-1}}_{B}\Big)^T \quad (\text{since } (AB)^T = B^T A^T) \\ &= X\left((X^T X)^{-1}\right)^T X^T \quad (\text{since } AB^T = B^T A^T \text{ and } \left(A^T\right)^T = A) \\ &= X(X^T X)^{-1} X^T \quad (\text{since } (X^T X)^{-1} \text{ is symmetric}). \end{aligned}$$

Since $P^T = P$ and $I = I^T$ we have $(I - P)^T = I^T - P^T = I - P$.

3.4 Systems of Equations

Problem 3.38 *[D***] Consider the following system of linear equations*

$$\begin{aligned} 4x_1 + 2x_2 &= 3 \\ 5x_1 + 6x_2 &= 7. \end{aligned}$$

Write these in matrix notation as $Ax = b$. Calculate $\det[A]$ and based on this determine how many solutions exist and whether A^{-1} exists. Calculate x_1, x_2 using Cramer's rule, and using $x = A^{-1}b$.

Answer: Here $Ax = b$ as

$$\begin{bmatrix} 4 & 2 \\ 5 & 6 \end{bmatrix} \begin{bmatrix} x_1 \\ x_2 \end{bmatrix} = \begin{bmatrix} 3 \\ 7 \end{bmatrix} \text{ with } \det[A] = \det\begin{bmatrix} 4 & 2 \\ 5 & 6 \end{bmatrix} = 14.$$

Since $\det[A] = 14 \neq 0$ it follows that exactly one solution exists and that A^{-1} exists. Using Cramer's rule we find that

$$x_1 = \frac{\det\begin{bmatrix} 3 & 2 \\ 7 & 6 \end{bmatrix}}{14} = \frac{4}{14} = \frac{2}{7}, \; x_2 = \frac{\det\begin{bmatrix} 4 & 3 \\ 5 & 7 \end{bmatrix}}{14} = \frac{13}{14}.$$

Now

$$\begin{aligned} A &= \begin{bmatrix} 4 & 2 \\ 5 & 6 \end{bmatrix} \\ &\Longrightarrow A^{-1} = \begin{bmatrix} 4 & 2 \\ 5 & 6 \end{bmatrix}^{-1} = \frac{1}{14}\begin{bmatrix} 6 & -2 \\ -5 & 4 \end{bmatrix} \end{aligned}$$

so that

$$x = \begin{bmatrix} x_1 \\ x_2 \end{bmatrix} = A^{-1}b = \frac{1}{14}\begin{bmatrix} 6 & -2 \\ -5 & 4 \end{bmatrix}\begin{bmatrix} 3 \\ 7 \end{bmatrix} = \begin{bmatrix} \frac{2}{7} \\ \frac{13}{14} \end{bmatrix}.$$

Problem 3.39 *[D***] Consider the following system of linear equations*

$$\begin{aligned} 8x_1 + 4x_2 &= 6 \\ 5x_1 + 3x_2 &= 15. \end{aligned}$$

*Write these in matrix notation as $Ax = b$. Calculate x_1, x_2 using Cramer's rule and using A^{-1}. [B***] Solve the following equations for y_1 and y_2*

$$\begin{aligned} e^{-6}\left(y_1^2 y_2\right)^4 &= 1 \\ e^{-15}\left(y_1\right)^5\left(y_2\right)^3 &= 1. \end{aligned}$$

Answer: We have $Ax = b$ as

$$\begin{bmatrix} 8 & 4 \\ 5 & 3 \end{bmatrix}\begin{bmatrix} x_1 \\ x_2 \end{bmatrix} = \begin{bmatrix} 6 \\ 15 \end{bmatrix} \text{ with } \det[A] = \det\begin{bmatrix} 8 & 4 \\ 5 & 3 \end{bmatrix} = 4.$$

Since $\det[A] = 4 \neq 0$ it follows that exactly one solution to these equations exists and that A^{-1} exists. Using Cramer's rule we find that

$$x_1 = \frac{\det\begin{bmatrix} 6 & 4 \\ 15 & 3 \end{bmatrix}}{4} = -\frac{42}{4} = -\frac{21}{2},\ x_2 = \frac{\det\begin{bmatrix} 8 & 6 \\ 5 & 15 \end{bmatrix}}{4} = \frac{90}{4} = \frac{45}{2}.$$

Or we have $x = A^{-1}b$ as

$$\begin{bmatrix} x_1 \\ x_2 \end{bmatrix} = \begin{bmatrix} 8 & 4 \\ 5 & 3 \end{bmatrix}^{-1}\begin{bmatrix} 6 \\ 15 \end{bmatrix} = \frac{1}{4}\begin{bmatrix} 3 & -4 \\ -5 & 8 \end{bmatrix}\begin{bmatrix} 6 \\ 15 \end{bmatrix} = \begin{bmatrix} -\frac{21}{2} \\ \frac{45}{2} \end{bmatrix}.$$

Taking $\ln(\,)$ of both sides of $e^{-6}\left(y_1^2 y_2\right)^4 = 1$ and $e^{-15}\left(y_1\right)^5\left(y_2\right)^3 = 1$ we find that

$$\begin{aligned} \ln\left(e^{-6}\left(y_1^2 y_2\right)^4\right) &= \ln(1) \\ &\implies 8\ln(y_1) + 4\ln(y_2) = 6 \\ \ln\left(e^{-15}\left(y_1\right)^5\left(y_2\right)^3\right) &= \ln(1) \\ &\implies 5\ln(y_1) + 3\ln(y_2) = 15. \end{aligned}$$

Setting $x_1 = \ln(y_1)$ and $x_2 = \ln(y_2)$ yields

$$\begin{aligned} 8x_1 + 4x_2 &= 6 \\ 5x_1 + 3x_2 &= 15 \end{aligned}$$

which is identical to the linear system of equations we just solved so that

$$\ln(y_1) = x_1 = -\frac{21}{2} \implies y_1 = e^{-\frac{21}{2}},\ \ln(y_2) = x_2 = \frac{45}{2} \implies y_2 = e^{\frac{45}{2}}.$$

Problem 3.40 *[D***] Consider the following system of linear equations*

$$\begin{aligned} 3x_1 + 4x_2 &= 6 \\ 5x_1 + 6x_2 &= 15. \end{aligned}$$

*Write these in matrix notation as $Ax = b$. Calculate x_1, x_2 using Cramer's rule and $x = A^{-1}b$. [B***] Solve*

$$\begin{aligned} (w_1)^3 (w_2)^4 &= e^6 \\ (w_1)^5 (w_2)^6 &= e^{15}. \end{aligned}$$

Answer: We have $Ax = b$ as

$$\begin{bmatrix} 3 & 4 \\ 5 & 6 \end{bmatrix} \begin{bmatrix} x_1 \\ x_2 \end{bmatrix} = \begin{bmatrix} 6 \\ 15 \end{bmatrix} \text{ with } \det[A] = \det \begin{bmatrix} 3 & 4 \\ 5 & 6 \end{bmatrix} = -2.$$

Since $\det[A] = -2 \neq 0$ it follows that exactly one solution to these equations exists and that A^{-1} exists. Using Cramer's rule we find that

$$\begin{aligned} x_1 &= \frac{\det \begin{bmatrix} 6 & 4 \\ 15 & 6 \end{bmatrix}}{-2} = \frac{-24}{-2} = 12 \\ x_2 &= \frac{\det \begin{bmatrix} 3 & 6 \\ 5 & 15 \end{bmatrix}}{-2} = \frac{15}{-2} = -\frac{15}{2}. \end{aligned}$$

Or we have $x = A^{-1}b$ as

$$\begin{aligned} x &= \begin{bmatrix} x_1 \\ x_2 \end{bmatrix} = A^{-1}b = \begin{bmatrix} 3 & 4 \\ 5 & 6 \end{bmatrix}^{-1} \begin{bmatrix} 6 \\ 15 \end{bmatrix} \\ &= \frac{1}{-2} \begin{bmatrix} 6 & -4 \\ -5 & 3 \end{bmatrix} \begin{bmatrix} 6 \\ 15 \end{bmatrix} = \begin{bmatrix} 12 \\ -\frac{15}{2} \end{bmatrix}. \end{aligned}$$

Now for the nonlinear system of equations we have

$$\begin{aligned} (w_1)^3 (w_2)^4 &= e^6 \Longrightarrow \ln\left((w_1)^3 (w_2)^4\right) = \ln\left(e^6\right) \\ &\Longrightarrow 3\ln(w_1) + 4\ln(w_2) = 6 \\ (w_1)^5 (w_2)^6 &= e^{15} \Longrightarrow \ln\left((w_1)^5 (w_2)^6\right) = \ln\left(e^{15}\right) \\ &\Longrightarrow 5\ln(w_1) + 6\ln(w_2) = 15. \end{aligned}$$

Setting $x_1 = \ln(w_1)$ and $x_2 = \ln(w_2)$ we find that

$$\begin{aligned} 3x_1 + 4x_2 &= 6 \\ 5x_1 + 6x_2 &= 15 \end{aligned}$$

which is identical to the linear system of equations we just solved so that

$$\begin{aligned} \ln(w_1) &= x_1 = 12 \Longrightarrow w_1 = e^{12} \\ \ln(w_2) &= x_2 = -\frac{15}{2} \Longrightarrow w_2 = e^{-\frac{15}{2}}. \end{aligned}$$

Problem 3.41 *[B***] Consider the following system of linear equations*

$$\begin{aligned} y_1^2 y_2^3 &= e^6 \\ y_1^5 y_2^3 &= e^{15}. \end{aligned}$$

Using the $\ln()$ *function, transform these into a system of two linear equations with* $x_1 = \ln(y_1)$ *and* $x_2 = \ln(y_2)$ *and solve for* x_1 *and* x_2 *using Cramer's rule. What are* y_1, y_2*?*

Answer: We have

$$\begin{aligned} \ln\left(y_1^2 y_2^3\right) &= \ln\left(e^6\right) \Longrightarrow 2\ln(y_1) + 3\ln(y_2) = 6 \Longrightarrow 2x_1 + 3x_2 = 6 \\ \ln\left(y_1^5 y_2^3\right) &= \ln\left(e^{15}\right) \Longrightarrow 5\ln(y_1) + 3\ln(y_2) = 15 \Longrightarrow 5x_1 + 3x_2 = 15 \end{aligned}$$

or in matrix notation $Ax = b$ as

$$\begin{bmatrix} 2 & 3 \\ 5 & 3 \end{bmatrix} \begin{bmatrix} x_1 \\ x_2 \end{bmatrix} = \begin{bmatrix} 6 \\ 15 \end{bmatrix}$$

with

$$\det[A] = \det \begin{bmatrix} 2 & 3 \\ 5 & 3 \end{bmatrix} = -9.$$

Solving using Cramer's rule we have

$$x_1 = \frac{\det \begin{bmatrix} 6 & 3 \\ 15 & 3 \end{bmatrix}}{-9} = \frac{-27}{-9} = 3,\ x_2 = \frac{\det \begin{bmatrix} 2 & 6 \\ 5 & 15 \end{bmatrix}}{-9} = \frac{0}{-9} = 0.$$

Therefore

$$y_1 = e^{x_1} = e^3 \approx 20.09,\ y_2 = e^{x_2} = e^0 = 1.$$

Problem 3.42 *[D***] Consider the following system of equations*

$$\begin{aligned} 3x_1 + 5x_2 &= 7 + 2x_3 \\ x_1 - 2x_2 + 3x_3 &= 4 + 2x_2 \\ x_1 - x_2 + x_3 &= 10. \end{aligned}$$

Writing this system as $Ax = b$*, what is* A*, what is* x *and what is* b*? Calculate* $\det[A]$*. Is* A *non-singular? Is it invertible? What is its rank? Solve* $Ax = 0$ *for* x *using* $x = A^{-1}b$*. Solve using Cramer's rule. [B***] Consider the system of equations* $yA^T = b^T$ *where* y *is an unknown* 1×3 *row vector and* A *and* b *are the same as above. What is* y*?*

Answer: Putting the $x's$ on the left-hand side we can rewrite the equations as

$$\begin{aligned} 3x_1 + 5x_2 - 2x_3 &= 7 \\ x_1 - 4x_2 + 3x_3 &= 4 \\ x_1 - x_2 + x_3 &= 10 \end{aligned}$$

or in matrix notation as

$$\begin{bmatrix} 3 & 5 & -2 \\ 1 & -4 & 3 \\ 1 & -1 & 1 \end{bmatrix} \begin{bmatrix} x_1 \\ x_2 \\ x_3 \end{bmatrix} = \begin{bmatrix} 7 \\ 4 \\ 10 \end{bmatrix}.$$

We have

$$\det \begin{bmatrix} 3 & 5 & -2 \\ 1 & -4 & 3 \\ 1 & -1 & 1 \end{bmatrix} = 3 \det \begin{bmatrix} -4 & 3 \\ -1 & 1 \end{bmatrix} - 5 \det \begin{bmatrix} 1 & 3 \\ 1 & 1 \end{bmatrix} + -2 \det \begin{bmatrix} 1 & -4 \\ 1 & -1 \end{bmatrix} = 1.$$

Since $\det[A] \neq 0$, it follows that A is invertible, has full rank, that is $rank\,[A] = 3$, and that $Ax = 0 \Longrightarrow x = 0$; that is $x_1 = 0, x_2 = 0, x_3 = 0$.

The adjoint of A is given by

$$adj \begin{bmatrix} 3 & 5 & -2 \\ 1 & -4 & 3 \\ 1 & -1 & 1 \end{bmatrix} = \begin{bmatrix} \det \begin{bmatrix} -4 & 3 \\ -1 & 1 \end{bmatrix} & -\det \begin{bmatrix} 1 & 3 \\ 1 & 1 \end{bmatrix} & \det \begin{bmatrix} 1 & -4 \\ 1 & -1 \end{bmatrix} \\ -\det \begin{bmatrix} 5 & -2 \\ -1 & 1 \end{bmatrix} & \det \begin{bmatrix} 3 & -2 \\ 1 & 1 \end{bmatrix} & -\det \begin{bmatrix} 3 & 5 \\ 1 & -1 \end{bmatrix} \\ \det \begin{bmatrix} 5 & -2 \\ -4 & 3 \end{bmatrix} & -\det \begin{bmatrix} 3 & -2 \\ 1 & 3 \end{bmatrix} & \det \begin{bmatrix} 3 & 5 \\ 1 & -4 \end{bmatrix} \end{bmatrix}^T$$

$$= \begin{bmatrix} -1 & -3 & 7 \\ 2 & 5 & -11 \\ 3 & 8 & -17 \end{bmatrix}.$$

Since $A^{-1} = \frac{1}{\det[A]} adj\,[A]$ and $\det[A] = 1$ it follows that for this matrix $A^{-1} = adj\,[A]$ so that

$$A^{-1} = \begin{bmatrix} -1 & -3 & 7 \\ 2 & 5 & -11 \\ 3 & 8 & -17 \end{bmatrix}$$

$$\Longrightarrow \quad x = A^{-1}b = \begin{bmatrix} -1 & -3 & 7 \\ 2 & 5 & -11 \\ 3 & 8 & -17 \end{bmatrix} \begin{bmatrix} 7 \\ 4 \\ 10 \end{bmatrix} = \begin{bmatrix} 51 \\ -76 \\ -117 \end{bmatrix}.$$

So $x_1 = 51, x_2 = -76, x_3 = -117$. Using Cramer's rule we find that

$$x_1 = \frac{\det \begin{bmatrix} 7 & 5 & -2 \\ 4 & -4 & 3 \\ 10 & -1 & 1 \end{bmatrix}}{1} = 51$$

$$x_2 = \frac{\det \begin{bmatrix} 3 & 7 & -2 \\ 1 & 4 & 3 \\ 1 & 10 & 1 \end{bmatrix}}{1} = -76$$

$$x_3 = \frac{\det \begin{bmatrix} 3 & 5 & 7 \\ 1 & -4 & 4 \\ 1 & -1 & 10 \end{bmatrix}}{1} = -117.$$

Taking transposes of both sides of $yA^T = b^T$ we obtain

$$\left(yA^T\right)^T = \left(b^T\right)^T \Longrightarrow Ay^T = b \Longrightarrow y^T = x \Longrightarrow y = \begin{bmatrix} 51 & -76 & -117 \end{bmatrix}.$$

Problem 3.43 *[D***] If $Ax = b$ where*

$$A = \begin{bmatrix} 1 & 2 & 4 \\ -2 & 3 & 2 \\ 3 & 1 & 3 \end{bmatrix}, \; x = \begin{bmatrix} x_1 \\ x_2 \\ x_3 \end{bmatrix}, \; b = \begin{bmatrix} 1 \\ 5 \\ 3 \end{bmatrix}$$

write this out as three linear equations with three unknowns. Calculate $x = A^{-1}b$. Now calculate x_1, x_2, x_3 using Cramer's rule.

Answer: We have

$$\begin{aligned} x_1 + 2x_2 + 4x_3 &= 1 \\ -2x_1 + 3x_2 + 2x_3 &= 5 \\ 3x_1 + x_2 + 3x_3 &= 3. \end{aligned}$$

We have

$$\det[A] = \det\begin{bmatrix} 1 & 2 & 4 \\ -2 & 3 & 2 \\ 3 & 1 & 3 \end{bmatrix} = -13 \text{ and } adj\,[A] = \begin{bmatrix} 7 & -2 & -8 \\ 12 & -9 & -10 \\ -11 & 5 & 7 \end{bmatrix}$$

so that

$$A^{-1} = \frac{1}{-13}\begin{bmatrix} 7 & -2 & -8 \\ 12 & -9 & -10 \\ -11 & 5 & 7 \end{bmatrix} = \begin{bmatrix} -\frac{7}{13} & \frac{2}{13} & \frac{8}{13} \\ -\frac{12}{13} & \frac{9}{13} & \frac{10}{13} \\ \frac{11}{13} & -\frac{5}{13} & -\frac{7}{13} \end{bmatrix}$$

and

$$\begin{bmatrix} x_1 \\ x_2 \\ x_3 \end{bmatrix} = \begin{bmatrix} -\frac{7}{13} & \frac{2}{13} & \frac{8}{13} \\ -\frac{12}{13} & \frac{9}{13} & \frac{10}{13} \\ \frac{11}{13} & -\frac{5}{13} & -\frac{7}{13} \end{bmatrix}\begin{bmatrix} 1 \\ 5 \\ 3 \end{bmatrix} = \begin{bmatrix} \frac{27}{13} \\ \frac{63}{13} \\ -\frac{35}{13} \end{bmatrix}.$$

So $x_1 = \frac{27}{13}, x_2 = \frac{63}{13}, x_3 = -\frac{35}{13}$. Using Cramer's rule obtain the same answers as

$$\begin{aligned} x_1 &= \frac{\det\begin{bmatrix} 1 & 2 & 4 \\ 5 & 3 & 2 \\ 3 & 1 & 3 \end{bmatrix}}{-13} = \frac{27}{13} \\ x_2 &= \frac{\det\begin{bmatrix} 1 & 1 & 4 \\ -2 & 5 & 2 \\ 3 & 3 & 3 \end{bmatrix}}{-13} = \frac{63}{13} \\ x_3 &= \frac{\det\begin{bmatrix} 1 & 2 & 1 \\ -2 & 3 & 5 \\ 3 & 1 & 3 \end{bmatrix}}{-13} = -\frac{35}{13}. \end{aligned}$$

Problem 3.44 *[D***] Consider the following system of linear equations*

$$\begin{aligned} 3x_1 + 4x_2 + 5x_3 &= 6 \\ -2x_1 + x_2 + 2x_3 &= 8 \\ 7x_1 + 3x_2 + 4x_3 &= 2. \end{aligned}$$

*If these are written as $Ax = b$, what are A, x, and b? Calculate $\det[A]$ using the Laplace expansion by going across the first row. Find the $2,3$ element (second row, third column) of A^{-1}. Find x_2 using Cramer's rule and [B**] $x = A^{-1}b$.*

Answer: We have

$$A = \begin{bmatrix} 3 & 4 & 5 \\ -2 & 1 & 2 \\ 7 & 3 & 4 \end{bmatrix}, \; x = \begin{bmatrix} x_1 \\ x_2 \\ x_3 \end{bmatrix}, \; b = \begin{bmatrix} 6 \\ 8 \\ 2 \end{bmatrix}$$

and

$$\det[A] = \det \begin{bmatrix} 3 & 4 & 5 \\ -2 & 1 & 2 \\ 7 & 3 & 4 \end{bmatrix} = 3\det\begin{bmatrix} 1 & 2 \\ 3 & 4 \end{bmatrix} - 4\det\begin{bmatrix} -2 & 2 \\ 7 & 4 \end{bmatrix} + 5\det\begin{bmatrix} -2 & 1 \\ 7 & 3 \end{bmatrix} = 17.$$

The $2,3$ element of A^{-1} is $\frac{c_{32}}{\det[A]} = \frac{-16}{17}$ since

$$c_{32} = \det\begin{bmatrix} 3 & 5 \\ -2 & 2 \end{bmatrix} \times -1^{3+2} = -16.$$

Using $A^{-1} = adj\,[A] \,/\, \det[A]$ we have

$$\begin{bmatrix} 3 & 4 & 5 \\ -2 & 1 & 2 \\ 7 & 3 & 4 \end{bmatrix}^{-1} = \frac{1}{17}\begin{bmatrix} -2 & -1 & 3 \\ 22 & -23 & -16 \\ -13 & 19 & 11 \end{bmatrix} = \begin{bmatrix} -\frac{2}{17} & -\frac{1}{17} & \frac{3}{17} \\ \frac{22}{17} & -\frac{23}{17} & -\frac{16}{17} \\ -\frac{13}{17} & \frac{19}{17} & \frac{11}{17} \end{bmatrix}.$$

Using Cramer's rule we have

$$x_2 = \frac{\det\begin{bmatrix} 3 & 6 & 5 \\ -2 & 8 & 2 \\ 7 & 2 & 4 \end{bmatrix}}{17} = -\frac{84}{17}.$$

Or directly we have

$$x = \begin{bmatrix} x_1 \\ x_2 \\ x_3 \end{bmatrix} = A^{-1}b = \begin{bmatrix} -\frac{2}{17} & -\frac{1}{17} & \frac{3}{17} \\ \frac{22}{17} & -\frac{23}{17} & -\frac{16}{17} \\ -\frac{13}{17} & \frac{19}{17} & \frac{11}{17} \end{bmatrix}\begin{bmatrix} 6 \\ 8 \\ 2 \end{bmatrix} = \begin{bmatrix} -\frac{14}{17} \\ -\frac{84}{17} \\ \frac{96}{17} \end{bmatrix}.$$

3.5 Idempotent Matrices

Problem 3.45 *[C***] A rule in matrix algebra is* $(AB)^{-1} = B^{-1}A^{-1}$. *Suppose* X *is an* $n \times p$ *matrix. Why then is* $\left(X^T X\right)^{-1} = X^{-1}\left(X^T\right)^{-1}$ *and*

$$P = X\left(X^T X\right)^{-1} X^T = XX^{-1}\left(X^T\right)^{-1} X^T = I \times I = I$$

incorrect? When would this be correct?

Answer: It is false to write $\left(X^T X\right)^{-1} = X^{-1}\left(X^T\right)^{-1}$ since if $n \neq p$ and X is not square, then X^{-1} does not exist. Only square matrices have inverses. The result $(AB)^{-1} = B^{-1}A^{-1}$ is only valid when A^{-1} and B^{-1} exist. If X is a square $n \times n$ matrix with $\det[X] \neq 0$ then $\left(X^T X\right)^{-1} = X^{-1}\left(X^T\right)^{-1}$ and $P = I$.

Problem 3.46 *[C***]Given*

$$X = \begin{bmatrix} 1 \\ 1 \\ 1 \end{bmatrix}$$

calculate $P = X\left(X^T X\right)^{-1} X^T$ *and verify that* $tr\,[P] = 1$, *the number of columns of* X. *Verify by direct multiplication that* P *is idempotent.*

Answer: We have

$$\begin{aligned} \left(X^T X\right)^{-1} &= \left(\begin{bmatrix} 1 & 1 & 1 \end{bmatrix}\begin{bmatrix} 1 \\ 1 \\ 1 \end{bmatrix}\right)^{-1} \\ &= \left(1^2 + 1^2 + 1^2\right)^{-1} = 3^{-1} = \frac{1}{3} \end{aligned}$$

so that

$$P = X\left(X^T X\right)^{-1} X^T = \frac{1}{3}\begin{bmatrix} 1 \\ 1 \\ 1 \end{bmatrix}\begin{bmatrix} 1 & 1 & 1 \end{bmatrix} = \begin{bmatrix} \frac{1}{3} & \frac{1}{3} & \frac{1}{3} \\ \frac{1}{3} & \frac{1}{3} & \frac{1}{3} \\ \frac{1}{3} & \frac{1}{3} & \frac{1}{3} \end{bmatrix}.$$

Thus

$$\begin{aligned} tr\,[P] &= tr\begin{bmatrix} \frac{1}{3} & \frac{1}{3} & \frac{1}{3} \\ \frac{1}{3} & \frac{1}{3} & \frac{1}{3} \\ \frac{1}{3} & \frac{1}{3} & \frac{1}{3} \end{bmatrix} = \frac{1}{3} + \frac{1}{3} + \frac{1}{3} = 1 \\ P \times P &= \begin{bmatrix} \frac{1}{3} & \frac{1}{3} & \frac{1}{3} \\ \frac{1}{3} & \frac{1}{3} & \frac{1}{3} \\ \frac{1}{3} & \frac{1}{3} & \frac{1}{3} \end{bmatrix}\begin{bmatrix} \frac{1}{3} & \frac{1}{3} & \frac{1}{3} \\ \frac{1}{3} & \frac{1}{3} & \frac{1}{3} \\ \frac{1}{3} & \frac{1}{3} & \frac{1}{3} \end{bmatrix} = \begin{bmatrix} \frac{1}{3} & \frac{1}{3} & \frac{1}{3} \\ \frac{1}{3} & \frac{1}{3} & \frac{1}{3} \\ \frac{1}{3} & \frac{1}{3} & \frac{1}{3} \end{bmatrix}. \end{aligned}$$

Problem 3.47 *[C**]Given the matrix*

$$X = \begin{bmatrix} 1 & 1 \\ 1 & 2 \\ 1 & 3 \end{bmatrix}$$

calculate $P = X\left(X^T X\right)^{-1} X^T$ *and verify that* $tr\,[P] = 2$*, the number of columns of* X*. Verify by direct multiplication that* P *is idempotent.*

Answer: We have

$$(X^T X)^{-1} = \left(\begin{bmatrix} 1 & 1 & 1 \\ 1 & 2 & 3 \end{bmatrix}\begin{bmatrix} 1 & 1 \\ 1 & 2 \\ 1 & 3 \end{bmatrix}\right)^{-1} = \begin{bmatrix} 3 & 6 \\ 6 & 14 \end{bmatrix}^{-1} = \begin{bmatrix} \frac{7}{3} & -1 \\ -1 & \frac{1}{2} \end{bmatrix}$$

so that

$$P = X\left(X^T X\right)^{-1} X^T = \begin{bmatrix} 1 & 1 \\ 1 & 2 \\ 1 & 3 \end{bmatrix}\begin{bmatrix} \frac{7}{3} & -1 \\ -1 & \frac{1}{2} \end{bmatrix}\begin{bmatrix} 1 & 1 & 1 \\ 1 & 2 & 3 \end{bmatrix} = \begin{bmatrix} \frac{5}{6} & \frac{1}{3} & -\frac{1}{6} \\ \frac{1}{3} & \frac{1}{3} & \frac{1}{3} \\ -\frac{1}{6} & \frac{1}{3} & \frac{5}{6} \end{bmatrix}.$$

Thus

$$\begin{aligned} tr\,[P] &= tr\begin{bmatrix} \frac{5}{6} & \frac{1}{3} & -\frac{1}{6} \\ \frac{1}{3} & \frac{1}{3} & \frac{1}{3} \\ -\frac{1}{6} & \frac{1}{3} & \frac{5}{6} \end{bmatrix} = \frac{5}{6} + \frac{1}{3} + \frac{5}{6} = 2 \\ P \times P &= \begin{bmatrix} \frac{5}{6} & \frac{1}{3} & -\frac{1}{6} \\ \frac{1}{3} & \frac{1}{3} & \frac{1}{3} \\ -\frac{1}{6} & \frac{1}{3} & \frac{5}{6} \end{bmatrix}\begin{bmatrix} \frac{5}{6} & \frac{1}{3} & -\frac{1}{6} \\ \frac{1}{3} & \frac{1}{3} & \frac{1}{3} \\ -\frac{1}{6} & \frac{1}{3} & \frac{5}{6} \end{bmatrix} \\ &= \begin{bmatrix} \frac{5}{6} & \frac{1}{3} & -\frac{1}{6} \\ \frac{1}{3} & \frac{1}{3} & \frac{1}{3} \\ -\frac{1}{6} & \frac{1}{3} & \frac{5}{6} \end{bmatrix}. \end{aligned}$$

Problem 3.48 *[A**] Suppose* X *is an* $n \times 1$ *column vector with* i^{th} *element denoted by* x_i*. Let* $P = X\left(X^T X\right)^{-1} X^T = [p_{ij}]$ *where* p_{ij} *is the* i, j^{th} *element of* P*. What is* p_{ij}*?*

Answer: We have

$$X^T X = \begin{bmatrix} x_1 & x_2 & \cdots & x_n \end{bmatrix}\begin{bmatrix} x_1 \\ x_2 \\ \vdots \\ x_n \end{bmatrix} = x_1^2 + x_2^2 + \cdots + x_n^2 \text{ a scalar.}$$

Therefore

$$
\begin{aligned}
X\underbrace{\left(X^TX\right)^{-1}}_{\frac{1}{x_1^2+x_2^2+\cdots+x_n^2}}X^T &= \frac{1}{x_1^2+x_2^2+\cdots+x_n^2}XX^T \\
&= \frac{1}{x_1^2+x_2^2+\cdots+x_n^2}\begin{bmatrix} x_1 \\ x_2 \\ \vdots \\ x_n \end{bmatrix}\begin{bmatrix} x_1 & x_2 & \cdots & x_n \end{bmatrix} \\
&= \frac{1}{x_1^2+x_2^2+\cdots+x_n^2}\begin{bmatrix} x_1^2 & x_1x_2 & \cdots & x_1x_n \\ x_1x_2 & x_2^2 & \cdots & x_2x_n \\ \vdots & \vdots & \ddots & \vdots \\ x_1x_n & x_2x_n & \cdots & x_n^2 \end{bmatrix}
\end{aligned}
$$

so that

$$p_{ij} = \frac{x_ix_j}{x_1^2+x_2^2+\cdots+x_n^2}.$$

Problem 3.49 *[B**] If X is any $n \times p$ matrix and $A = X^TBX$ where B is symmetric, show that A is symmetric and that $P = X\left(X^TBX\right)^{-1}X^TB$ is idempotent.*

Answer: Since B is symmetric we have $B^T = B$ so that

$$A^T = \left(X^TBX\right)^T = X^TB^T\left(X^T\right)^T = X^TBX = A$$

and A is symmetric.

The matrix P is idempotent as

$$
\begin{aligned}
PP &= X\left(X^TBX\right)^{-1}X^TB \times X\left(X^TBX\right)^{-1}X^TB \\
&= X\underbrace{\left(X^TBX\right)^{-1}X^TBX}_{I}\left(X^TBX\right)^{-1}X^TB \\
&= XI\left(X^TBX\right)^{-1}X^TB = X\left(X^TBX\right)^{-1}X^TB = P.
\end{aligned}
$$

Problem 3.50 *[B**] Suppose R is an $r \times p$ matrix, X is a $n \times p$ matrix and Σ is an $n \times n$ symmetric matrix. Consider the matrices*

$$
\begin{aligned}
A &= X(X^T\Sigma^{-1}X)^{-1}X^T\Sigma^{-1}, B = X(X^T\Sigma^{-1}X)^{-1}X^T \\
C &= (RX^T\Sigma^{-1}XR^T)^{-1},\ D = (R^TX^T\Sigma^{-1}XR)^{-1}.
\end{aligned}
$$

Find the dimensions of the matrices A, B, and C. What can you say about the matrix D? Which matrices are symmetric? Which are idempotent? Determine AB, BA, B^TA, AB^T, $\left(B^TA^T\right)^T$, and $\left(A^TB^T\right)^T$. Make any possible simplifications , and if possible avoid any unnecessary work in your derivations.

Answer: A is $n \times n$, B is $n \times n$, C is $r \times r$, and D is not defined since XR and $R^T X^T$ are not defined. The matrices B and C are symmetric while the matrix A is idempotent.

We have

$$\begin{aligned}
AB &= X(X^T\Sigma^{-1}X)^{-1}\underbrace{X^T\Sigma^{-1}X(X^T\Sigma^{-1}X)^{-1}}_{I}X^T = X(X^T\Sigma^{-1}X)^{-1}X^T = B \\
BA &= X(X^T\Sigma^{-1}X)^{-1}X^TX(X^T\Sigma^{-1}X)^{-1}X^T\Sigma^{-1} \\
B^TA &= BA \text{ and } AB^T = AB = B \text{ (since } B^T = B) \\
\left(B^TA^T\right)^T &= AB = B \text{ since } \left(B^TA^T\right)^T = \left(A^T\right)^T\left(B^T\right)^T = AB \\
\left(A^TB^T\right)^T &= BA \text{ (since } \left(A^TB^T\right)^T = \left(B^T\right)^T\left(A^T\right)^T = BA).
\end{aligned}$$

Problem 3.51 *[A **] Recall from statistics that the sample mean and variance are given by* $\bar{Y} = \frac{1}{n}\sum_{i=1}^{n} Y_i$ *and* $s^2 = \frac{1}{n-1}\sum_{i=1}^{n}\left(Y_i - \bar{Y}\right)^2$. *Let* $Y = [Y_i]$ *be an* $n \times 1$ *vector and let* ι *be an* $n \times 1$ *vector of* $1's$. *Let* $P = \frac{1}{n}\,\iota\iota^T$. *Show that* P *and* $I - P$ *are symmetric and idempotent, that* $\bar{Y} = \frac{1}{n}\iota^T Y$, *and that* $s^2 = \frac{Y^T(I-P)Y}{n-1}$.

Answer: First note that $\iota^T\iota = n$ since

$$\iota^T\iota = \begin{bmatrix} 1 & 1 & \cdots & 1 \end{bmatrix}\begin{bmatrix} 1 \\ 1 \\ \vdots \\ 1 \end{bmatrix} = 1^2 + 1^2 + \cdots + 1^2 = n.$$

Multiplying P by itself we find that

$$PP = \left(\frac{1}{n}\iota\iota^T\right)\left(\frac{1}{n}\iota\iota^T\right) = \frac{1}{n^2}\iota\underbrace{\iota^T\iota}_{n}\iota^T = \frac{n}{n^2}\iota\iota^T = \frac{1}{n}\iota\iota^T = P$$

and so P idempotent. Now multiplying $I - P$ by itself and using $PP = P$ we find that

$$(I-P)(I-P) = II - IP - PI + PP = I - P - P + P = I - P \Longrightarrow I - P \text{ is idempotent.}$$

To show that $\bar{Y} = \frac{1}{n}\iota^T Y$ note that

$$\begin{aligned}
\frac{1}{n}\iota^T Y &= \frac{1}{n}\begin{bmatrix} 1 & 1 & \cdots & 1 \end{bmatrix}\begin{bmatrix} Y_1 \\ Y_2 \\ \vdots \\ Y_n \end{bmatrix} \\
&= \frac{1}{n}(1 \times Y_1 + 1 \times Y_2 + \cdots + 1 \times Y_n) \\
&= \frac{1}{n}(Y_1 + Y_2 + \cdots + Y_n) = \bar{Y}.
\end{aligned}$$

Furthermore

$$(I-P)Y = Y - \frac{1}{n}\iota\iota^T Y = Y - \iota\underbrace{\left(\frac{1}{n}\iota^T Y\right)}_{\bar{Y}} = Y - \iota\bar{Y} = \begin{bmatrix} Y_1 - \bar{Y} \\ Y_2 - \bar{Y} \\ \vdots \\ Y_n - \bar{Y} \end{bmatrix}.$$

Therefore

$$\begin{aligned} ((I-P)Y)^T((I-P))Y &= \begin{bmatrix} Y_1 - \bar{Y} & Y_2 - \bar{Y} & \cdots & Y_n - \bar{Y} \end{bmatrix} \begin{bmatrix} Y_1 - \bar{Y} \\ Y_2 - \bar{Y} \\ \vdots \\ Y_n - \bar{Y} \end{bmatrix} \\ &= \left(Y_1 - \bar{Y}\right)^2 + \left(Y_2 - \bar{Y}\right)^2 + \cdots + \left(Y_n - \bar{Y}\right)^2 \\ &= (n-1)s^2. \end{aligned}$$

But using the fact that $I - P$ is symmetric and idempotent we have

$$\begin{aligned} ((I-P)Y)^T((I-P))Y &= Y^T(I-P)^T(I-P)Y \\ &= Y^T(I-P)(I-P)Y = Y^T(I-P)Y \end{aligned}$$

so that

$$Y^T(I-P)Y = (n-1)s^2 \Longrightarrow s^2 = \frac{Y^T(I-P)Y}{n-1}.$$

3.6 Eigenvalues and Eigenvectors

Problem 3.52 *[D***] Consider the matrix*

$$A = \begin{bmatrix} 2 & -1 \\ -1 & 3 \end{bmatrix}.$$

Calculate the eigenvalues from $\det[A - \lambda I] = 0$. *Verify that the determinant is the product of the eigenvalues and the trace is the sum of the eigenvalues.*

Answer: We have

$$\begin{aligned} \det[A - \lambda I] &= \det\left[\begin{bmatrix} 2 & -1 \\ -1 & 3 \end{bmatrix} - \lambda\begin{bmatrix} 1 & 0 \\ 0 & 1 \end{bmatrix}\right] = \det\begin{bmatrix} 2-\lambda & -1 \\ -1 & 3-\lambda \end{bmatrix} \\ &= (2-\lambda)(3-\lambda) - (-1)\times(-1) = \lambda^2 - 5\lambda + 5 \end{aligned}$$

and so the eigenvalues satisfy

$$\begin{aligned} \lambda^2 - 5\lambda + 5 &= 0 \\ \Longrightarrow \quad \lambda &= \frac{-(-5) \pm \sqrt{(-5)^2 - 4\times 5}}{2} = \frac{5 \pm \sqrt{5}}{2} \\ \Longrightarrow \quad \lambda_1 &= \frac{5+\sqrt{5}}{2} \approx 3.62, \lambda_2 = \frac{5-\sqrt{5}}{2} \approx 1.38. \end{aligned}$$

We then have

$$\begin{aligned} \det[A] &= \lambda_1 \times \lambda_2 = \frac{5+\sqrt{5}}{2} \times \frac{5-\sqrt{5}}{2} = \frac{5^2 - \left(\sqrt{5}\right)^2}{4} = 5 \\ tr[A] &= \lambda_1 + \lambda_2 = \frac{5+\sqrt{5}}{2} + \frac{5-\sqrt{5}}{2} = 5. \end{aligned}$$

Problem 3.53 *[D***] Consider the matrix*

$$A = \begin{bmatrix} 5 & -2 \\ -2 & 5 \end{bmatrix}.$$

Calculate the eigenvalues from $\det[A - \lambda I] = 0$. *Find the eigenvectors of* A *that satisfy* $\|x\| = 1$. *[A**] Show that* A *can be written as* $A = C\Lambda C^T$ *where* $C^T = C^{-1}$ *and* Λ *is a diagonal matrix with the eigenvalues along the diagonal. Calculate* A^2 *as* $A^2 = C\Lambda^2 C^T$ *and* $A^2 = A \times A$ *and show that they are the same. Calculate* A^{-1} *as* $A^{-1} = C\Lambda^{-1}C^T$ *and* $A^{-1} = \frac{1}{|A|} adj[A]$ *and show that they are the same.*

Answer: The eigenvalues are the roots of

$$\det\left[\begin{bmatrix} 5 & -2 \\ -2 & 5 \end{bmatrix} - \lambda \begin{bmatrix} 1 & 0 \\ 0 & 1 \end{bmatrix}\right] = \lambda^2 - 10\lambda + 21 = 0$$

so that

$$\begin{aligned} \lambda &= \frac{-(-10) \pm \sqrt{(-10)^2 - 4 \times 21}}{2} \\ &\Longrightarrow \lambda_1 = 3, \lambda_2 = 7. \end{aligned}$$

An eigenvector associated with $\lambda_1 = 3$ is any non-zero column of the adjoint of $A - \lambda_1 I$ given by

$$A - \lambda_1 I = \begin{bmatrix} 5 & -2 \\ -2 & 5 \end{bmatrix} - \lambda_1 \begin{bmatrix} 1 & 0 \\ 0 & 1 \end{bmatrix} = \begin{bmatrix} 5 & -2 \\ -2 & 5 \end{bmatrix} - 3 \begin{bmatrix} 1 & 0 \\ 0 & 1 \end{bmatrix} = \begin{bmatrix} 2 & -2 \\ -2 & 2 \end{bmatrix}$$

which is

$$\begin{aligned} adj \begin{bmatrix} 2 & -2 \\ -2 & 2 \end{bmatrix} &= \begin{bmatrix} 2 & 2 \\ 2 & 2 \end{bmatrix} \\ &\Longrightarrow x_1 = \begin{bmatrix} 2 \\ 2 \end{bmatrix}. \end{aligned}$$

The length of x_1 is

$$\|x_1\| = \sqrt{x_1^T x_1} = \sqrt{2^2 + 2^2} = 2\sqrt{2}$$

so that normalizing x_1 to have a length of 1 yields

$$x_1 = \begin{bmatrix} 2 \\ 2 \end{bmatrix} \Longrightarrow x_1 = \frac{1}{2\sqrt{2}} \begin{bmatrix} 2 \\ 2 \end{bmatrix} = \begin{bmatrix} \frac{1}{\sqrt{2}} \\ \frac{1}{\sqrt{2}} \end{bmatrix}.$$

An eigenvector associated with $\lambda_2 = 7$ is any non-zero column of the adjoint matrix of

$$\begin{bmatrix} 5 & -2 \\ -2 & 5 \end{bmatrix} - \lambda_2 \begin{bmatrix} 1 & 0 \\ 0 & 1 \end{bmatrix} = \begin{bmatrix} 5 & -2 \\ -2 & 5 \end{bmatrix} - 7 \begin{bmatrix} 1 & 0 \\ 0 & 1 \end{bmatrix} = \begin{bmatrix} -2 & -2 \\ -2 & -2 \end{bmatrix}$$

which is

$$adj \begin{bmatrix} -2 & -2 \\ -2 & -2 \end{bmatrix} = \begin{bmatrix} -2 & 2 \\ 2 & -2 \end{bmatrix}$$

so that taking the first column

$$x_2 = \begin{bmatrix} -2 \\ 2 \end{bmatrix}.$$

The length of x_2 is

$$\|x_2\| = \sqrt{x_2^T x_2} = \sqrt{(-2)^2 + 2^2} = 2\sqrt{2}$$

so that normalizing x_2 to have a length of 1 yields

$$x_2 = \begin{bmatrix} -2 \\ 2 \end{bmatrix} \Longrightarrow x_2 = \frac{1}{2\sqrt{2}} \begin{bmatrix} -2 \\ 2 \end{bmatrix} = \begin{bmatrix} -\frac{1}{\sqrt{2}} \\ \frac{1}{\sqrt{2}} \end{bmatrix}.$$

Putting x_1 and x_2 in the columns of C then yields

$$C = \begin{bmatrix} \frac{1}{\sqrt{2}} & -\frac{1}{\sqrt{2}} \\ \frac{1}{\sqrt{2}} & \frac{1}{\sqrt{2}} \end{bmatrix}$$

which is an orthogonal matrix with $C^{-1} = C^T$.

Now

$$\begin{aligned} AC &= \begin{bmatrix} 5 & -2 \\ -2 & 5 \end{bmatrix} \begin{bmatrix} \frac{1}{\sqrt{2}} & -\frac{1}{\sqrt{2}} \\ \frac{1}{\sqrt{2}} & \frac{1}{\sqrt{2}} \end{bmatrix} = \begin{bmatrix} \frac{3}{2}\sqrt{2} & -\frac{7}{2}\sqrt{2} \\ \frac{3}{2}\sqrt{2} & \frac{7}{2}\sqrt{2} \end{bmatrix} \\ &= \begin{bmatrix} \frac{1}{\sqrt{2}} & -\frac{1}{\sqrt{2}} \\ \frac{1}{\sqrt{2}} & \frac{1}{\sqrt{2}} \end{bmatrix} \begin{bmatrix} 3 & 0 \\ 0 & 7 \end{bmatrix} = C\Lambda \end{aligned}$$

and so

$$AC = C\Lambda \Longrightarrow A = C\Lambda C^{-1} = C\Lambda C^T$$

or

$$A = \begin{bmatrix} 5 & -2 \\ -2 & 5 \end{bmatrix} = \begin{bmatrix} \frac{1}{\sqrt{2}} & -\frac{1}{\sqrt{2}} \\ \frac{1}{\sqrt{2}} & \frac{1}{\sqrt{2}} \end{bmatrix} \begin{bmatrix} 3 & 0 \\ 0 & 7 \end{bmatrix} \begin{bmatrix} \frac{1}{\sqrt{2}} & \frac{1}{\sqrt{2}} \\ -\frac{1}{\sqrt{2}} & \frac{1}{\sqrt{2}} \end{bmatrix}.$$

Now

$$A^2 = AA = \begin{bmatrix} 5 & -2 \\ -2 & 5 \end{bmatrix} \begin{bmatrix} 5 & -2 \\ -2 & 5 \end{bmatrix} = \begin{bmatrix} 29 & -20 \\ -20 & 29 \end{bmatrix}$$

and

$$\begin{aligned} A^2 &= C\Lambda^2 C^T = \begin{bmatrix} \frac{1}{\sqrt{2}} & -\frac{1}{\sqrt{2}} \\ \frac{1}{\sqrt{2}} & \frac{1}{\sqrt{2}} \end{bmatrix} \begin{bmatrix} 3^2 & 0 \\ 0 & 7^2 \end{bmatrix} \begin{bmatrix} \frac{1}{\sqrt{2}} & \frac{1}{\sqrt{2}} \\ -\frac{1}{\sqrt{2}} & \frac{1}{\sqrt{2}} \end{bmatrix} \\ &= \begin{bmatrix} 29 & -20 \\ -20 & 29 \end{bmatrix}. \end{aligned}$$

For A^{-1} we have

$$A^{-1} = \frac{1}{|A|} adj\,[A] = \frac{1}{21} \begin{bmatrix} 5 & 2 \\ 2 & 5 \end{bmatrix} = \begin{bmatrix} \frac{5}{21} & \frac{2}{21} \\ \frac{2}{21} & \frac{5}{21} \end{bmatrix}$$

and

$$\begin{aligned} A^{-1} &= C\Lambda^{-1} C^T = \begin{bmatrix} \frac{1}{\sqrt{2}} & -\frac{1}{\sqrt{2}} \\ \frac{1}{\sqrt{2}} & \frac{1}{\sqrt{2}} \end{bmatrix} \begin{bmatrix} 3^{-1} & 0 \\ 0 & 7^{-1} \end{bmatrix} \begin{bmatrix} \frac{1}{\sqrt{2}} & \frac{1}{\sqrt{2}} \\ -\frac{1}{\sqrt{2}} & \frac{1}{\sqrt{2}} \end{bmatrix} \\ &= \begin{bmatrix} \frac{5}{21} & \frac{2}{21} \\ \frac{2}{21} & \frac{5}{21} \end{bmatrix}. \end{aligned}$$

Problem 3.54 *[D***] Given the two matrices*

$$A = \begin{bmatrix} 1 & 1 \\ 2 & 1 \\ 3 & 1 \end{bmatrix}, \; B = \begin{bmatrix} 3 & 3 & 3 \\ 1 & 2 & 3 \end{bmatrix}$$

calculate the eigenvalues of BA, $(BA)^{-1}$, and $A^T B^T$.

Answer: We have

$$BA = \begin{bmatrix} 3 & 3 & 3 \\ 1 & 2 & 3 \end{bmatrix} \begin{bmatrix} 1 & 1 \\ 2 & 1 \\ 3 & 1 \end{bmatrix} = \begin{bmatrix} 18 & 9 \\ 14 & 6 \end{bmatrix}.$$

The eigenvalues of BA are calculated as

$$\begin{aligned} \det \left[\begin{bmatrix} 18 & 9 \\ 14 & 6 \end{bmatrix} - \lambda \begin{bmatrix} 1 & 0 \\ 0 & 1 \end{bmatrix} \right] &= \lambda^2 - 24\lambda - 18 = 0 \\ &\Longrightarrow \lambda_1 = 12 + 9\sqrt{2}, \lambda_2 = 12 - 9\sqrt{2}. \end{aligned}$$

The eigenvalues of $(BA)^{-1}$ are

$$\lambda_1 = \left(12 + 9\sqrt{2}\right)^{-1} = \frac{1}{12 + 9\sqrt{2}} \text{ and } \lambda_2 = \left(12 - 9\sqrt{2}\right)^{-1} = \frac{1}{12 - 9\sqrt{2}}.$$

Since $A^T B^T = (BA)^T$ the eigenvalues of $A^T B^T$ are $\lambda_1 = 12 + 9\sqrt{2}, \lambda_2 = 12 - 9\sqrt{2}$

Problem 3.55 *[D***] Verify for*

$$A = \begin{bmatrix} 5 & 1 \\ 2 & 4 \end{bmatrix}$$

that the eigenvalues of A and A^{-1} are inverses of each other.

Answer: The eigenvalues found as

$$\lambda^2 - 9\lambda + 18 = 0 \Longrightarrow \lambda = \frac{-(-9) \pm \sqrt{(-9)^2 - 4 \times 18}}{2} = \frac{9 \pm 3}{2} \Longrightarrow \lambda_1 = 3, \lambda_2 = 6.$$

The inverse of A is

$$A^{-1} = \begin{bmatrix} 5 & 1 \\ 2 & 4 \end{bmatrix}^{-1} = \frac{1}{18}\begin{bmatrix} 4 & -1 \\ -2 & 5 \end{bmatrix} = \begin{bmatrix} \frac{4}{18} & -\frac{1}{18} \\ -\frac{2}{18} & \frac{5}{18} \end{bmatrix}.$$

The eigenvalues of A^{-1} are the roots of

$$\lambda^2 - \left(\frac{4}{18} + \frac{5}{18}\right)\lambda + \frac{1}{18} = \lambda^2 - \frac{1}{2}\lambda + \frac{1}{18} = 0$$

so that

$$\begin{aligned} \lambda &= \frac{-\left(-\frac{1}{2}\right) \pm \sqrt{\left(-\frac{1}{2}\right)^2 - 4 \times \frac{1}{18}}}{2} = \frac{\frac{1}{2} \pm \frac{1}{6}}{2} \\ &\Longrightarrow \lambda_1 = \frac{1}{3}, \lambda_2 = \frac{1}{6}. \end{aligned}$$

Problem 3.56 *[A**] Prove that the eigenvalues of A and A^T are identical.*

Answer: The eigenvalues of A are the roots of the equation

$$\det[A - \lambda I] = 0.$$

From $\det[B] = \det\left[B^T\right]$, $(A+B)^T = A^T + B^T$, and $I = I^T$ we have

$$\det[A - \lambda I] = \det\left[(A - \lambda I)^T\right] = \det\left[A^T - \lambda I^T\right] = \det\left[A^T - \lambda I\right]$$

so that

$$\det[A - \lambda I] = 0 \Longleftrightarrow \det\left[A^T - \lambda I\right] = 0.$$

So λ is an eigenvalue of A if and only if λ is an eigenvalue of A^T. (Note A and A^T will not in general have the same eigenvectors.)

Problem 3.57 *[C***] Calculate the eigenvalues from* $\det[A - \lambda I] = 0$ *where*

$$A = \begin{bmatrix} 4 & 0 & 0 \\ 0 & 5 & 0 \\ 0 & 0 & 6 \end{bmatrix}.$$

Answer: We have

$$\begin{aligned}
\det\left[A-\lambda I\right] &= \det\left[\begin{bmatrix} 4 & 0 & 0 \\ 0 & 5 & 0 \\ 0 & 0 & 6 \end{bmatrix} - \lambda\begin{bmatrix} 1 & 0 & 0 \\ 0 & 1 & 0 \\ 0 & 0 & 1 \end{bmatrix}\right] \\
&= \det\begin{bmatrix} 4-\lambda & 0 & 0 \\ 0 & 5-\lambda & 0 \\ 0 & 0 & 6-\lambda \end{bmatrix} \\
&= (4-\lambda)(5-\lambda)(6-\lambda).
\end{aligned}$$

So the eigenvalues satisfy

$$(4-\lambda)(5-\lambda)(6-\lambda)=0 \Longrightarrow \lambda_1=4, \lambda_2=5, \lambda_3=6.$$

Problem 3.58 *[A **] Suppose that A is an $n\times n$ symmetric matrix with eigenvalues ordered from largest to smallest as $\lambda_1 \geq \lambda_2 \geq \cdots \geq \lambda_n$, and with corresponding eigenvectors $x_1, x_2, \ldots, x_n$ normalized so that $x_i^T x_i = 1$. Consider the problem of maximizing $f(x) = x^T A x$ subject to the constraint that $x^T x = 1$. Show that the solution is $x^* = x_1$. What is the $f(x^*)$ then? Show that for minimizing $x^T A x$ subject to the constraint that $x^T x = 1$ the solution is $x^* = x_n$.*

Answer: Let x be any vector that satisfies $x^T x = 1$. Writing A as $A = C\Lambda C^T$ where $C^T = C^{-1}$ is orthogonal and Λ is a diagonal matrix with eigenvalues along the diagonal, we have for the $n\times 1$ vector $y = C^T x$ and $1\times n$ vector $y^T = x^T C$ that

$$f(x) = x^T A x = x^T C\Lambda C^T x = y^T \Lambda y = y_1^2\lambda_1 + y_2^2\lambda_2 + \cdots + y_n^2\lambda_n.$$

Since $y_i^2 \geq 0$, if we replace the λ_i 's with the largest value λ_1 this must make the expression larger and so

$$f(x) = y_1^2\lambda_1 + y_2^2\lambda_2 + \cdots + y_n^2\lambda_n \leq y_1^2\lambda_1 + y_2^2\lambda_1 + \cdots + y_n^2\lambda_1 = \lambda_1\left(y_1^2 + y_2^2 + \cdots + y_n^2\right) = \lambda_1$$

where we use

$$y^T y = x^T C C^T x = x^T x = 1 \Longrightarrow y_1^2 + y_2^2 + \cdots + y_n^2 = 1.$$

Thus $f(x) \leq \lambda_1$ for all x that satisfy $x^T x = 1$. Now $f(x_1) = \lambda_1$ since

$$Ax_1 = \lambda_1 x_1 \Longrightarrow f(x_1) = x_1^T A x_1 = \lambda_1 x_1^T x_1 = \lambda_1$$

and so $f(x) \leq f(x_1)$. It follows that $x^* = x_1$ maximizes $f(x)$ subject to the constraint that $x^T x = 1$, and that $f(x^*) = \lambda_1$.

To minimize $f(x)$ subject to the constraint that $x^T x = 1$ we have using the same type of argument that

$$f(x) = y_1^2\lambda_1 + y_2^2\lambda_2 + \cdots + y_n^2\lambda_n \geq \lambda_n\left(y_1^2 + y_2^2 + \cdots + y_n^2\right) = \lambda_n = f(x_n)$$

and so $f(x) \geq f(x_n)$. It follows that $x^* = x_n$ maximizes $f(x)$ subject to the constraint that $x^T x = 1$, and that $f(x^*) = \lambda_n$.

3.7 Quadratic Forms

Problem 3.59 *[C***] Consider the matrix*

$$A = \begin{bmatrix} 2 & 0 \\ 0 & 3 \end{bmatrix}.$$

Show that this matrix is positive definite using: 1) the quadratic form, 2) leading principal minors, and 3) eigenvalues.

Answer: The quadratic form is

$$\begin{aligned} x^T Ax &= \begin{bmatrix} x_1 & x_2 \end{bmatrix} \begin{bmatrix} 2 & 0 \\ 0 & 3 \end{bmatrix} \begin{bmatrix} x_1 \\ x_2 \end{bmatrix} \\ &= 2x_1^2 + 3x_2^2 \geq 0 \end{aligned}$$

since $x_1^2 \geq 0, x_2^2 \geq 0$ (so that A is positive semi-definite). Also

$$\begin{aligned} x^T Ax &= 0 \Longrightarrow 2x_1^2 + 3x_2^2 = 0 \\ &\Longrightarrow x_1 = x_2 = 0 \Longrightarrow x = 0 \end{aligned}$$

so that A is positive definite.

Using leading principal minors we have

$$M_1 = 2 > 0, M_2 = \det \begin{bmatrix} 2 & 0 \\ 0 & 3 \end{bmatrix} = 6 > 0$$

and so A is positive definite.

Since A is a diagonal matrix the eigenvalues of A are the diagonal elements so that $\lambda_1 = 2 > 0, \lambda_2 = 3 > 0$ and so A is positive definite.

Problem 3.60 *[C***] Consider the matrix*

$$A = \begin{bmatrix} -2 & 0 \\ 0 & -3 \end{bmatrix}.$$

Show that this matrix is negative definite using: 1) the quadratic form, 2) leading principal minors, and 3) eigenvalues. How do the conditions involving leading principal minors and eigenvalues differ?

Answer: The quadratic form is

$$x^T Ax = \begin{bmatrix} x_1 & x_2 \end{bmatrix} \begin{bmatrix} -2 & 0 \\ 0 & -3 \end{bmatrix} \begin{bmatrix} x_1 \\ x_2 \end{bmatrix} = (-2)\, x_1^2 + (-3)\, x_2^2 \leq 0$$

since $x_1^2 \geq 0, x_2^2 \geq 0$. Also

$$x^T Ax = 0 \Longrightarrow -2x_1^2 + -3x_2^2 = 0 \Longrightarrow x_1 = x_2 = 0 \Longrightarrow x = 0$$

so that A is negative definite.

Using leading principal minors we have

$$M_1 = -2 < 0, M_2 = \det \begin{bmatrix} -2 & 0 \\ 0 & -3 \end{bmatrix} = 6 > 0.$$

Since $M_1 < 0$, $M_2 > 0$ the matrix A is negative definite. Since A is a diagonal matrix the eigenvalues of A are the diagonal elements so that $\lambda_1 = -2 < 0, \lambda_2 = -3 < 0$ and so A is negative definite.

Notice that while both eigenvalues are negative, the leading principal minors alternate in sign with $M_2 > 0$.

Problem 3.61 *[C***] Consider the matrix*

$$A = \begin{bmatrix} -9 & 3 \\ 3 & -5 \end{bmatrix}.$$

*Write out the quadratic form $x^T Ax$. Find the eigenvalues of A and use these to determine whether A is positive definite or negative definite. [D***] Now use leading principal minors to determine if A is positive definite or negative definite.*

Answer: We have

$$\begin{aligned} x^T Ax &= \begin{bmatrix} x_1 & x_2 \end{bmatrix} \begin{bmatrix} -9 & 3 \\ 3 & -5 \end{bmatrix} \begin{bmatrix} x_1 \\ x_2 \end{bmatrix} \\ &= -9x_1^2 + 6x_1x_2 - 5x_2^2. \end{aligned}$$

To find the eigenvalues of A note that

$$\begin{aligned} \det [A - \lambda I] &= \det \left[\begin{bmatrix} -9 & 3 \\ 3 & -5 \end{bmatrix} - \lambda \begin{bmatrix} 1 & 0 \\ 0 & 1 \end{bmatrix} \right] \\ &= 36 + 14\lambda + \lambda^2 = 0 \\ &\implies \lambda_1 = -7 + \sqrt{13} \approx -3.39 < 0, \ \lambda_2 = -7 - \sqrt{13} \approx -10.61 < 0. \end{aligned}$$

Since $\lambda_1 < 0, \lambda_2 < 0$ it follows that A is negative definite. Since A is negative definite it follows that it is also negative semi-definite.

Using leading principal minors we find that

$$\begin{aligned} M_1 &= \det [-9] = -9 < 0 \\ M_2 &= \det \begin{bmatrix} -9 & 3 \\ 3 & -5 \end{bmatrix} = 36 > 0. \end{aligned}$$

Since they alternate in sign with $M_1 < 0$, it follows that A is negative definite.

Problem 3.62 *[C***] If A is given by*

$$A = \begin{bmatrix} 5 & -4 \\ -4 & 2 \end{bmatrix}$$

show that A is positive definite by 1) using leading principal minors and 2) calculating the eigenvalues of A.

Answer: The eigenvalues are the solutions to

$$\det\left(\begin{bmatrix} 5 & -4 \\ -4 & 2 \end{bmatrix} - \lambda \begin{bmatrix} 1 & 0 \\ 0 & 1 \end{bmatrix}\right) = 0 \Longrightarrow \lambda^2 - 7\lambda - 6 = 0$$

so that

$$\lambda_1 = \frac{7}{2} + \frac{1}{2}\sqrt{73} \approx 7.77 > 0, \lambda_2 = \frac{7}{2} - \frac{1}{2}\sqrt{73} \approx -0.772 < 0.$$

Since $\lambda_2 < 0$ it follows that A is **not** positive definite.

Using leading principal minors we find that

$$M_1 = 5 > 0, M_2 = \det\begin{bmatrix} 5 & -4 \\ -4 & 2 \end{bmatrix} = -6 < 0.$$

Since $M_2 < 0$ it follows that A is **not** positive definite.

Problem 3.63 *[B***] Consider the matrix*

$$A = \begin{bmatrix} -3 & 0 & 0 \\ 0 & -4 & 0 \\ 0 & 0 & -5 \end{bmatrix}.$$

Show that this matrix is negative definite using: 1) the quadratic form, 2) leading principal minors, and 3) eigenvalues.

Answer: The quadratic form is

$$\begin{aligned} x^T A x &= \begin{bmatrix} x_1 & x_2 & x_3 \end{bmatrix} \begin{bmatrix} -3 & 0 & 0 \\ 0 & -4 & 0 \\ 0 & 0 & -5 \end{bmatrix} \begin{bmatrix} x_1 \\ x_2 \\ x_3 \end{bmatrix} \\ &= (-3)\, x_1^2 + (-4)\, x_2^2 + (-5)\, x_3^2 \le 0 \end{aligned}$$

since $x_1^2 \ge 0, x_2^2 \ge 0, x_3^2 \ge 0$ (so that A is negative semi-definite). Also

$$x^T A x = 0 \Longrightarrow -3x_1^2 + -4x_2^2 + -5x_3^2 = 0 \Longrightarrow x_1 = x_2 = x_3 = 0 \Longrightarrow x = 0$$

so that A is negative definite.

The matrix A is negative definite since the leading principal minors alternate in sign with $M_1 < 0$ as

$$\begin{aligned} M_1 &= -3 < 0 \\ M_2 &= \det\begin{bmatrix} -3 & 0 \\ 0 & -4 \end{bmatrix} = 12 > 0 \\ M_3 &= \det\begin{bmatrix} -3 & 0 & 0 \\ 0 & -4 & 0 \\ 0 & 0 & -5 \end{bmatrix} = -60 < 0. \end{aligned}$$

Since A is a diagonal matrix the eigenvalues of A are the diagonal elements so that $\lambda_1 = -3 < 0, \lambda_2 = -4 < 0, \lambda_3 = -5 < 0$ and so A is negative definite.

Problem 3.64 *[C***] Consider*

$$A = \begin{bmatrix} 3 & -2 & 1 \\ -2 & 4 & 1 \\ 1 & 1 & 5 \end{bmatrix}.$$

Show that A is positive definite.

Answer: The matrix A is positive definite since all leading principal minors are positive as

$$\begin{aligned} M_1 &= 3 > 0 \\ M_2 &= \det \begin{bmatrix} 3 & -2 \\ -2 & 4 \end{bmatrix} = 8 > 0 \\ M_3 &= \det \begin{bmatrix} 3 & -2 & 1 \\ -2 & 4 & 1 \\ 1 & 1 & 5 \end{bmatrix} = 29 > 0. \end{aligned}$$

Problem 3.65 *Consider the matrix*

$$A = \begin{bmatrix} 1 & 1 & 1 \\ 1 & 2 & 2 \\ 1 & 2 & 3 \end{bmatrix}.$$

*Show that A is positive definite using 1) [A***] the quadratic form, 2) [D***] leading principal minors.*

Answer: We can rewrite the matrix as $A = A_1 + A_2 + A_3$ as

$$\begin{bmatrix} 1 & 1 & 1 \\ 1 & 2 & 2 \\ 1 & 2 & 3 \end{bmatrix} = \begin{bmatrix} 1 & 1 & 1 \\ 1 & 1 & 1 \\ 1 & 1 & 1 \end{bmatrix} + \begin{bmatrix} 0 & 0 & 0 \\ 0 & 1 & 1 \\ 0 & 1 & 1 \end{bmatrix} + \begin{bmatrix} 0 & 0 & 0 \\ 0 & 0 & 0 \\ 0 & 0 & 1 \end{bmatrix}$$

so that the quadratic form becomes

$$\begin{aligned} x^T A x &= x^T (A_1 + A_2 + A_3)\, x = x^T A_1 x + x^T A_2 x + x^T A_3 x \\ &= (x_1 + x_2 + x_3)^2 + (x_2 + x_3)^2 + x_3^2. \end{aligned}$$

If $x^T A x = 0$ then we have $x = 0$ as

$$\begin{aligned} x_3^2 &= 0 \Longrightarrow x_3 = 0 \\ (x_2 + x_3)^2 &= x_2^2 = 0 \Longrightarrow x_2 = 0 \\ (x_1 + x_2 + x_3)^2 &= 0 \Longrightarrow x_1^2 = 0 \Longrightarrow x_1 = 0. \end{aligned}$$

Therefore A is positive definite. The matrix A is also positive definite since the leading principal minors satisfy

$$M_1 = 1 > 0, M_2 = \det \begin{bmatrix} 1 & 1 \\ 1 & 2 \end{bmatrix} = 1 > 0, M_3 = \det \begin{bmatrix} 1 & 1 & 1 \\ 1 & 2 & 2 \\ 1 & 2 & 3 \end{bmatrix} = 1 > 0.$$

Problem 3.66 *[C***] Consider*

$$A = \begin{bmatrix} -3 & -2 & -1 \\ -2 & -4 & 1 \\ -1 & 1 & -5 \end{bmatrix}.$$

Show that A is negative definite.

Answer: The matrix A is negative definite since the leading principal minors alternate in sign with $M_1 < 0$ as

$$\begin{aligned} M_1 &= -3 < 0 \\ M_2 &= \det \begin{bmatrix} -3 & -2 \\ -2 & -4 \end{bmatrix} = 8 > 0 \\ M_3 &= \det \begin{bmatrix} -3 & -2 & -1 \\ -2 & -4 & 1 \\ -1 & 1 & -5 \end{bmatrix} = -29 < 0. \end{aligned}$$

Problem 3.67 *[D***] If*

$$X = \begin{bmatrix} 1 & -1 \\ 1 & -2 \\ 1 & -3 \end{bmatrix}$$

*then calculate $A = X^T X$. Calculate the eigenvalues of A and show from these eigenvalues that A is positive definite. Use leading principal minors to show that A is positive definite. [A***] Now show that A is positive definite using the quadratic form.*

Answer: We have

$$A = X^T X = \begin{bmatrix} 1 & 1 & 1 \\ -1 & -2 & -3 \end{bmatrix} \begin{bmatrix} 1 & -1 \\ 1 & -2 \\ 1 & -3 \end{bmatrix} = \begin{bmatrix} 3 & -6 \\ -6 & 14 \end{bmatrix}.$$

The eigenvalues of A are found from

$$\begin{aligned} \lambda^2 - 17\lambda + 6 &= 0 \\ \Longrightarrow \quad \lambda_1 &= \frac{17}{2} + \frac{1}{2}\sqrt{265} \approx 16.6, \lambda_2 = \frac{17}{2} - \frac{1}{2}\sqrt{265} \approx 0.36. \end{aligned}$$

Since $\lambda_1 > 0$ and $\lambda_2 > 0$ it follows that A is positive definite. Alternatively A is positive definite as

$$M_1 = 3 > 0, M_2 = \det \begin{bmatrix} 3 & -6 \\ -6 & 14 \end{bmatrix} = 6 > 0.$$

Using the quadratic form with 2×1 vector y we have

$$\begin{aligned} y^T A y &= \begin{bmatrix} y_1 & y_2 \end{bmatrix} \begin{bmatrix} 1 & 1 & 1 \\ -1 & -2 & -3 \end{bmatrix} \begin{bmatrix} 1 & -1 \\ 1 & -2 \\ 1 & -3 \end{bmatrix} \begin{bmatrix} y_1 \\ y_2 \end{bmatrix} \\ &= \begin{bmatrix} y_1 - y_2 & y_1 - 2y_2 & y_1 - 3y_2 \end{bmatrix} \begin{bmatrix} y_1 - y_2 \\ y_1 - 2y_2 \\ y_1 - 3y_2 \end{bmatrix} \\ &= (y_1 - y_2)^2 + (y_1 - 2y_2)^2 + (y_1 - 3y_2)^2 \geq 0 \end{aligned}$$

and so A is positive semi-definite. If $y^T A y = 0$ then

$$y_1 - y_2 = 0 \Longrightarrow y_1 = y_2 \text{ and } y_1 - 2y_2 = 0 \Longrightarrow y_1 = 2y_2 \text{ and } y_1 - 3y_2 = 0 \Longrightarrow y_1 = 3y_2$$

which only has a solution $y_1 = y_2 = 0$. Thus A is positive definite.

Problem 3.68 *[A***] Consider $A = X^T X$ where X is an $n \times p$ matrix. Show that $A = X^T X$ is positive semi-definite. Show that $A^{-1} = \left(X^T X\right)^{-1}$ does not necessarily exist. Suppose that the column vectors of X are all linearly independent, or rank $[X] = p$, or $Xy = 0 \Longrightarrow y = 0$ where y is a $p \times 1$ vector. Show that $X^T X$ is then positive definite. Does A^{-1} necessarily exist then?*

Answer: Let y be any $p \times 1$ vector. Then

$$y^T A y = y^T X^T X y = z^T z$$

where $z = Xy$ is an $n \times 1$ column vector and $z^T = (Xy)^T = y^T X^T$.We have

$$\begin{aligned} y^T A y &= z^T z = \begin{bmatrix} z_1 & z_2 & \cdots & z_n \end{bmatrix} \begin{bmatrix} z_1 \\ z_2 \\ \vdots \\ z_n \end{bmatrix} \\ &= z_1^2 + z_2^2 + \cdots + z_n^2 \geq 0 \end{aligned}$$

since $z_i^2 \geq 0$ (a squared number is always non-negative). Since $y^T A y \geq 0$ for all y it follows that A is positive semi-definite.

A positive semi-definite matrix that is not positive definite does not have an inverse, and so A^{-1} does not necessarily exist. But if we make the assumption that *rank* $[X] = p$, or that $Xy = 0 \Longrightarrow y = 0$ where y is a $p \times 1$ vector, then $A = X^T X$ is positive definite and $A^{-1} = \left(X^T X\right)^{-1}$ does exist. In order to show A is positive definite we need to show that

$$y^T A y = 0 \Longrightarrow y = 0.$$

We have

$$y^T A y = 0 \Longrightarrow y^T X^T X y = 0 \Longrightarrow z^T z = 0$$

where $z = Xy$. But $z = Xy = 0 \Longrightarrow y = 0$ and so $A = X^T X$ is positive definite and $A^{-1} = \left(X^T X\right)^{-1}$ exists.

Problem 3.69 *[A***] The matrix $A = X^T X$ is always positive semi-definite and $A = X^T X = X^T I X$ where I is positive definite, and hence positive semi-definite. Consider generalizing this result. Replace I with B and assume that B is positive semi-definite. Show that $A = X^T BX$ is always positive semi-definite. Now assume in addition that rank $[X] = p$ so that*

$$Xy = 0 \Longrightarrow y = 0.$$

If B is positive definite then show that $X^T BX$ is positive definite.

Answer: If y is any $p \times 1$ vector then

$$y^T Ay = y^T X^T BXy = z^T Bz \geq 0$$

where $z = Xy$ and $z^T = (Xy)^T = y^T X^T$ and where the inequality follows from the fact that B is positive semi-definite. Since $y^T Ay \geq 0$ it follows that A is positive semi-definite.

Now suppose that $rank\,[X] = p$. Then $z = Xy = 0 \Longrightarrow y = 0$. We then have

$$y^T Ay = 0 \Longrightarrow z^T Bz = 0 \Longrightarrow z = 0 \Longrightarrow Xy = 0 \Longrightarrow y = 0.$$

It follows then that $y^T Ay > 0$ for $y \neq 0$ and so A is positive definite.

Problem 3.70 *[B**] If A is negative definite and B is negative definite, show $A + B$ is negative definite.*

Answer: If A is negative definite then for any $x \neq 0$

$$x^T Ax < 0, x^T Bx \leq 0 \Longrightarrow x^T (A + B)\, x = \underbrace{x^T Ax}_{-} + \underbrace{x^T Bx}_{-} < 0.$$

Since $x^T (A + B)\, x < 0$ for $x \neq 0$ it follows that $A + B$ is negative definite.

Problem 3.71 *[A**] If tr $[A] < 0$ can A be positive definite?*

Answer: Since $tr\,[A]$ is the sum of the eigenvalues or

$$tr\,[A] = \lambda_1 + \lambda_2 + \cdots + \lambda_n < 0$$

it follows that at least one eigenvalue must be negative and hence A cannot be positive definite.

Problem 3.72 *[A**] Suppose that A is symmetric and positive semi-definite. Since $A \geq 0$, and since positive numbers have a real square root, it is natural to consider the problem of calculating the square root of A, or to find a matrix B such that $A = BB$. (This turns out to be an important problem in econometrics.) Show that B exists, and show that B is also positive semi-definite.*

Answer: Since A is symmetric we can write A as $A = C\Lambda C^T$ where C is orthogonal or $C^T = C^{-1}$ and Λ is diagonal with the eigenvalues of A along the diagonal as

$$\Lambda = \begin{bmatrix} \lambda_1 & 0 & \cdots & 0 \\ 0 & \lambda_2 & \cdots & 0 \\ \vdots & \vdots & \ddots & \vdots \\ 0 & 0 & 0 & \lambda_n \end{bmatrix}.$$

Since A is positive semi-definite $\lambda_i \geq 0$ for $i = 1, 2, \ldots, n$. Since $\lambda_i \geq 0$ we can take the positive square root of each of the eigenvalues. Therefore define B by $B = C\Lambda^{\frac{1}{2}}C^T$ where

$$\Lambda^{\frac{1}{2}} = \begin{bmatrix} \sqrt{\lambda_1} & 0 & \cdots & 0 \\ 0 & \sqrt{\lambda_2} & \cdots & 0 \\ \vdots & \ddots & \ddots & \vdots \\ 0 & 0 & 0 & \sqrt{\lambda_n} \end{bmatrix}.$$

You can then verify that B is symmetric, positive semi-definite, and that using $C^TC = C^{-1}C = I$ we have

$$BB = C\Lambda^{\frac{1}{2}}C^TC\Lambda^{\frac{1}{2}}C^T = C\Lambda^{\frac{1}{2}}\Lambda^{\frac{1}{2}}C^T = C\Lambda C^T = A.$$

The eigenvalues of B are $\sqrt{\lambda_i} \geq 0$ for $i = 1, 2, \ldots, n$ and so B is positive semi-definite as well.

Problem 3.73 *[B**] For the matrix A below the representation $A = C\Lambda C^T$ is*

$$A = \begin{bmatrix} 5 & -2 \\ -2 & 5 \end{bmatrix} = \begin{bmatrix} \frac{1}{\sqrt{2}} & -\frac{1}{\sqrt{2}} \\ \frac{1}{\sqrt{2}} & \frac{1}{\sqrt{2}} \end{bmatrix} \begin{bmatrix} 3 & 0 \\ 0 & 7 \end{bmatrix} \begin{bmatrix} \frac{1}{\sqrt{2}} & \frac{1}{\sqrt{2}} \\ -\frac{1}{\sqrt{2}} & \frac{1}{\sqrt{2}} \end{bmatrix}.$$

Calculate the square root of A, that is that B for which $A = BB$. Verify that $BB = A$ by carrying out the multiplication.

Answer: We have

$$\begin{aligned} B &= \begin{bmatrix} \frac{1}{\sqrt{2}} & -\frac{1}{\sqrt{2}} \\ \frac{1}{\sqrt{2}} & \frac{1}{\sqrt{2}} \end{bmatrix} \begin{bmatrix} \sqrt{3} & 0 \\ 0 & \sqrt{7} \end{bmatrix} \begin{bmatrix} \frac{1}{\sqrt{2}} & \frac{1}{\sqrt{2}} \\ -\frac{1}{\sqrt{2}} & \frac{1}{\sqrt{2}} \end{bmatrix} \\ &= \begin{bmatrix} \frac{\sqrt{3}+\sqrt{7}}{2} & \frac{\sqrt{3}-\sqrt{7}}{2} \\ \frac{\sqrt{3}-\sqrt{7}}{2} & \frac{\sqrt{3}+\sqrt{7}}{2} \end{bmatrix}. \end{aligned}$$

Multiplying we have

$$BB = \begin{bmatrix} \frac{\sqrt{3}+\sqrt{7}}{2} & \frac{\sqrt{3}-\sqrt{7}}{2} \\ \frac{\sqrt{3}-\sqrt{7}}{2} & \frac{\sqrt{3}+\sqrt{7}}{2} \end{bmatrix} \begin{bmatrix} \frac{\sqrt{3}+\sqrt{7}}{2} & \frac{\sqrt{3}-\sqrt{7}}{2} \\ \frac{\sqrt{3}-\sqrt{7}}{2} & \frac{\sqrt{3}+\sqrt{7}}{2} \end{bmatrix} = \begin{bmatrix} 5 & -2 \\ -2 & 5 \end{bmatrix}.$$

Problem 3.74 *[A**] Suppose that A is positive semi-definite or $A \geq 0$. Consider the problem of calculating the square root of A but now we find a matrix B such that $A = BB^T$. Show that such a matrix B exists.*

Answer: Since A is symmetric we can write A as $A = C\Lambda C^T$ where C is orthogonal or $C^T = C^{-1}$ and Λ is diagonal with the eigenvalues of A along the diagonal or

$$\Lambda = \begin{bmatrix} \lambda_1 & 0 & \cdots & 0 \\ 0 & \lambda_2 & \cdots & 0 \\ \vdots & \vdots & \ddots & \vdots \\ 0 & 0 & 0 & \lambda_n \end{bmatrix}$$

and where $\lambda_i \geq 0$ since A is positive semi-definite. Since $\lambda_i \geq 0$ we can take the positive square root of each of the eigenvalues. Define B by $B = C\Lambda^{\frac{1}{2}}$ from which it follows that $B^T = \Lambda^{\frac{1}{2}}C^T$. It follows then that

$$BB^T = C\Lambda^{\frac{1}{2}}\Lambda^{\frac{1}{2}}C^T = C\Lambda C^T = A.$$

Problem 3.75 *[A**] For the matrix A below the representation $A = C\Lambda C^T$ is*

$$A = \begin{bmatrix} 5 & -2 \\ -2 & 5 \end{bmatrix} = \begin{bmatrix} \frac{1}{\sqrt{2}} & -\frac{1}{\sqrt{2}} \\ \frac{1}{\sqrt{2}} & \frac{1}{\sqrt{2}} \end{bmatrix} \begin{bmatrix} 3 & 0 \\ 0 & 7 \end{bmatrix} \begin{bmatrix} \frac{1}{\sqrt{2}} & \frac{1}{\sqrt{2}} \\ -\frac{1}{\sqrt{2}} & \frac{1}{\sqrt{2}} \end{bmatrix}.$$

Calculate the square root of A, that is that B for which $A = BB^T$. Verify that $BB^T = A$ by carrying out the multiplication.

Answer: We have

$$B = \begin{bmatrix} \frac{1}{\sqrt{2}} & -\frac{1}{\sqrt{2}} \\ \frac{1}{\sqrt{2}} & \frac{1}{\sqrt{2}} \end{bmatrix} \begin{bmatrix} \sqrt{3} & 0 \\ 0 & \sqrt{7} \end{bmatrix} = \begin{bmatrix} \frac{\sqrt{3}}{\sqrt{2}} & -\frac{\sqrt{7}}{\sqrt{2}} \\ \frac{\sqrt{3}}{\sqrt{2}} & \frac{\sqrt{7}}{\sqrt{2}} \end{bmatrix}.$$

Multiplying we have

$$BB^T = \begin{bmatrix} \frac{\sqrt{3}}{\sqrt{2}} & -\frac{\sqrt{7}}{\sqrt{2}} \\ \frac{\sqrt{3}}{\sqrt{2}} & \frac{\sqrt{7}}{\sqrt{2}} \end{bmatrix} \begin{bmatrix} \frac{\sqrt{3}}{\sqrt{2}} & \frac{\sqrt{3}}{\sqrt{2}} \\ -\frac{\sqrt{7}}{\sqrt{2}} & \frac{\sqrt{7}}{\sqrt{2}} \end{bmatrix} = \begin{bmatrix} 5 & -2 \\ -2 & 5 \end{bmatrix}.$$

Problem 3.76 *[A**] For scalars $a > 0$ and $b > 0$ we have*

$$a - b \geq 0 \Longleftrightarrow \frac{1}{b} - \frac{1}{a} \geq 0.$$

For example if $a = 5$ and $b = 3$ then

$$a - b = 5 - 3 = 2 \geq 0 \Longleftrightarrow \frac{1}{3} - \frac{1}{5} = \frac{2}{15} \geq 0.$$

Now consider generalizing this to matrices (this turns out to be important for econometrics). Suppose that $A > 0$ and $B > 0$, that is both A and B are positive definite. Prove that $A - B \geq 0$ if and only if $B^{-1} - A^{-1} \geq 0$; that is: $A - B$ is positive semi-definite if and only if $B^{-1} - A^{-1}$ is positive semi-definite. Hint: prove that $A - B$ is positive semi-definite if and only if $\det[A - \lambda B] = 0 \Longrightarrow \lambda \geq 1$.

Answer: Since $B = C\Lambda C^T$ is positive definite we can write $B = DD^T$ where $D = C\Lambda^{\frac{1}{2}}$. We then have

$$\begin{aligned}
\det[A - \lambda B] &= \det[A - B - (\lambda - 1) B] = \det\left[A - B - (\lambda - 1) DD^T\right] \\
&= \det\left[D\left(D^{-1}(A - B)\left(D^T\right)^{-1} - (\lambda - 1) I\right) D^T\right] \\
&= \det[D] \det\left[D^{-1}(A - B)\left(D^{-1}\right)^T - (\lambda - 1) I\right] \det\left[D^T\right] \\
&= \det\left[DD^T\right] \det\left[D^{-1}(A - B)\left(D^{-1}\right)^T - (\lambda - 1) I\right] \\
&= \det[B] \det\left[D^{-1}(A - B)\left(D^{-1}\right)^T - (\lambda - 1) I\right]
\end{aligned}$$

where $\det[B] > 0$. Therefore $\det[A - \lambda B] = 0$ if and only if $\lambda - 1$ is an eigenvalue of the matrix $E = D^{-1}(A - B)\left(D^{-1}\right)^T$.

Now suppose that $A - B$ is positive semi-definite. It follows then that $E = D^{-1}(A - B)\left(D^{-1}\right)^T$ is positive semi-definite since E has the form $C\Omega C^T$ where $\Omega = A - B$ is positive semi-definite and $C = D^{-1}$. Therefore the eigenvalues of E are non-negative so that $\det[A - \lambda B] = 0$ implies that $\lambda - 1 \geq 0$ or $\lambda \geq 1$.

Now suppose that $\det[A - \lambda B] = 0$ implies that $\lambda \geq 1$. It follows that E is positive semi-definite. Therefore

$$x^T(A - B)x = x^T DD^{-1}(A - B)\left(D^{-1}\right)^T D^T x = y^T E y \geq 0$$

where $y = D^T x$. It follows then that $A - B$ is positive semi-definite.

We then have

$$\det[A - \lambda B] = \det\left[A\left(B^{-1} - \lambda A^{-1}\right) B\right] = \det[A] \det\left[B^{-1} - \lambda A^{-1}\right] \det[B]$$

so that $\det[A - \lambda B] = 0 \Longleftrightarrow \det\left[B^{-1} - \lambda A^{-1}\right] = 0$. Therefore

$$\begin{aligned}
A - B \text{ is positive semi-definite} &\Longleftrightarrow \det[A - \lambda B] = 0 \Longrightarrow \lambda \geq 1 \\
&\Longleftrightarrow \det\left[B^{-1} - \lambda A^{-1}\right] \Longrightarrow \lambda \geq 1 \\
&\Longleftrightarrow B^{-1} - A^{-1} \text{ is positive semi-definite.}
\end{aligned}$$

Problem 3.77 *[A**] The multivariate generalization of the quadratic $ax^2 + bx + c$ is*

$$f(x) = x^T A x + b^T x + c$$

where x and b are $n \times 1$ vectors, A is a non-singular, symmetric $n \times n$ matrix, and c is a scalar. In Problem 2.10 on page 18 we showed how to optimize scalar quadratics by completing the square. In a similar spirit show that

$$f(x) = (x - d)^T A(x - d) + e$$

where $d = -\frac{1}{2}A^{-1}b$ and $e = c - \frac{b^T A^{-1} b}{4}$. Use this result to show that if A is negative definite then $f(x)$ has a global maximum at $x^* = -\frac{1}{2}A^{-1}b$.

Answer: We have

$$(x-d)^T A(x-d)+e=x^T Ax-d^T Ax-x^T Ad+d^T Ad+e.$$

Now the two terms $d^T Ax$ and $x^T Ad$ are identical since they are both scalars, and hence equal to their transpose, and since

$$\left(x^T Ad\right)^T=d^T A^T x=d^T Ax.$$

We therefore have

$$(x-d)^T A(x-d)+e=x^T Ax-2d^T Ax-d^T Ad+e=x^T Ax+b^T x+c$$

from which we conclude that

$$\begin{aligned} b^T &= -2d^T A \Longrightarrow b=-2Ad \\ &\Longrightarrow d=-\frac{1}{2}A^{-1}b \\ -d^T Ad+e &= c \\ &\Longrightarrow e=c-d^T Ad=c-\left(-\frac{1}{2}A^{-1}b\right)^T A\left(-\frac{1}{2}A^{-1}b\right)=c-\frac{b^T A^{-1}b}{4}. \end{aligned}$$

If A is negative definite then

$$\begin{aligned} (x-d)^T A(x-d) &< 0 \text{ for } x\neq d \\ (x-d)^T A(x-d) &= 0 \text{ for } x=d \end{aligned}$$

and so $f(x)\leq f(d)$ as

$$f(x)=x^T Ax+b^T x+c=(x-d)^T A(x-d)+e\leq e=f(d).$$

It follows then that $x^*=d=-\frac{1}{2}A^{-1}b$ is a global maximizer.

Chapter 4

Multivariate Calculus

4.1 Partial Derivatives

Problem 4.1 *[D***] Suppose the demand curves for two goods Q_1, Q_2 are*

$$Q_1^* = \frac{Y}{4P_1}, Q_2^* = \frac{3Y}{4P_2}$$

where P_1, P_2 are the respective prices, and Y is household income. Using partial derivatives show that both goods have downward sloping demand curves. Calculate the own price elasticities of demand from the definition of elasticity. Show that both goods are normal. Are the two goods substitutes or complements?

Answer: Both demand curves slope downwards since

$$\begin{aligned} Q_1^* &= \frac{Y}{4P_1} \\ &\Longrightarrow \frac{\partial Q_1^*}{\partial P_1} = -\frac{Y}{4P_1^2} < 0 \\ Q_2^* &= \frac{3Y}{4P_2} \\ &\Longrightarrow \frac{\partial Q_2^*}{\partial P_2} = -\frac{3Y}{4P_2^2} < 0. \end{aligned}$$

The elasticity of demand for both goods is -1 as

$$\frac{\partial Q_1^*}{\partial P_1}\frac{P_1}{Q_1^*} = -\frac{Y}{4P_1^2}\frac{P_1}{\frac{Y}{4P_1}} = -1, \frac{\partial Q_2^*}{\partial P_2}\frac{P_2}{Q_2^*} = -\frac{3Y}{4P_2^2}\frac{P_2}{\frac{3Y}{4P_2}} = -1.$$

Both goods are normal since

$$\frac{\partial Q_1^*}{\partial Y} = \frac{1}{4P_1} > 0, \frac{\partial Q_2^*}{\partial Y} = \frac{3}{4P_2} > 0$$

or because they both have unit income elasticities as

$$\frac{\partial Q_1^*}{\partial Y}\frac{Y}{Q_1^*} = \frac{1}{4P_1}\frac{Y}{\frac{Y}{4P_1}} = 1, \quad \frac{\partial Q_2^*}{\partial Y}\frac{Y}{Q_2^*} = \frac{3}{4P_2}\frac{Y}{\frac{3Y}{4P_2}} = 1.$$

The goods are neither substitutes $\frac{\partial Q_1^*}{\partial P_2} > 0$ nor complements $\frac{\partial Q_1^*}{\partial P_2} < 0$ since $\frac{\partial Q_1^*}{\partial P_2} = \frac{\partial Q_2^*}{\partial P_1} = 0$.

Problem 4.2 *[D***] Suppose the demand curve for Q_1 is*

$$Q_1^* = \frac{YP_1^{-\frac{3}{2}}}{P_1^{-\frac{1}{2}} + P_2^{-\frac{1}{2}}}.$$

Show the demand curve slopes downwards and that the own price elasticity of demand will depend on P_1, P_2, but not on Y. Show that demand is elastic. Show that Q_1 is normal and calculate the income elasticity. Are Q_1 and Q_2 substitutes or complements? Calculate the cross elasticity between Q_1 and Q_2, and show it's less than $\frac{1}{2}$.

Answer: We can make our things easier by getting all $P_1's$ in the denominator as

$$Q_1^* = \frac{YP_1^{-\frac{3}{2}}}{P_1^{-\frac{1}{2}} + P_2^{-\frac{1}{2}}} = \frac{Y}{P_1^{\frac{3}{2}}\left(P_1^{-\frac{1}{2}} + P_2^{-\frac{1}{2}}\right)} = \frac{Y}{P_1 + P_1^{\frac{3}{2}}P_2^{-\frac{1}{2}}}$$

so that the demand for Q_1 slopes downwards as

$$\begin{aligned} Q_1^* &= \frac{Y}{P_1 + P_1^{\frac{3}{2}}P_2^{-\frac{1}{2}}} \\ \Longrightarrow \frac{\partial Q_1^*}{\partial P_1} &= -\frac{Y\left(1 + \frac{3}{2}P_1^{\frac{1}{2}}P_2^{-\frac{1}{2}}\right)}{\left(P_1 + P_1^{\frac{3}{2}}P_2^{-\frac{1}{2}}\right)^2} < 0. \end{aligned}$$

The own price elasticity of demand for Q_1 is

$$\begin{aligned} \frac{\partial Q_1^*}{\partial P_1}\frac{P_1}{Q_1^*} &= -\frac{Y\left(1 + \frac{3}{2}P_1^{\frac{1}{2}}P_2^{-\frac{1}{2}}\right)}{\left(P_1 + P_1^{\frac{3}{2}}P_2^{-\frac{1}{2}}\right)^2}\frac{P_1}{\frac{Y}{P_1 + P_1^{\frac{3}{2}}P_2^{-\frac{1}{2}}}} = -\frac{P_1 + \frac{3}{2}P_1^{\frac{3}{2}}P_2^{-\frac{1}{2}}}{P_1 + P_1^{\frac{3}{2}}P_2^{-\frac{1}{2}}} \\ &= -\left(\frac{P_1 + P_1^{\frac{3}{2}}P_2^{-\frac{1}{2}}}{P_1 + P_1^{\frac{3}{2}}P_2^{-\frac{1}{2}}} + \frac{\frac{1}{2}P_1^{\frac{3}{2}}P_2^{-\frac{1}{2}}}{P_1 + P_1^{\frac{3}{2}}P_2^{-\frac{1}{2}}}\right) = -\left(1 + \frac{\frac{1}{2}P_1^{\frac{3}{2}}P_2^{-\frac{1}{2}}}{P_1 + P_1^{\frac{3}{2}}P_2^{-\frac{1}{2}}}\right) < -1 \end{aligned}$$

and so demand is elastic. Notice that the elasticity depends on P_1, P_2 but not on Y. The good is normal since

$$\frac{\partial Q_1^*}{\partial Y} = \frac{P_1^{-\frac{3}{2}}}{P_1^{-\frac{1}{2}} + P_2^{-\frac{1}{2}}} > 0$$

with an income elasticity of 1, the exponent on Y, or from the definition of elasticity as

$$\frac{\partial Q_1^*}{\partial Y}\frac{Y}{Q_1^*} = \frac{P_1^{-\frac{3}{2}}}{P_1^{-\frac{1}{2}} + P_2^{-\frac{1}{2}}}\frac{Y}{\frac{YP_1^{-\frac{3}{2}}}{P_1^{-\frac{1}{2}}+P_2^{-\frac{1}{2}}}} = 1.$$

The goods Q_1 and Q_2 are substitutes since

$$\frac{\partial Q_1^*}{\partial P_2} = -\frac{YP_1^{-\frac{3}{2}}\left(-\frac{1}{2}P_2^{-\frac{3}{2}}\right)}{\left(P_1^{-\frac{1}{2}} + P_2^{-\frac{1}{2}}\right)^2} = \frac{YP_1^{-\frac{3}{2}}P_2^{-\frac{3}{2}}}{2\left(P_1^{-\frac{1}{2}} + P_2^{-\frac{1}{2}}\right)^2} > 0.$$

Since $0 < \frac{P_2^{-\frac{1}{2}}}{P_1^{-\frac{1}{2}}+P_2^{-\frac{1}{2}}} < 1$ the cross elasticity is less than $\frac{1}{2}$ as

$$\frac{\partial Q_1^*}{\partial P_2}\frac{P_2}{Q_1^*} = \frac{YP_1^{-\frac{3}{2}}P_2^{-\frac{3}{2}}}{2\left(P_1^{-\frac{1}{2}} + P_2^{-\frac{1}{2}}\right)^2}\frac{P_2}{\frac{YP_1^{-\frac{3}{2}}}{P_1^{-\frac{1}{2}}+P_2^{-\frac{1}{2}}}} = \frac{1}{2}\frac{P_2^{-\frac{1}{2}}}{P_1^{-\frac{1}{2}} + P_2^{-\frac{1}{2}}} < \frac{1}{2}.$$

Problem 4.3 *[D***] Consider the function*

$$f(x_1, x_2) = \frac{4}{5}\ln(x_1) + \frac{1}{5}\ln(x_2) - 5x_1 - 4x_2.$$

Find the first and second-order partial derivatives, construct the Hessian, and determine if it is globally convex or concave.

Answer: We have

$$\begin{aligned}
\frac{\partial f(x_1,x_2)}{\partial x_1} &= \frac{4}{5}\frac{1}{x_1} - 5, \frac{\partial f(x_1,x_2)}{\partial x_2} = \frac{1}{5}\frac{1}{x_2} - 4 \\
\frac{\partial^2 f(x_1,x_2)}{\partial x_1^2} &= \frac{\partial}{\partial x_1}\left(\frac{4}{5}\frac{1}{x_1} - 5\right) = -\frac{4}{5}\frac{1}{x_1^2}, \frac{\partial^2 f(x_1,x_2)}{\partial x_1 \partial x_2} = \frac{\partial}{\partial x_2}\left(\frac{4}{5}\frac{1}{x_1} - 5\right) = 0 \\
\frac{\partial^2 f(x_1,x_2)}{\partial x_2^2} &= \frac{\partial}{\partial x_2}\left(\frac{1}{5}\frac{1}{x_2} - 4\right) = -\frac{1}{5}\frac{1}{x_2^2}, \frac{\partial^2 f(x_1,x_2)}{\partial x_2 \partial x_1} = \frac{\partial}{\partial x_1}\left(\frac{1}{5}\frac{1}{x_2} - 4\right) = 0.
\end{aligned}$$

Since the Hessian

$$H(x_1, x_2) = \begin{bmatrix} -\frac{4}{5x_1^2} & 0 \\ 0 & -\frac{1}{5x_2^2} \end{bmatrix}$$

is a diagonal matrix with negative elements on the diagonal for all x_1, x_2, it follows that $f(x_1, x_2)$ is globally concave.

Problem 4.4 *[D***] For*

$$f(x_1, x_2) = \ln\left(e^{3\ln(x_1)}e^{5\ln(x_2)}\left(e^0 - e^{\ln(x_1)} - x_2\right)^6\right)$$

find the first and second-order partial derivatives. Determine if the function is globally concave or convex.

Answer: Simplifying we have

$$\begin{aligned} f(x_1,x_2) &= \ln\left(e^{3\ln(x_1)}e^{5\ln(x_2)}\left(e^0 - e^{\ln(x_1)} - x_2\right)^6\right) \\ &= \ln\left(e^{3\ln(x_1)}e^{5\ln(x_2)}\right) + \ln\left((1-x_1-x_2)^6\right) \\ &= 3\ln(x_1) + 5\ln(x_2) + 6\ln(1-x_1-x_2). \end{aligned}$$

We have

$$\frac{\partial f(x_1,x_2)}{\partial x_1} = \frac{3}{x_1} - \frac{6}{1-x_1-x_2}, \quad \frac{\partial f(x_1,x_2)}{\partial x_2} = \frac{5}{x_2} - \frac{6}{1-x_1-x_2}.$$

The second-order partial derivatives are

$$\begin{aligned} \frac{\partial^2 f(x_1,x_2)}{\partial x_1^2} &= -\frac{3}{x_1^2} - \frac{6}{(1-x_1-x_2)^2}, \quad \frac{\partial^2 f(x_1,x_2)}{\partial x_1 \partial x_2} = -\frac{6}{(1-x_1-x_2)^2} \\ \frac{\partial^2 f(x_1,x_2)}{\partial x_2^2} &= -\frac{5}{x_2^2} - \frac{6}{(1-x_1-x_2)^2} \end{aligned}$$

so that the Hessian is

$$H(x_1,x_2) = \begin{bmatrix} -\frac{3}{x_1^2} - \frac{6}{(1-x_1-x_2)^2} & -\frac{6}{(1-x_1-x_2)^2} \\ -\frac{6}{(1-x_1-x_2)^2} & -\frac{5}{x_2^2} - \frac{6}{(1-x_1-x_2)^2} \end{bmatrix}.$$

We have

$$\begin{aligned} M_1 &= \det\left[-\frac{3}{x_1^2} - \frac{6}{(1-x_1-x_2)^2}\right] \\ &= -\frac{3}{x_1^2} - \frac{6}{(1-x_1-x_2)^2} < 0 \\ M_2 &= \det\begin{bmatrix} -\frac{3}{x_1^2} - \frac{6}{(1-x_1-x_2)^2} & -\frac{6}{(1-x_1-x_2)^2} \\ -\frac{6}{(1-x_1-x_2)^2} & -\frac{5}{x_2^2} - \frac{6}{(1-x_1-x_2)^2} \end{bmatrix} \\ &= \left(-\frac{3}{x_1^2} - \frac{6}{(1-x_1-x_2)^2}\right)\left(-\frac{5}{x_2^2} - \frac{6}{(1-x_1-x_2)^2}\right) - \left(-\frac{6}{(1-x_1-x_2)^2}\right)^2 \\ &= \frac{15}{x_1^2 x_2^2} + \frac{18}{x_1^2}\frac{1}{(1-x_1-x_2)^2} + \frac{30}{x_2^2}\frac{1}{(1-x_1-x_2)^2} > 0. \end{aligned}$$

Since $M_1 < 0, M_2 > 0$ for all x_1, x_2, the Hessian is negative definite and hence $f(x_1,x_2)$ is globally concave.

Problem 4.5 *[B**] Consider*

$$f(x_1,x_2) = -\ln(x_1) - \frac{x_2^2 - x_2 + 1}{x_1} \text{ for } 0 < x_1 < \frac{3}{2}.$$

*Calculate the first and second-order partial derivatives, and construct the Hessian. Determine if the function is locally concave or convex at $x_1 = 1, x_2 = 1$. [A**] Is the function globally concave or globally convex?*

Answer: We have

$$\frac{\partial f(x_1,x_2)}{\partial x_1} = -\frac{1}{x_1} + \frac{x_2^2 - x_2 + 1}{x_1^2}, \quad \frac{\partial f(x_1,x_2)}{\partial x_2} = -\frac{2x_2 - 1}{x_1}.$$

The second derivatives are

$$\frac{\partial^2 f(x_1,x_2)}{\partial x_1^2} = \frac{1}{x_1^2} - 2\frac{x_2^2 - x_2 + 1}{x_1^3}, \quad \frac{\partial^2 f(x_1,x_2)}{\partial x_1 \partial x_2} = \frac{2x_2 - 1}{x_1^2}, \quad \frac{\partial^2 f(x_1,x_2)}{\partial x_2^2} = -\frac{2}{x_1}$$

so the Hessian is

$$H(x_1,x_2) = \begin{bmatrix} \frac{1}{x_1^2} - 2\frac{x_2^2 - x_2 + 1}{x_1^3} & \frac{2x_2-1}{x_1^2} \\ \frac{2x_2-1}{x_1^2} & -\frac{2}{x_1} \end{bmatrix}.$$

At $x_1 = 1, x_2 = 1$ we have

$$\begin{aligned} H(1,1) &= \begin{bmatrix} -1 & 1 \\ 1 & -2 \end{bmatrix} \\ &\implies M_1 = -1 < 0, M_2 = \det \begin{bmatrix} -1 & 1 \\ 1 & -2 \end{bmatrix} = 1 > 0 \end{aligned}$$

and so $H(1,1)$ is negative definite. It follows that $f(x_1,x_2)$ is locally concave at $x_1 = 1, x_2 = 1$.

To show that the function is globally concave we have first that

$$\begin{aligned} M_1 &= \frac{1}{x_1^2} - 2\frac{x_2^2 - x_2 + 1}{x_1^3} = \frac{1}{x_1^3}\left(x_1 - 2\left(x_2^2 - x_2 + 1\right)\right) \\ &= \frac{1}{x_1^3}\left(x_1 - 2\left(\left(x_2 - \frac{1}{2}\right)^2 + \frac{3}{4}\right)\right) = -\frac{1}{x_1^3}\left(\frac{3}{2} - x_1\right) - \frac{2\left(x_2 - \frac{1}{2}\right)^2}{x_1^3} < 0 \end{aligned}$$

since $x_1 < \frac{3}{2}$ and by completing the square as

$$x_2^2 - x_2 + 1 = \left(x_2 - \frac{1}{2}\right)^2 + \frac{3}{4}.$$

Also

$$M_2 = \det \begin{bmatrix} \frac{1}{x_1^2} - 2\frac{x_2^2 - x_2 + 1}{x_1^3} & \frac{2x_2-1}{x_1^2} \\ \frac{2x_2-1}{x_1^2} & -\frac{2}{x_1} \end{bmatrix} = -\frac{2x_1 - 3}{x_1^4} = \frac{2\left(\frac{3}{2} - x_1\right)}{x_1^4} > 0$$

since $x_1 < \frac{3}{2}$. Since $M_1 < 0, M_2 > 0$ for all x_1, x_2 it follows that the function is globally concave.

Problem 4.6 *Suppose that a firm has a production function given by*

$$Q = 5L^{\frac{2}{5}}K^{\frac{1}{5}}.$$

*[D***] Show there is a positive and diminishing marginal product of labour, and a positive and diminishing marginal product of capital. [C***] Show that increasing K leads to an increase in the marginal product of labour, and increasing L leads to an increase in the marginal product of capital. [B***] Show that the production function is globally concave.*

Answer: We have positive marginal products as

$$MP_L(L,K) \equiv \frac{\partial F(L,K)}{\partial L} = 2L^{-\frac{3}{5}}K^{\frac{1}{5}} > 0, MP_K(L,K) \equiv \frac{\partial F(L,K)}{\partial K} = L^{\frac{2}{5}}K^{-\frac{4}{5}} > 0.$$

We have diminishing marginal products since

$$\frac{\partial MP_L(L,K)}{\partial L} = -\frac{6}{5}L^{-\frac{8}{5}}K^{\frac{1}{5}} < 0 \text{ and } \frac{\partial MP_K(L,K)}{\partial K} = -\frac{4}{5}L^{\frac{2}{5}}K^{-\frac{9}{5}} < 0.$$

By Young's theorem

$$\frac{\partial MP_K(L,K)}{\partial L} = \frac{\partial MP_L(L,K)}{\partial K} = \frac{\partial^2 F(L,K)}{\partial L \partial K} = \frac{2}{5}L^{-\frac{3}{5}}K^{-\frac{4}{5}} > 0$$

and so an increase in K increases the marginal product of labour MP_L, and an increase in L increases the marginal product of capital MP_K.

The Hessian of $Q = 5L^{\frac{2}{5}}K^{\frac{1}{5}}$ is

$$H(L,K) = \begin{bmatrix} -\frac{6}{5}L^{-\frac{8}{5}}K^{\frac{1}{5}} & \frac{2}{5}L^{-\frac{3}{5}}K^{-\frac{4}{5}} \\ \frac{2}{5}L^{-\frac{3}{5}}K^{-\frac{4}{5}} & -\frac{4}{5}L^{\frac{2}{5}}K^{-\frac{9}{5}} \end{bmatrix}.$$

Using principal minors we find that

$$\begin{aligned} M_1 &= -\frac{6}{5}L^{-\frac{8}{5}}K^{\frac{1}{5}} < 0 \\ M_2 &= \left(-\frac{6}{5}L^{-\frac{8}{5}}K^{\frac{1}{5}}\right)\left(-\frac{4}{5}L^{\frac{2}{5}}K^{-\frac{9}{5}}\right) - \left(\frac{2}{5}L^{-\frac{3}{5}}K^{-\frac{4}{5}}\right)^2 = \frac{20}{25}L^{-\frac{6}{5}}K^{-\frac{8}{5}} > 0 \end{aligned}$$

so $H(L,K)$ is negative definite for all L, K, and so $Q = 5L^{\frac{2}{5}}K^{\frac{1}{5}}$ is globally concave.

Problem 4.7 *[D***] Suppose the demand for labour and capital are given by*

$$L^* = 4w^{-2}r^{-\frac{1}{2}}, K^* = 2w^{-1}r^{-\frac{3}{2}}.$$

Show that both demand curves slope downwards. Show that $\frac{\partial L^*}{\partial r} = \frac{\partial K^*}{\partial w} < 0$. Find the elasticity of labour demand directly from the definition (without reading off exponents).

Answer: Both demand curves slope downwards as

$$\begin{aligned} L^* &= 4w^{-2}r^{-\frac{1}{2}} \\ &\Longrightarrow \frac{\partial L^*}{\partial w} = -8w^{-3}r^{-\frac{1}{2}} < 0 \\ K^* &= 2w^{-1}r^{-\frac{3}{2}} \\ &\Longrightarrow \frac{\partial K^*}{\partial r} = -3w^{-1}r^{-\frac{5}{2}} < 0. \end{aligned}$$

We have

$$\frac{\partial L^*}{\partial r} = \frac{\partial K^*}{\partial w} = -2w^{-2}r^{-\frac{3}{2}} < 0.$$

The elasticity of the demand for labour is

$$\frac{\partial L^*}{\partial w}\frac{w}{L^*} = -8w^{-3}r^{-\frac{1}{2}}\frac{w}{4w^{-2}r^{-\frac{1}{2}}} = -2w^{-2}r^{-\frac{1}{2}}\frac{1}{w^{-2}r^{-\frac{1}{2}}} = -2$$

which is the exponent on w in $L^* = 4w^{-2}r^{-\frac{1}{2}}$.

Problem 4.8 *Suppose that a perfectly competitive firm has a production function given by*

$$Q = L^{\alpha}K^{\beta}$$

*where $0 < \alpha < 1$, $0 < \beta < 1$, and $\alpha + \beta < 1$. [C***] Calculate the marginal product of labour and capital and show they are positive. Show that there is a positive and diminishing marginal product of labour and capital. Show that increasing K increases the marginal product of labour, and increasing L increases the marginal product of capital. [B***] Show that the production function is globally concave.*

Answer: Since $\alpha > 0, \beta > 0$ we have

$$MP_L(L,K) = \frac{\partial F(L,K)}{\partial L} = \alpha L^{\alpha-1}K^{\beta} > 0 \text{ and } MP_K(L,K) = \frac{\partial F(L,K)}{\partial K} = \beta L^{\alpha}K^{\beta-1} > 0.$$

Furthermore since $\alpha < 1, \beta < 1$ we have

$$\frac{\partial MP_L(L,K)}{\partial L} = \alpha(\alpha-1)L^{\alpha-2}K^{\beta} < 0 \text{ and } \frac{\partial MP_K(L,K)}{\partial K} = \beta(\beta-1)L^{\alpha}K^{\beta-2} < 0$$

and so there is a diminishing marginal product of labour and capital.

By Young's theorem

$$\frac{\partial MP_K(L,K)}{\partial L} = \frac{\partial MP_L(L,K)}{\partial K} = \frac{\partial^2 F(L,K)}{\partial L \partial K} = \alpha\beta L^{\alpha-1}K^{\beta-1} > 0$$

so that an increase in capital K increases the marginal product of labour $MP_L(L,K)$, and an increase in labour L increases the marginal product of capital $MP_K(L,K)$.

The Hessian of $Q = L^{\alpha}K^{\beta}$ is

$$H(L,K) = \begin{bmatrix} \alpha(\alpha-1)L^{\alpha-2}K^{\beta} & \alpha\beta L^{\alpha-1}K^{\beta-1} \\ \alpha\beta L^{\alpha-1}K^{\beta-1} & \beta(\beta-1)L^{\alpha}K^{\beta-2} \end{bmatrix}.$$

Using principal minors we find for all L, K that

$$\begin{aligned}
M_1 &= \alpha(\alpha-1)L^{\alpha-2}K^{\beta} < 0 \text{ since } 0<\alpha<1 \\
M_2 &= \left(\alpha(\alpha-1)L^{\alpha-2}K^{\beta}\right)\left(\beta(\beta-1)L^{\alpha}K^{\beta-2}\right) - \left(\alpha\beta L^{\alpha-1}K^{\beta-1}\right)^2 \\
&= \alpha(\alpha-1)\beta(\beta-1)L^{2\alpha-2}K^{2\beta-2} - (\alpha\beta)^2 L^{2\alpha-2}K^{2\beta-2} \\
&= L^{2\alpha-2}K^{2\beta-2}\left(\alpha(\alpha-1)\beta(\beta-1) - (\alpha\beta)^2\right) \\
&= \alpha\beta L^{2\alpha-2}K^{2\beta-2}\left((\alpha-1)(\beta-1) - \alpha\beta\right) \\
&= \alpha\beta L^{2\alpha-2}K^{2\beta-2}\underbrace{(1-(\alpha+\beta))}_{+\text{ since }\alpha+\beta<1} > 0
\end{aligned}$$

so that $Q = L^{\alpha}K^{\beta}$ is globally concave.

Problem 4.9 *Suppose the firm's production function is given by*

$$Q = F(L,K) = 4L^{\frac{1}{2}}K^{\frac{3}{4}}.$$

*[C***] Show that the marginal products of labour and capital are positive and diminishing. Show that increasing K increases the marginal product of labour, and increasing L increases the marginal product of capital. [B***] Is this production function concave?*

Answer: Since $L^{-\frac{1}{2}} > 0, K^{\frac{3}{4}} > 0$ we have

$$MP_L(L,K) = 2L^{-\frac{1}{2}}K^{\frac{3}{4}} > 0, MP_K(L,K) = 3L^{\frac{1}{2}}K^{-\frac{1}{4}} > 0.$$

There is a diminishing marginal product of labour and capital since

$$\frac{\partial MP_L(L,K)}{\partial L} = -L^{-\frac{3}{2}}K^{\frac{3}{4}} < 0 \text{ and } \frac{\partial MP_K(L,K)}{\partial K} = -\frac{3}{4}L^{\frac{1}{2}}K^{-\frac{5}{4}} < 0.$$

Increasing K increases the marginal product of labour, and increasing L increases the marginal product of capital since

$$\frac{\partial MP_K(L,K)}{\partial L} = \frac{\partial MP_L(L,K)}{\partial K} = \frac{3}{2}L^{-\frac{1}{2}}K^{-\frac{1}{4}} > 0.$$

The Hessian of the production function is

$$H(L,K) = \begin{bmatrix} -L^{-\frac{3}{2}}K^{\frac{3}{4}} & \frac{3}{2}L^{-\frac{1}{2}}K^{-\frac{1}{4}} \\ \frac{3}{2}L^{-\frac{1}{2}}K^{-\frac{1}{4}} & -\frac{3}{4}L^{\frac{1}{2}}K^{-\frac{5}{4}} \end{bmatrix}.$$

Using principal minors we find that

$$\begin{aligned}
M_1 &= -L^{-\frac{3}{2}}K^{\frac{3}{4}} < 0 \\
M_2 &= \left(-L^{-\frac{3}{2}}K^{\frac{3}{4}}\right)\left(-\frac{3}{4}L^{\frac{1}{2}}K^{-\frac{5}{4}}\right) - \left(\frac{3}{2}L^{-\frac{1}{2}}K^{-\frac{1}{4}}\right)^2 \\
&= \frac{3}{4}L^{-1}K^{-\frac{1}{2}} - \frac{9}{4}L^{-1}K^{-\frac{1}{2}} = -\frac{3}{2}L^{-1}K^{-\frac{1}{2}} < 0.
\end{aligned}$$

Since $M_2 < 0$ (and not $M_2 > 0$ as required) the Hessian $H(L,K)$ is not negative definite, and so the production function is *not* concave. This is because there are increasing returns to scale: the sum of the exponents on labour and capital is greater than 1 as $\frac{1}{2} + \frac{3}{4} = \frac{5}{4} > 1$.

Problem 4.10 *Suppose that the firm's production function is given by*

$$Q = F(L,K) = 3L^{\frac{1}{3}} + 6K^{\frac{1}{6}}.$$

*[D***] Show that the marginal products of labour and capital are positive and diminishing. What happens to the marginal product of labour when K is increased? What happens to the marginal product of capital when L is increased? [B***] Show that the production function is globally concave.*

Answer: We have

$$MP_L(L,K) = \frac{\partial F(L,K)}{\partial L} = L^{-\frac{2}{3}} > 0, MP_K(L,K) = \frac{\partial F(L,K)}{\partial K} = K^{-\frac{5}{6}} > 0.$$

There is a diminishing marginal product of labour and capital since

$$\frac{\partial MP_L(L,K)}{\partial L} = -\frac{2}{3}L^{-\frac{5}{3}} < 0 \text{ and } \frac{\partial MP_K(L,K)}{\partial K} = -\frac{5}{6}K^{-\frac{11}{6}} < 0.$$

Since

$$\frac{\partial MP_L(L,K)}{\partial K} = \frac{\partial MP_K(L,K)}{\partial L} = \frac{\partial}{\partial K}\left(L^{-\frac{2}{3}}\right) = \frac{\partial}{\partial L}\left(K^{-\frac{5}{6}}\right) = 0$$

an increase in capital has no effect on the marginal product of labour, and an increase in labour has no effect on the marginal product of capital.

The Hessian is

$$H(L,K) = \begin{bmatrix} -\frac{2}{3}L^{-\frac{5}{3}} & 0 \\ 0 & -\frac{5}{6}K^{-\frac{11}{6}} \end{bmatrix}.$$

Since this is a diagonal matrix with negative diagonal elements, it follows that the Hessian is negative definite for all L, K, so that $F(L,K)$ is globally concave.

Problem 4.11 *Consider a special case of the CES (constant elasticity of substitution) production function*

$$Q = F(L,K) = 2\left(L^{\frac{3}{4}} + K^{\frac{3}{4}}\right)^{\frac{2}{3}}.$$

*[C***] Show that the marginal products of labour and capital are positive and diminishing. Show that increasing K decreases the marginal product of labour, and increasing L decreases the marginal product of capital. (Normally we imagine that an increase in K increases the marginal product of labour. Here we have a counter-example.) [A***] Show that the production function is globally concave.*

Answer: Using the chain rule we have positive marginal products of labour and capital since

$$\begin{aligned} MP_L(L,K) &= 2\times\frac{2}{3}\left(L^{\frac{3}{4}}+K^{\frac{3}{4}}\right)^{\frac{2}{3}-1}\frac{3}{4}L^{\frac{3}{4}-1}=\left(L^{\frac{3}{4}}+K^{\frac{3}{4}}\right)^{-\frac{1}{3}}L^{-\frac{1}{4}}>0 \\ MP_K(L,K) &= 2\times\frac{2}{3}\left(L^{\frac{3}{4}}+K^{\frac{3}{4}}\right)^{\frac{2}{3}-1}\frac{3}{4}K^{\frac{3}{4}-1}=\left(L^{\frac{3}{4}}+K^{\frac{3}{4}}\right)^{-\frac{1}{3}}K^{-\frac{1}{4}}>0. \end{aligned}$$

Now

$$\frac{\partial^2 F(L,K)}{\partial L\partial K}=-\frac{1}{3}\left(L^{\frac{3}{4}}+K^{\frac{3}{4}}\right)^{-\frac{1}{3}-1}L^{-\frac{1}{4}}\frac{3}{4}K^{\frac{3}{4}-1}=-\frac{1}{4}\left(L^{\frac{3}{4}}+K^{\frac{3}{4}}\right)^{-\frac{4}{3}}L^{-\frac{1}{4}}K^{-\frac{1}{4}}<0$$

and so by Young's theorem we have

$$\frac{\partial MP_L(L,K)}{\partial K}=\frac{\partial MP_K(L,K)}{\partial L}=\frac{\partial^2 F(L,K)}{\partial L\partial K}=-\frac{1}{4}\left(L^{\frac{3}{4}}+K^{\frac{3}{4}}\right)^{-\frac{4}{3}}L^{-\frac{1}{4}}K^{-\frac{1}{4}}<0.$$

We have diminishing marginal products of labour and capital since

$$\begin{aligned} \frac{\partial MP_L(L,K)}{\partial L} &= \frac{\partial^2 F(L,K)}{\partial L^2}=\frac{\partial}{\partial L}\left(\left(L^{\frac{3}{4}}+K^{\frac{3}{4}}\right)^{-\frac{1}{3}}L^{-\frac{1}{4}}\right) \\ &= -\frac{1}{3}\left(L^{\frac{3}{4}}+K^{\frac{3}{4}}\right)^{-\frac{1}{3}-1}\frac{3}{4}L^{-\frac{1}{4}}L^{-\frac{1}{4}}-\frac{1}{4}\left(L^{\frac{3}{4}}+K^{\frac{3}{4}}\right)^{-\frac{1}{3}}L^{-\frac{1}{4}-1} \\ &= -\frac{1}{4}\left(L^{\frac{3}{4}}+K^{\frac{3}{4}}\right)^{-\frac{4}{3}}L^{-\frac{1}{2}}-\frac{1}{4}\left(L^{\frac{3}{4}}+K^{\frac{3}{4}}\right)^{-\frac{1}{3}}L^{-\frac{5}{4}} \\ &= -\frac{1}{4}\left(L^{\frac{3}{4}}+K^{\frac{3}{4}}\right)^{-\frac{4}{3}}\left(L^{-\frac{1}{2}}+\left(L^{\frac{3}{4}}+K^{\frac{3}{4}}\right)L^{-\frac{5}{4}}\right)<0 \\ \frac{\partial MP_K(L,K)}{\partial K} &= \frac{\partial^2 F(L,K)}{\partial K^2}=\frac{\partial}{\partial K}\left(\left(L^{\frac{3}{4}}+K^{\frac{3}{4}}\right)^{-\frac{1}{3}}K^{-\frac{1}{4}}\right) \\ &= -\frac{1}{4}\left(L^{\frac{3}{4}}+K^{\frac{3}{4}}\right)^{-\frac{4}{3}}K^{-\frac{1}{2}}-\frac{1}{4}\left(L^{\frac{3}{4}}+K^{\frac{3}{4}}\right)^{-\frac{1}{3}}K^{-\frac{5}{4}} \\ &= -\frac{1}{4}\left(L^{\frac{3}{4}}+K^{\frac{3}{4}}\right)^{-\frac{4}{3}}\left(K^{-\frac{1}{2}}+\left(L^{\frac{3}{4}}+K^{\frac{3}{4}}\right)K^{-\frac{5}{4}}\right)<0. \end{aligned}$$

To show concavity we construct the Hessian

$$\begin{aligned} H(L,K) &= \begin{bmatrix} \frac{\partial^2 F(L,K)}{\partial L^2} & \frac{\partial^2 F(L,K)}{\partial L\partial K} \\ \frac{\partial^2 F(L,K)}{\partial L\partial K} & \frac{\partial^2 F(L,K)}{\partial K^2} \end{bmatrix} \\ &= -\frac{1}{4}\left(L^{\frac{3}{4}}+K^{\frac{3}{4}}\right)^{-\frac{4}{3}} \\ &\quad\times\begin{bmatrix} L^{-\frac{1}{2}}+\left(L^{\frac{3}{4}}+K^{\frac{3}{4}}\right)L^{-\frac{5}{4}} & L^{-\frac{1}{4}}K^{-\frac{1}{4}} \\ L^{-\frac{1}{4}}K^{-\frac{1}{4}} & K^{-\frac{1}{2}}+\left(L^{\frac{3}{4}}+K^{\frac{3}{4}}\right)K^{-\frac{5}{4}} \end{bmatrix}. \end{aligned}$$

We have already shown that

$$M_1=\frac{\partial^2 F(L,K)}{\partial L^2}<0$$

so all we need to show that $H(L,K)$ is negative definite is to show that $M_2 > 0$. Here we have

$$\begin{aligned}
M_2 &= \det\left[H(L,K)\right] = \frac{\partial^2 F(L,K)}{\partial L^2}\frac{\partial^2 F(L,K)}{\partial K^2} - \left(\frac{\partial^2 F(L,K)}{\partial L \partial K}\right)^2 \\
&= \left(-\frac{1}{4}\left(L^{\frac{3}{4}}+K^{\frac{3}{4}}\right)^{-\frac{4}{3}}\right)^2 \\
&\quad \times \left(\left(L^{-\frac{1}{2}}+\left(L^{\frac{3}{4}}+K^{\frac{3}{4}}\right)L^{-\frac{5}{4}}\right)\left(K^{-\frac{1}{2}}+\left(L^{\frac{3}{4}}+K^{\frac{3}{4}}\right)K^{-\frac{5}{4}}\right)-\left(L^{-\frac{1}{4}}K^{-\frac{1}{4}}\right)^2\right) \\
&= \frac{\left(L^{\frac{3}{4}}+K^{\frac{3}{4}}\right)^{-\frac{8}{3}}}{16}\left(\left(L^{-\frac{1}{2}}+\left(L^{\frac{3}{4}}+K^{\frac{3}{4}}\right)L^{-\frac{5}{4}}\right)\left(K^{-\frac{1}{2}}+\left(L^{\frac{3}{4}}+K^{\frac{3}{4}}\right)K^{-\frac{5}{4}}\right)-\left(L^{-\frac{1}{4}}K^{-\frac{1}{4}}\right)^2\right) \\
&= \frac{\left(L^{\frac{3}{4}}+K^{\frac{3}{4}}\right)^{-\frac{8}{3}}}{16}\left(\begin{array}{c} L^{-\frac{1}{2}}K^{-\frac{1}{2}}+\left(L^{\frac{3}{4}}+K^{\frac{3}{4}}\right)L^{-\frac{1}{2}}K^{-\frac{5}{4}} \\ +\left(L^{\frac{3}{4}}+K^{\frac{3}{4}}\right)L^{-\frac{5}{4}}K^{-\frac{1}{2}}+\left(L^{\frac{3}{4}}+K^{\frac{3}{4}}\right)^2 L^{-\frac{5}{4}}K^{-\frac{5}{4}}-L^{-\frac{1}{2}}K^{-\frac{1}{2}} \end{array}\right) \\
&= \frac{\left(L^{\frac{3}{4}}+K^{\frac{3}{4}}\right)^{-\frac{8}{3}}}{16}\left(L^{\frac{3}{4}}+K^{\frac{3}{4}}\right)\left(L^{-\frac{1}{2}}K^{-\frac{5}{4}}+L^{-\frac{5}{4}}K^{-\frac{1}{2}}+\left(L^{\frac{3}{4}}+K^{\frac{3}{4}}\right)L^{-\frac{5}{4}}K^{-\frac{5}{4}}\right) \\
&= \frac{\left(L^{\frac{3}{4}}+K^{\frac{3}{4}}\right)^{-\frac{4}{3}}\left(L^{-\frac{1}{2}}K^{-\frac{5}{4}}+L^{-\frac{5}{4}}K^{-\frac{1}{2}}\right)}{8} > 0.
\end{aligned}$$

Since $M_1 < 0$ and $M_2 > 0$ for all L, K, it follows that $H(L,K)$ is negative definite for all L, K and hence the production function is globally concave.

Problem 4.12 *Consider the CES production function*

$$Q = F(L,K) = \frac{1}{\gamma}\left(L^{\rho}+K^{\rho}\right)^{\frac{\gamma}{\rho}}$$

*where $\rho < 1$, and $0 < \gamma < 1$. The parameter ρ determines the degree of substitutability between labour and capital while γ determines the returns to scale. The assumption $0 < \gamma < 1$ is equivalent to assuming decreasing returns to scale. [C***] Show that the marginal products of labour and capital are positive and diminishing. [B***]When is $\frac{\partial^2 F(L,K)}{\partial L \partial K} < 0$? [A***]Show that the CES production function is globally concave.*

Answer: Using the chain rule we have positive marginal products since

$$\begin{aligned}
MP_L(L,K) &= \left(L^{\rho}+K^{\rho}\right)^{\frac{\gamma}{\rho}-1}L^{\rho-1} = \left(L^{\rho}+K^{\rho}\right)^{\frac{\gamma-\rho}{\rho}}L^{\rho-1} > 0 \\
MP_K(L,K) &= \left(L^{\rho}+K^{\rho}\right)^{\frac{\gamma}{\rho}-1}K^{\rho-1} = \left(L^{\rho}+K^{\rho}\right)^{\frac{\gamma-\rho}{\rho}}K^{\rho-1} > 0.
\end{aligned}$$

We have diminishing marginal products of labour and capital since

$$\begin{aligned}
\frac{\partial MP_L(L,K)}{\partial L} &= \frac{\partial^2 F(L,K)}{\partial L^2} = \frac{\partial}{\partial L}\left(\left(L^\rho + K^\rho\right)^{\frac{\gamma-\rho}{\rho}} L^{\rho-1}\right) \\
&= (\gamma-\rho)\left(L^\rho + K^\rho\right)^{\frac{\gamma-2\rho}{\rho}} L^{2(\rho-1)} + (\rho-1)\left(L^\rho + K^\rho\right)^{\frac{\gamma-\rho}{\rho}} L^{\rho-2} \\
&= \left(L^\rho + K^\rho\right)^{\frac{\gamma-2\rho}{\rho}} L^{\rho-2}\left((\gamma-\rho)L^\rho + (\rho-1)\left(L^\rho + K^\rho\right)\right) \\
&= \left(L^\rho + K^\rho\right)^{\frac{\gamma-2\rho}{\rho}} L^{\rho-2}\left((\gamma-\rho+\rho-1)L^\rho + (\rho-1)K^\rho\right) \\
&= \left(L^\rho + K^\rho\right)^{\frac{\gamma-2\rho}{\rho}} L^{\rho-2}\left(\underbrace{(\gamma-1)}_{-}L^\rho + \underbrace{(\rho-1)}_{-}K^\rho\right) < 0
\end{aligned}$$

and similarly

$$\begin{aligned}
\frac{\partial MP_K(L,K)}{\partial K} &= \frac{\partial^2 F(L,K)}{\partial K^2} = \frac{\partial}{\partial K}\left(\left(L^\rho + K^\rho\right)^{\frac{\gamma-\rho}{\rho}} K^{\rho-1}\right) \\
&= \left(L^\rho + K^\rho\right)^{\frac{\gamma-2\rho}{\rho}} K^{\rho-2}\left(\underbrace{(\gamma-1)}_{-}K^\rho + \underbrace{(\rho-1)}_{-}L^\rho\right) < 0.
\end{aligned}$$

We have

$$\begin{aligned}
\frac{\partial^2 F(L,K)}{\partial L \partial K} &= \frac{\partial MP_L(L,K)}{\partial K} = \frac{\partial}{\partial K}\left(\left(L^\rho + K^\rho\right)^{\frac{\gamma-\rho}{\rho}} L^{\rho-1}\right) \\
&= (\gamma-\rho)\left(L^\rho + K^\rho\right)^{\frac{\gamma-2\rho}{\rho}} L^{\rho-1}K^{\rho-1}.
\end{aligned}$$

If $\gamma > \rho$ then $\frac{\partial^2 F(L,K)}{\partial L \partial K} > 0$ while if $\gamma < \rho$ then $\frac{\partial^2 F(L,K)}{\partial L \partial K} < 0$. For example when $\rho = \frac{3}{4}$ and $\gamma = \frac{1}{2}$ we have $\gamma - \rho = -\frac{1}{4} < 0$ and so $\frac{\partial^2 F(L,K)}{\partial L \partial K} < 0$.

To show concavity we construct the Hessian

$$H(L,K) = \begin{bmatrix} \frac{\partial^2 F(L,K)}{\partial L^2} & \frac{\partial^2 F(L,K)}{\partial L \partial K} \\ \frac{\partial^2 F(L,K)}{\partial L \partial K} & \frac{\partial^2 F(L,K)}{\partial K^2} \end{bmatrix}.$$

We have

$$M_1 = \frac{\partial^2 F(L,K)}{\partial L^2} = \left(L^\rho + K^\rho\right)^{\frac{\gamma-2\rho}{\rho}} L^{\rho-2}\left(\underbrace{(\gamma-1)}_{-}L^\rho + \underbrace{(\rho-1)}_{-}K^\rho\right) < 0.$$

The real work is to show $M_2 > 0$ as

$$\begin{aligned}
M_2 &= \det\left[H(L,K)\right] = \frac{\partial^2 F(L,K)}{\partial L^2}\frac{\partial^2 F(L,K)}{\partial K^2} - \left(\frac{\partial^2 F(L,K)}{\partial L \partial K}\right)^2 \\
&= \left(\left(L^\rho + K^\rho\right)^{\frac{\gamma-2\rho}{\rho}} L^{\rho-2}\left((\gamma-1)L^\rho + (\rho-1)K^\rho\right)\right) \\
&\quad \times \left(\left(L^\rho + K^\rho\right)^{\frac{\gamma-2\rho}{\rho}} K^{\rho-2}\left((\gamma-1)K^\rho + (\rho-1)L^\rho\right)\right) \\
&\quad - \left((\gamma-\rho)\left(L^\rho + K^\rho\right)^{\frac{\gamma-2\rho}{\rho}} L^{\rho-1}K^{\rho-1}\right)^2.
\end{aligned}$$

To make our life easier let's factor out some positive terms from M_2 as

$$M_2 = \underbrace{\left(\left(L^\rho + K^\rho\right)^{\frac{\gamma-2\rho}{\rho}}\right)^2}_{+} \tilde{M}_2$$

where to show $M_2 > 0$ we need only show $\tilde{M}_2 > 0$ where

$$\begin{aligned}
\tilde{M}_2 &= \left((\gamma-1) L^{2\rho-2} + (\rho-1) K^\rho L^{\rho-2}\right) \\
&\quad \times \left((\gamma-1) K^{2\rho-2} + (\rho-1) L^\rho K^{\rho-2}\right) - (\gamma-\rho)^2 L^{2\rho-2} K^{2\rho-2} \\
&= \left((\gamma-1)^2 + (\rho-1)^2 - (\gamma-\rho)^2\right) L^{2\rho-2} K^{2\rho-2} \\
&\quad + (\gamma-1)(\rho-1)\left(K^{3\rho-2} L^{\rho-2} + K^{\rho-2} L^{3\rho-2}\right).
\end{aligned}$$

You can verify that

$$\begin{aligned}
\left((\gamma-1)^2 + (\rho-1)^2 - (\gamma-\rho)^2\right) &= \left((\gamma-1)^2 + (\rho-1)^2 - ((\gamma-1)-(\rho-1))^2\right) \\
&= 2(\gamma-1)(\rho-1)
\end{aligned}$$

so that

$$\begin{aligned}
\tilde{M}_2 &= (\gamma-1)(\rho-1)\left(2L^{2\rho-2}K^{2\rho-2} + K^{3\rho-2}L^{\rho-2} + K^{\rho-2}L^{3\rho-2}\right) \\
&= (\gamma-1)(\rho-1) L^{\rho-2} K^{\rho-2}\left(L^{2\rho} + 2K^\rho L^\rho + K^{2\rho}\right) \\
&= (\gamma-1)(\rho-1) L^{\rho-2} K^{\rho-2}\left(L^\rho + K^\rho\right)^2 > 0
\end{aligned}$$

where $(\gamma-1)(\rho-1) > 0$ since $\gamma < 1$ and $\rho < 1$ by assumption. Thus $M_2 > 0$ and the production function is globally concave. In fact we can write M_2 in quite a simple form as

$$\begin{aligned}
M_2 &= \left(\gamma\left(L^\rho + K^\rho\right)^{\frac{\gamma-2\rho}{\rho}}\right)^2 (\gamma-1)(\rho-1) L^{\rho-2} K^{\rho-2}\left(L^\rho + K^\rho\right)^2 \\
&= (\gamma-1)(\rho-1)\gamma^2\left(L^\rho + K^\rho\right)^{\frac{\gamma-\rho}{\rho}} L^{\rho-2} K^{\rho-2} > 0.
\end{aligned}$$

Problem 4.13 *Consider a more general form of the CES production function*

$$Q = F(L, K) = \frac{1}{\gamma}\left(\alpha L^\rho + (1-\alpha) K^\rho\right)^{\frac{\gamma}{\rho}}$$

*where $\rho < 1$, $0 < \alpha < 1$ and $0 < \gamma < 1$. [A**] Show that this reduces to a Cobb-Douglas when $\rho = 0$.*

Answer: We have

$$\begin{aligned}
Q &= \frac{1}{\gamma}\left(\alpha L^\rho + (1-\alpha) K^\rho\right)^{\frac{\gamma}{\rho}} \\
&\Longrightarrow \left((\gamma Q)^{\frac{1}{\gamma}}\right)^\rho = \alpha L^\rho + (1-\alpha) K^\rho \\
&\Longrightarrow \frac{\left((\gamma Q)^{\frac{1}{\gamma}}\right)^\rho - 1}{\rho} = \alpha\frac{L^\rho - 1}{\rho} + (1-\alpha)\frac{K^\rho - 1}{\rho}.
\end{aligned}$$

Note that all terms are of the form $\frac{x^\rho-1}{\rho}$ for some x. Using L'Hôpital's rule (see Problem 2.155 on page 120) we have

$$\lim_{\rho\to 0}\frac{x^\rho-1}{\rho}=\ln(x)$$

and so as $\rho\to 0$ the production function becomes

$$\begin{aligned}\ln\left((\gamma Q)^{\frac{1}{\gamma}}\right) &= \alpha\ln(L)+(1-\alpha)\ln(K)\\ &\implies (\gamma Q)^{\frac{1}{\gamma}}=L^\alpha K^{1-\alpha}\\ &\implies Q=\gamma^{-\frac{1}{\gamma}}L^{\alpha\gamma}K^{(1-\alpha)\gamma}\end{aligned}$$

which is Cobb-Douglas functional form.

Problem 4.14 *Consider a household with utility function*

$$U(Q_1,Q_2)=4Q_1^{\frac{1}{4}}+4Q_2^{\frac{1}{4}}.$$

*[D***] Find the marginal utility of Q_1 and Q_2. Show that there is positive and diminishing marginal utility for both goods. If Q_2 increases, then what happens to the marginal utility of Q_1? [C***] Show that the utility function is concave.*

Answer: We have positive and diminishing marginal utility as

$$\begin{aligned}MU_1(Q_1,Q_2) &= Q_1^{-\frac{3}{4}}>0,\ \frac{\partial MU_1(Q_1,Q_2)}{\partial Q_1}=-\frac{3}{4}Q_1^{-\frac{7}{4}}<0\\ MU_2(Q_1,Q_2) &= Q_2^{-\frac{3}{4}}>0,\ \frac{\partial MU_2(Q_1,Q_2)}{\partial Q_2}=-\frac{3}{4}Q_2^{-\frac{7}{4}}<0.\end{aligned}$$

Since

$$\frac{\partial MU_1(Q_1,Q_2)}{\partial Q_2}=\frac{\partial MU_2(Q_1,Q_2)}{\partial Q_1}=\frac{\partial^2 U(Q_1,Q_2)}{\partial Q_1\partial Q_2}=0$$

nothing happens to $MU_1(Q_1,Q_2)$ when Q_2 changes, nor does anything happen to $MU_2(Q_1,Q_2)$ when Q_1 changes. The Hessian of $U(Q_1,Q_2)$ is

$$H(Q_1,Q_2)=\begin{bmatrix}-\frac{3}{4}Q_1^{-\frac{7}{4}} & 0\\ 0 & -\frac{3}{4}Q_2^{-\frac{7}{4}}\end{bmatrix}$$

which is a diagonal matrix with negative elements along the diagonal and hence is negative definite. Therefore $U(Q_1,Q_2)$ is concave.

4.2 Unconstrained Optimization

Problem 4.15 *Consider*

$$f(x_1,x_2)=3x_1^2-6x_1x_2+5x_2^2-4x_1-2x_2+8.$$

*[D***] Find the first-order conditions and solve using Cramer's rule. [B***] Determine if your solution is a global minimizer or maximizer.*

Answer: We have

$$\begin{aligned}\frac{\partial f(x_1,x_2)}{\partial x_1} &= 6x_1-6x_2-4\\ \frac{\partial f(x_1,x_2)}{\partial x_2} &= -6x_1+10x_2-2.\end{aligned}$$

The first-order conditions then are

$$\begin{aligned}6x_1^*-6x_2^*-4 &= 0\\ -6x_1^*+10x_2^*-2 &= 0\end{aligned}$$

so that

$$\begin{bmatrix}6 & -6\\ -6 & 10\end{bmatrix}\begin{bmatrix}x_1^*\\ x_2^*\end{bmatrix}=\begin{bmatrix}4\\ 2\end{bmatrix} \text{ and } \det\begin{bmatrix}6 & -6\\ -6 & 10\end{bmatrix}=24.$$

Using Cramer's rule we find that

$$x_1^*=\frac{\det\begin{bmatrix}4 & -6\\ 2 & 10\end{bmatrix}}{24}=\frac{52}{24}=\frac{13}{6},\ x_2^*=\frac{\det\begin{bmatrix}6 & 4\\ -6 & 2\end{bmatrix}}{24}=\frac{36}{24}=\frac{3}{2}.$$

Now

$$\begin{aligned}\frac{\partial^2 f(x_1,x_2)}{\partial x_1^2} &= \frac{\partial}{\partial x_1}(6x_1-6x_2-4)=6\\ \frac{\partial^2 f(x_1,x_2)}{\partial x_1\partial x_2} &= \frac{\partial}{\partial x_2}(6x_1-6x_2-4)=-6\\ \frac{\partial^2 f(x_1,x_2)}{\partial x_2^2} &= \frac{\partial}{\partial x_2}(-6x_1+10x_2-2)=10\\ \frac{\partial^2 f(x_1,x_2)}{\partial x_2\partial x_1} &= \frac{\partial}{\partial x_1}(-6x_1+10x_2-2)=-6.\end{aligned}$$

The Hessian $H(x_1,x_2)$ is positive definite for all x_1,x_2 since

$$H(x_1,x_2) = \begin{bmatrix}6 & -6\\ -6 & 10\end{bmatrix}$$

$$\Longrightarrow M_1=\det[6]=6>0, M_2=\det\begin{bmatrix}6 & -6\\ -6 & 10\end{bmatrix}=24>0.$$

It follows that $f(x_1,x_2)$ is globally convex and $x_1^*=\frac{13}{6}, x_2^*=\frac{3}{2}$ is a global minimizer.

Problem 4.16 *Consider the function*

$$f(x_1,x_2)=5x_1^2-6x_1x_2+4x_2^2-2x_1-3x_2.$$

*[B**] Write this function using matrix notation as $f(x)=a^Tx+x^TAx$ where x and a are 2×1 vectors and A is a symmetric 2×2 matrix. [D***] Find the first-order conditions and solve using Cramer's rule. [B***] Determine if your solution is a global minimizer or maximizer.*

Answer: We can write $f(x)$ as

$$f(x) = a^T x + x^T A x = \begin{bmatrix} -2 & -3 \end{bmatrix} \begin{bmatrix} x_1 \\ x_2 \end{bmatrix} + \begin{bmatrix} x_1 & x_2 \end{bmatrix} \begin{bmatrix} 5 & -3 \\ -3 & 4 \end{bmatrix} \begin{bmatrix} x_1 \\ x_2 \end{bmatrix}.$$

Now

$$\frac{\partial f(x_1, x_2)}{\partial x_1} = 10x_1 - 6x_2 - 2, \frac{\partial f(x_1, x_2)}{\partial x_2} = -6x_1 + 8x_2 - 3.$$

The first-order conditions then are

$$\begin{aligned} 10x_1^* - 6x_2^* - 2 &= 0 \\ -6x_1^* + 8x_2^* - 3 &= 0. \end{aligned}$$

In matrix notation

$$\begin{bmatrix} 10 & -6 \\ -6 & 8 \end{bmatrix} \begin{bmatrix} x_1^* \\ x_2^* \end{bmatrix} = \begin{bmatrix} 2 \\ 3 \end{bmatrix} \text{ and } \det \begin{bmatrix} 10 & -6 \\ -6 & 8 \end{bmatrix} = 44.$$

Using Cramer's rule we have

$$\begin{aligned} x_1^* &= \frac{\det \begin{bmatrix} 2 & -6 \\ 3 & 8 \end{bmatrix}}{44} = \frac{34}{44} = \frac{17}{22} \\ x_2^* &= \frac{\det \begin{bmatrix} 10 & 2 \\ -6 & 3 \end{bmatrix}}{44} = \frac{42}{44} = \frac{21}{22}. \end{aligned}$$

We have

$$\begin{aligned} \frac{\partial^2 f(x_1, x_2)}{\partial x_1^2} &= \frac{\partial}{\partial x_1}(10x_1 - 6x_2 - 2) = 10 \\ \frac{\partial^2 f(x_1, x_2)}{\partial x_2 \partial x_1} &= \frac{\partial}{\partial x_2}(10x_1 - 6x_2 - 2) = -6 \\ \frac{\partial^2 f(x_1, x_2)}{\partial x_2^2} &= \frac{\partial}{\partial x_2}(-6x_1 + 8x_2 - 3) = 8 \\ \frac{\partial^2 f(x_1, x_2)}{\partial x_1 \partial x_2} &= \frac{\partial}{\partial x_1}(-6x_1 + 8x_2 - 3) = -6. \end{aligned}$$

The Hessian $H(x_1, x_2)$ is positive definite for all x_1, x_2 since

$$\begin{aligned} H(x_1, x_2) &= \begin{bmatrix} 10 & -6 \\ -6 & 8 \end{bmatrix} \\ &\implies M_1 = \det[10] = 10 > 0, M_2 = \det \begin{bmatrix} 10 & -6 \\ -6 & 8 \end{bmatrix} = 44 > 0. \end{aligned}$$

It follows that $f(x_1, x_2)$ is globally convex and $x_1^* = \frac{17}{22}, x_2^* = \frac{21}{22}$ is a global minimizer.

Problem 4.17 *Consider*

$$f(x_1, x_2) = -4x_1^2 - 6x_2^2 + 4x_1x_2 + 3x_1 + 2x_2.$$

*[D***] Find the first-order conditions and solve using Cramer's rule. [B***] Determine if your solution is a global minimizer or maximizer using both leading principal minors and eigenvalues.*

Answer: We have

$$\begin{aligned}\frac{\partial f(x_1, x_2)}{\partial x_1} &= -8x_1 + 4x_2 + 3 \\ \frac{\partial f(x_1, x_2)}{\partial x_2} &= 4x_1 - 12x_2 + 2.\end{aligned}$$

The first-order conditions then are

$$\begin{aligned}-8x_1^* + 4x_2^* + 3 &= 0 \\ 4x_1^* - 12x_2^* + 2 &= 0\end{aligned}$$

so that

$$\begin{bmatrix} -8 & 4 \\ 4 & -12 \end{bmatrix} \begin{bmatrix} x_1^* \\ x_2^* \end{bmatrix} = \begin{bmatrix} -3 \\ -2 \end{bmatrix} \text{ and } \det \begin{bmatrix} -8 & 4 \\ 4 & -12 \end{bmatrix} = 80.$$

Using Cramer's rule it follows that

$$x_1^* = \frac{\det \begin{bmatrix} -3 & 4 \\ -2 & -12 \end{bmatrix}}{80} = \frac{44}{80} = \frac{11}{20}, \; x_2^* = \frac{\det \begin{bmatrix} -8 & -3 \\ 4 & -2 \end{bmatrix}}{80} = \frac{28}{80} = \frac{7}{20}.$$

Since

$$\begin{aligned}\frac{\partial^2 f(x_1, x_2)}{\partial x_1^2} &= \frac{\partial}{\partial x_1}(-8x_1 + 4x_2 + 3) = -8 \\ \frac{\partial^2 f(x_1, x_2)}{\partial x_1 \partial x_2} &= \frac{\partial}{\partial x_2}(-8x_1 + 4x_2 + 3) = 4 \\ \frac{\partial^2 f(x_1, x_2)}{\partial x_2^2} &= \frac{\partial}{\partial x_2}(3 + 4x_1 - 12x_2) = -12 \\ \frac{\partial^2 f(x_1, x_2)}{\partial x_2 \partial x_1} &= \frac{\partial}{\partial x_1}(3 + 4x_1 - 12x_2) = 4\end{aligned}$$

the Hessian is negative definite as

$$\begin{aligned}H(x_1, x_2) &= \begin{bmatrix} -8 & 4 \\ 4 & -12 \end{bmatrix} \\ &\Longrightarrow M_1 = \det[-8] = -8 < 0, M_2 = \det \begin{bmatrix} -8 & 4 \\ 4 & -12 \end{bmatrix} = 80 > 0.\end{aligned}$$

It follows that $f(x_1, x_2)$ is globally concave and $x_1^* = \frac{11}{20}, x_2^* = \frac{7}{20}$ is a global maximizer.

Alternative we can show that $H(x_1, x_2)$ is negative definite by calculating eigenvalues from

$$\begin{aligned} \lambda^2 + 20\lambda + 80 &= 0 \\ &\implies \lambda_1 = -10 + 2\sqrt{5} \approx -5.5, \lambda_2 = -10 - 2\sqrt{5} \approx -14.5. \end{aligned}$$

Since $\lambda_1 < 0, \lambda_2 < 0$ for all x_1, x_2 it follows that $f(x_1, x_2)$ is globally concave. Since $f(x_1, x_2)$ is globally concave, it follows that $x_1^* = \frac{11}{20}, x_2^* = \frac{7}{20}$ is a global maximizer.

Problem 4.18 *Consider*

$$f(x_1, x_2) = 3x_1^2 + 2x_2^2 - 4x_1x_2 - 2x_1 - 3x_2 + 7.$$

*[D***] Find the first-order conditions and solve. [B***] Determine if the solution to the first-order conditions is a global maximizer or minimizer. [B**] Now write this function as*

$$f(x) = x^T Ax + b^T x + c$$

*and find A, b, c. [A**] Show that your solution to the first-order conditions can be written as $x^* = -\frac{1}{2}A^{-1}b$. Now generalize and suppose A is a positive definite $n \times n$ matrix. [A**] Consider the problem of minimizing*

$$f(x) = x^T Ax + b^T x + c$$

where x and b are $n \times 1$ vectors and c is a 1×1 scalar. Show that the global minimum of $f(x)$ is given by $x^* = -\frac{1}{2}A^{-1}b$.

Answer: We have

$$\begin{aligned} \frac{\partial f(x_1, x_2)}{\partial x_1} &= 6x_1 - 4x_2 - 2 \\ \frac{\partial f(x_1, x_2)}{\partial x_2} &= -4x_1 + 4x_2 - 3 \end{aligned}$$

so the first-order conditions are

$$\begin{aligned} 6x_1^* - 4x_2^* - 2 &= 0 \\ -4x_1^* + 4x_2^* - 3 &= 0 \end{aligned}$$

so that in matrix notation we have

$$\begin{bmatrix} 6 & -4 \\ -4 & 4 \end{bmatrix} \begin{bmatrix} x_1^* \\ x_2^* \end{bmatrix} = \begin{bmatrix} 2 \\ 3 \end{bmatrix} \text{ and } \det \begin{bmatrix} 6 & -4 \\ -4 & 4 \end{bmatrix} = 8.$$

Using Cramer's rule yields

$$x_1^* = \frac{\det \begin{bmatrix} 2 & -4 \\ 3 & 4 \end{bmatrix}}{8} = \frac{20}{8} = \frac{5}{2}, \; x_2^* = \frac{\det \begin{bmatrix} 6 & 2 \\ -4 & 3 \end{bmatrix}}{8} = \frac{26}{8} = \frac{13}{4}.$$

Since

$$\frac{\partial^2 f(x_1,x_2)}{\partial x_1^2}=6,\ \frac{\partial^2 f(x_1,x_2)}{\partial x_2^2}=4,\ \frac{\partial^2 f(x_1,x_2)}{\partial x_2\partial x_1}=-4.$$

the Hessian is given by

$$H(x_1,x_2)=\begin{bmatrix} 6 & -4 \\ -4 & 4 \end{bmatrix}.$$

Since

$$M_1=\det[6]=6>0, M_2=\det\begin{bmatrix} 6 & -4 \\ -4 & 4 \end{bmatrix}=8>0$$

it follows that $H(x_1,x_2)$ is positive definite for all x_1,x_2, $f(x_1,x_2)$ is globally convex, and $x_1^*=\frac{5}{2}, x_2^*=\frac{13}{4}$ is a global minimizer.

We can re-write $f(x_1,x_2)$ in matrix notation as

$$\begin{aligned} f(x_1,x_2) &= 3x_1^2+2x_2^2-4x_1x_2-2x_1-3x_2+7=x^TAx+b^Tx+c \\ &= \begin{bmatrix} x_1 & x_2 \end{bmatrix}\begin{bmatrix} 3 & -2 \\ -2 & 2 \end{bmatrix}\begin{bmatrix} x_1 \\ x_2 \end{bmatrix}+\begin{bmatrix} -2 & -3 \end{bmatrix}\begin{bmatrix} x_1 \\ x_2 \end{bmatrix}+7 \end{aligned}$$

so that

$$A=\begin{bmatrix} 3 & -2 \\ -2 & 2 \end{bmatrix}, b=\begin{bmatrix} -2 \\ -3 \end{bmatrix}, c=7.$$

Now

$$x^*=\begin{bmatrix} x_1^* \\ x_2^* \end{bmatrix}=-\frac{1}{2}A^{-1}b=-\frac{1}{2}\begin{bmatrix} 3 & -2 \\ -2 & 2 \end{bmatrix}^{-1}\begin{bmatrix} -2 \\ -3 \end{bmatrix}=\begin{bmatrix} \frac{5}{2} \\ \frac{13}{4} \end{bmatrix}$$

or $x_1^*=\frac{5}{2}, x_2^*=\frac{13}{4}$ as we computed above.

Now in general we have

$$f(x)=x^TAx+b^Tx+c \Longrightarrow \frac{\partial f(x)}{\partial x}=2Ax+b$$

and so the first-order condition $\frac{\partial f(x^*)}{\partial x}=0$ yields

$$2Ax^*+b=0 \Longrightarrow x^*=-\frac{1}{2}A^{-1}b.$$

This corresponds to a global minimum since the Hessian is positive definite as

$$H(x)=\frac{\partial f(x)}{\partial x\partial x^T}=2A>0$$

since $A>0$ is positive definite.

Problem 4.19 *[C***] Consider*

$$f(x_1,x_2)=-3e^{2\ln(x_1)}+2e^{\ln\left(2x_1x_2^2\right)-\ln(x_2)}-2x_2^2+8\ln\left(e^{x_1}\right)+16e^{\ln(x_2)}.$$

Find the first-order conditions and solve using Cramer's rule. Using eigenvalues show that your solution is a global maximizer.

Answer: Simplifying we have

$$\begin{aligned} f(x_1 x_2) &= -3e^{2\ln(x_1)} + 2e^{\ln\left(2x_1x_2^2\right)-\ln(x_2)} - 2x_2^2 + 8\ln\left(e^{x_1}\right) + 16e^{\ln(x_2)} \\ &= -3x_1^2 + 4x_1x_2 - 2x_2^2 + 8x_1 + 16x_2 \end{aligned}$$

so that

$$\frac{\partial f(x_1,x_2)}{\partial x_1} = -6x_1 + 4x_2 + 8, \quad \frac{\partial f(x_1,x_2)}{\partial x_2} = 4x_1 - 4x_2 + 16$$

and the first-order conditions are

$$\begin{aligned} -6x_1^* + 4x_2^* + 8 &= 0 \\ 4x_1^* - 4x_2^* + 16 &= 0. \end{aligned}$$

In matrix notation

$$\begin{bmatrix} -6 & 4 \\ 4 & -4 \end{bmatrix}\begin{bmatrix} x_1^* \\ x_2^* \end{bmatrix} = \begin{bmatrix} -8 \\ -16 \end{bmatrix} \text{ and } \det\begin{bmatrix} -6 & 4 \\ 4 & -4 \end{bmatrix} = 8.$$

Cramer's rule then gives

$$x_1^* = \frac{\det\begin{bmatrix} -8 & 4 \\ -16 & -4 \end{bmatrix}}{8} = \frac{96}{8} = 12, \; x_2^* = \frac{\det\begin{bmatrix} -6 & -8 \\ 4 & -16 \end{bmatrix}}{8} = \frac{128}{8} = 16.$$

The second derivatives are

$$\frac{\partial^2 f(x_1,x_2)}{\partial x_1^2} = -6, \; \frac{\partial^2 f(x_1,x_2)}{\partial x_2 \partial x_1} = 4, \; \frac{\partial^2 f(x_1,x_2)}{\partial x_2^2} = -4$$

so the Hessian is

$$H(x_1,x_2) = \begin{bmatrix} -6 & 4 \\ 4 & -4 \end{bmatrix}.$$

Using eigenvalues we have

$$\begin{aligned} \lambda^2 + 10\lambda + 8 &= 0 \Rightarrow \lambda = \frac{-10 \pm \sqrt{10^2 - 4\times 8}}{2} = -5 \pm \sqrt{17} \\ &\Longrightarrow \lambda_1 = -5 + \sqrt{17} \approx -0.88 < 0 \text{ and } \lambda_2 = -5 - \sqrt{17} \approx -9.1 < 0 \end{aligned}$$

Since $\lambda_1 < 0, \lambda_2 < 0$ for all x_1, x_2, it follows that the Hessian is negative definite for all x_1, x_2, and so $f(x_1, x_2)$ is globally concave. Thus $x_1^* = 12, x_2^* = 16$ is a global maximizer.

Problem 4.20 *Consider*

$$f(x_1,x_2) = 2\ln(x_1) + 3\ln(x_2) + \ln(1 - x_1 - x_2).$$

*[C***] Find the first-order conditions and solve. [B**] Show that your solution is a local maximizer. [B***] Now show that is a global maximizer.*

Answer: We have

$$\begin{aligned}\frac{\partial f(x_1,x_2)}{\partial x_1} &= \frac{2}{x_1} - \frac{1}{1-x_1-x_2}\\ \frac{\partial f(x_1,x_2)}{\partial x_2} &= \frac{3}{x_2} - \frac{1}{1-x_1-x_2}\end{aligned}$$

so that the first-order conditions yield

$$\begin{aligned}\frac{2}{x_1^*} - \frac{1}{1-x_1^*-x_2^*} &= 0\\ &\implies 2(1-x_1^*-x_2^*) = x_1^*\\ &\implies 3x_1^* + 2x_2^* = 2\\ \frac{3}{x_2^*} - \frac{1}{1-x_1^*-x_2^*} &= 0\\ &\implies 3(1-x_1^*-x_2^*) = x_2^*\\ &\implies 3x_1^* + 4x_2^* = 3.\end{aligned}$$

In matrix notation we have

$$\begin{bmatrix}3 & 2\\ 3 & 4\end{bmatrix}\begin{bmatrix}x_1^*\\ x_2^*\end{bmatrix} = \begin{bmatrix}2\\ 3\end{bmatrix} \text{ and } \det\begin{bmatrix}3 & 2\\ 3 & 4\end{bmatrix} = 6.$$

Using Cramer's rule we find that

$$x_1^* = \frac{\det\begin{bmatrix}2 & 2\\ 3 & 4\end{bmatrix}}{6} = \frac{2}{6} = \frac{1}{3},\ x_2^* = \frac{\det\begin{bmatrix}3 & 2\\ 3 & 3\end{bmatrix}}{6} = \frac{3}{6} = \frac{1}{2}.$$

The Hessian is found from the second derivatives

$$\begin{aligned}\frac{\partial^2 f(x_1,x_2)}{\partial x_1^2} &= \frac{\partial}{\partial x_1}\left(\frac{2}{x_1} - \frac{1}{1-x_1-x_2}\right)\\ &= -\frac{2}{x_1^2} - \frac{1}{(1-x_1-x_2)^2}\\ \frac{\partial^2 f(x_1,x_2)}{\partial x_2^2} &= \frac{\partial}{\partial x_2}\left(\frac{3}{x_2} - \frac{1}{1-x_1-x_2}\right)\\ &= -\frac{3}{x_2^2} - \frac{1}{(1-x_1-x_2)^2}\\ \frac{\partial^2 f(x_1,x_2)}{\partial x_1 \partial x_2} &= \frac{\partial}{\partial x_2}\left(\frac{2}{x_1} - \frac{1}{1-x_1-x_2}\right)\\ &= -\frac{1}{(1-x_1-x_2)^2}\end{aligned}$$

so that

$$H(x_1,x_2) = \begin{bmatrix}-\frac{2}{x_1^2} - \frac{1}{(1-x_1-x_2)^2} & -\frac{1}{(1-x_1-x_2)^2}\\ -\frac{1}{(1-x_1-x_2)^2} & -\frac{3}{x_2^2} - \frac{1}{(1-x_1-x_2)^2}\end{bmatrix}.$$

At $x_1^* = \frac{1}{3}, x_2^* = \frac{1}{2}$ we have

$$H\left(\frac{1}{3}, \frac{1}{2}\right) = \begin{bmatrix} -\frac{2}{\left(\frac{1}{3}\right)^2} - \frac{1}{\left(1-\frac{1}{3}-\frac{1}{2}\right)^2} & -\frac{1}{\left(1-\frac{1}{3}-\frac{1}{2}\right)^2} \\ -\frac{1}{\left(1-\frac{1}{3}-\frac{1}{2}\right)^2} & -\frac{3}{\left(\frac{1}{2}\right)^2} - \frac{1}{\left(1-\frac{1}{3}-\frac{1}{2}\right)^2} \end{bmatrix} = \begin{bmatrix} -54 & -36 \\ -36 & -48 \end{bmatrix}$$

so that the function is locally concave and as a local maximum at $x_1^* = \frac{1}{3}, x_2^* = \frac{1}{2}$ as

$$M_1 = -54 < 0 \text{ and } M_2 = \det \begin{bmatrix} -54 & -36 \\ -36 & -48 \end{bmatrix} = 1296 > 0.$$

Now to show that $x_1^* = \frac{1}{3}, x_2^* = \frac{1}{2}$ corresponds to a global maximum. We have

$$\begin{aligned} M_1 &= \det \left[-\frac{2}{x_1^2} - \frac{1}{(1-x_1-x_2)^2} \right] \\ &= -\frac{2}{x_1^2} - \frac{1}{(1-x_1-x_2)^2} < 0 \\ M_2 &= \det \begin{bmatrix} -\frac{2}{x_1^2} - \frac{1}{(1-x_1-x_2)^2} & -\frac{1}{(1-x_1-x_2)^2} \\ -\frac{1}{(1-x_1-x_2)^2} & -\frac{3}{x_2^2} - \frac{1}{(1-x_1-x_2)^2} \end{bmatrix} \\ &= \left(-\frac{2}{x_1^2} - \frac{1}{(1-x_1-x_2)^2}\right)\left(-\frac{3}{x_2^2} - \frac{1}{(1-x_1-x_2)^2}\right) - \left(\frac{1}{(1-x_1-x_2)^2}\right)^2 \\ &= \frac{6}{x_1^2 x_2^2} + \frac{2}{x_1^2} \frac{1}{(1-x_1-x_2)^2} + \frac{3}{x_2^2} \frac{1}{(1-x_1-x_2)^2} > 0. \end{aligned}$$

Therefore $H(x_1, x_2)$ is negative definite for all x, $f(x_1, x_2)$ is globally concave, and $x_1^* = \frac{1}{3}, x_2^* = \frac{1}{2}$ is a global maximizer.

Problem 4.21 *Consider the function*

$$f(x_1, x_2) = 4x_1^{\frac{1}{4}} x_2^{\frac{1}{2}} - x_1 - x_2$$

where $x_1 > 0, x_2 > 0$. *[D***] Find the first-order conditions. [B***] Using the* ln( ) *function, convert the first-order conditions into a system of two linear equations with unknowns* $\ln(x_1^*), \ln(x_2^*)$. *Solve this system of equations using Cramer's rule and find* x_1^*, x_2^*. *[B**] By substituting* x_1^*, x_2^* *into the Hessian show that* x_1^*, x_2^* *is a local maximizer. [B***] Show that* x_1^*, x_2^* *is a global maximizer. [A ***] Consider*

$$g(x_1, x_2) = 4\exp\left(\frac{1}{4}\ln(x_1) + \frac{1}{2}\ln(x_2)\right) - e^{\ln(x_1)} - e^{\ln(x_2)}.$$

Simplify the function so there are no $\exp'$ s *or* $\ln'$ s. *Find the global maximum without doing any additional calculus.*

Answer: We have

$$\begin{aligned}\frac{\partial f(x_1,x_2)}{\partial x_1} &= x_1^{-\frac{3}{4}}x_2^{\frac{1}{2}}-1\\ \frac{\partial f(x_1,x_2)}{\partial x_2} &= 2x_1^{\frac{1}{4}}x_2^{-\frac{1}{2}}-1\end{aligned}$$

so that the first-order conditions yield

$$\begin{aligned}(x_1^*)^{-\frac{3}{4}}(x_2^*)^{\frac{1}{2}}-1 &= 0\\ &\implies (x_1^*)^{-\frac{3}{4}}(x_2^*)^{\frac{1}{2}}=1\\ &\implies -\frac{3}{4}\ln(x_1^*)+\frac{1}{2}\ln(x_2^*)=\ln(1)=0\\ 2(x_1^*)^{\frac{1}{4}}(x_2^*)^{-\frac{1}{2}}-1 &= 0\\ &\implies (x_1^*)^{\frac{1}{4}}(x_2^*)^{-\frac{1}{2}}=\frac{1}{2}\\ &\implies \frac{1}{4}\ln(x_1^*)-\frac{1}{2}\ln(x_2^*)=\ln\left(\frac{1}{2}\right)=-\ln(2)\end{aligned}$$

or in matrix notation

$$\begin{bmatrix}-\frac{3}{4} & \frac{1}{2}\\ \frac{1}{4} & -\frac{1}{2}\end{bmatrix}\begin{bmatrix}\ln(x_1^*)\\ \ln(x_2^*)\end{bmatrix}=\begin{bmatrix}0\\ -\ln(2)\end{bmatrix}\text{ and }\det[A]=\det\begin{bmatrix}-\frac{3}{4} & \frac{1}{2}\\ \frac{1}{4} & -\frac{1}{2}\end{bmatrix}=\frac{1}{4}.$$

From Cramer's rule we have

$$\begin{aligned}\ln(x_1^*) &= \frac{\det\begin{bmatrix}0 & \frac{1}{2}\\ -\ln(2) & -\frac{1}{2}\end{bmatrix}}{\frac{1}{4}}=2\ln(2)\\ &\implies x_1^*=e^{y_1^*}=e^{2\ln(2)}=4\\ \ln(x_2^*) &= \frac{\det\begin{bmatrix}-\frac{3}{4} & 0\\ \frac{1}{4} & -\ln(2)\end{bmatrix}}{\frac{1}{4}}=3\ln(2)\\ &\implies x_2^*=e^{y_2^*}=e^{3\ln(2)}=8.\end{aligned}$$

To find the Hessian we need to calculate second derivatives as

$$\frac{\partial^2 f(x_1,x_2)}{\partial x_1^2}=-\frac{3}{4}x_1^{-\frac{7}{4}}x_2^{\frac{1}{2}},\ \frac{\partial^2 f(x_1,x_2)}{\partial x_2^2}=-x_1^{\frac{1}{4}}x_2^{-\frac{3}{2}},\ \frac{\partial^2 f(x_1,x_2)}{\partial x_1\partial x_2}=\frac{1}{2}x_1^{-\frac{3}{4}}x_2^{-\frac{1}{2}}$$

so that the Hessian is given by

$$H(x_1,x_2)=\begin{bmatrix}-\frac{3}{4}x_1^{-\frac{7}{4}}x_2^{\frac{1}{2}} & \frac{1}{2}x_1^{-\frac{3}{4}}x_2^{-\frac{1}{2}}\\ \frac{1}{2}x_1^{-\frac{3}{4}}x_2^{-\frac{1}{2}} & -x_1^{\frac{1}{4}}x_2^{-\frac{3}{2}}\end{bmatrix}.$$

Substituting $x_1^* = 4, x_2^* = 8$ into $H(x_1, x_2)$ we find that

$$\begin{aligned} H(x_1^*, x_2^*) &= H(4,8) = \begin{bmatrix} -\frac{3}{4}4^{-\frac{7}{4}}8^{\frac{1}{2}} & \frac{1}{2}4^{-\frac{3}{4}}8^{-\frac{1}{2}} \\ \frac{1}{2}4^{-\frac{3}{4}}8^{-\frac{1}{2}} & -4^{\frac{1}{4}}8^{-\frac{3}{2}} \end{bmatrix} \\ &= \begin{bmatrix} -\frac{3}{16} & \frac{1}{16} \\ \frac{1}{16} & -\frac{1}{16} \end{bmatrix}. \end{aligned}$$

We have

$$M_1 = -\frac{3}{16} < 0, M_2 = \det \begin{bmatrix} -\frac{3}{16} & \frac{1}{16} \\ \frac{1}{16} & -\frac{1}{16} \end{bmatrix} = \frac{1}{128} > 0$$

so that $f(x_1, x_2)$ is locally concave at $x_1^* = 4, x_2^* = 8$, and so $x_1^* = 4, x_2^* = 8$ is a local maximizer. (Alternatively the eigenvalues of $H(4,8)$ are the roots of

$$\begin{aligned} \lambda^2 + \frac{1}{4}\lambda + \frac{1}{128} &= 0 \\ \implies \lambda_1 = -\frac{1}{8} + \frac{1}{16}\sqrt{2} \approx -0.036 < 0, \lambda_2 &= -\frac{1}{8} - \frac{1}{16}\sqrt{2} \approx -0.21 < 0 \end{aligned}$$

so that $H(4,8)$ is negative definite and hence $x_1^* = 4, x_2^* = 8$ is a local maximizer.)

We can in fact prove that $H(x_1, x_2)$ is negative definite for all x_1, x_2 since

$$\begin{aligned} M_1 &= -\frac{3}{4}x_1^{-\frac{7}{4}}x_2^{\frac{1}{2}} < 0 \\ M_2 &= \det \begin{bmatrix} -\frac{3}{4}x_1^{-\frac{7}{4}}x_2^{\frac{1}{2}} & \frac{1}{2}x_1^{-\frac{3}{4}}x_2^{-\frac{1}{2}} \\ \frac{1}{2}x_1^{-\frac{3}{4}}x_2^{-\frac{1}{2}} & -x_1^{\frac{1}{4}}x_2^{-\frac{3}{2}} \end{bmatrix} = \frac{3}{4}x_1^{-\frac{6}{4}}x_2^{-\frac{2}{2}} - \frac{1}{4}x_1^{-\frac{6}{4}}x_2^{-\frac{2}{2}} = \frac{2}{4}x_1^{-\frac{6}{4}}x_2^{-\frac{2}{2}} > 0 \end{aligned}$$

so that $f(x_1, x_2)$ is globally concave so that $x_1^* = 4, x_2^* = 8$ is a global maximizer.

Now for

$$g(x_1, x_2) = 4\exp\left(\frac{1}{4}\ln(x_1) + \frac{1}{2}\ln(x_2)\right) - e^{\ln(x_1)} - e^{\ln(x_2)}$$

we have by simplifying that

$$\begin{aligned} g(x_1, x_2) &= 4\exp\left(\frac{1}{4}\ln(x_1) + \frac{1}{2}\ln(x_2)\right) - e^{\ln(x_1)} - e^{\ln(x_2)} \\ &= 4\exp\left(\frac{1}{4}\ln(x_1)\right)\exp\left(\frac{1}{2}\ln(x_2)\right) - x_1 - x_2 \\ &= 4\exp\left(\ln\left(x_1^{\frac{1}{4}}\right)\right)\exp\left(\ln\left(x_2^{\frac{1}{2}}\right)\right) - x_1 - x_2 = 4x_1^{\frac{1}{4}}x_2^{\frac{1}{2}} - x_1 - x_2 = f(x_1, x_2) \end{aligned}$$

which is the same function as before and so $x_1^* = 4, x_2^* = 8$.

Problem 4.22 *Consider*

$$f(x_1, x_2) = \frac{4}{5}\ln(x_1) + \frac{1}{5}\ln(x_2) - 5x_1 - 4x_2.$$

*[C***] Find the first-order conditions and solve. [B***] Show that your solution is a global maximizer.*

Answer: We have

$$\begin{aligned}\frac{\partial f(x_1,x_2)}{\partial x_1} &= \frac{4}{5}\frac{1}{x_1}-5\\ \frac{\partial f(x_1,x_2)}{\partial x_2} &= \frac{1}{5}\frac{1}{x_2}-4\end{aligned}$$

and so the first-order conditions yield

$$\begin{aligned}\frac{4}{5}\frac{1}{x_1^*}-5 &= 0\\ &\implies x_1^*=\frac{4}{25}\end{aligned}$$

so that

$$\begin{aligned}\frac{1}{5}\frac{1}{x_2^*}-4 &= 0\\ &\implies x_2^*=\frac{1}{20}.\end{aligned}$$

Now

$$\begin{aligned}\frac{\partial^2 f(x_1,x_2)}{\partial x_1^2} &= \frac{\partial}{\partial x_1}\left(\frac{4}{5}\frac{1}{x_1}-5\right)=-\frac{4}{5}\frac{1}{x_1^2},\ \frac{\partial^2 f(x_1,x_2)}{\partial x_1\partial x_2}=\frac{\partial}{\partial x_2}\left(\frac{4}{5}\frac{1}{x_1}-5\right)=0\\ \frac{\partial^2 f(x_1,x_2)}{\partial x_2^2} &= \frac{\partial}{\partial x_2}\left(\frac{1}{5}\frac{1}{x_2}-4\right)=-\frac{1}{5}\frac{1}{x_2^2},\ \frac{\partial^2 f(x_1,x_2)}{\partial x_2\partial x_1}=\frac{\partial}{\partial x_1}\left(\frac{1}{5}\frac{1}{x_2}-4\right)=0\end{aligned}$$

and so the Hessian is

$$H(x_1,x_2)=\begin{bmatrix}-\frac{4}{5}\frac{1}{x_1^2} & 0\\ 0 & -\frac{1}{5}\frac{1}{x_2^2}\end{bmatrix}.$$

This is negative definite since it is a diagonal matrix with negative elements on the diagonal. Thus $f(x_1,x_2)$ is globally concave so that $x_1^*=\frac{4}{25}, x_2^*=\frac{1}{20}$ is a global maximizer.

Problem 4.23 *Consider*

$$f(x_1,x_2)=-\ln(x_1)-\frac{x_2^2-x_2+1}{x_1}\ \textit{for}\ 0<x_1<\frac{3}{2}.$$

*[B***] Find the first-order conditions and solve. Determine if the function is locally concave or convex at the solution to the first-order conditions and whether this solution is a local maximizer or minimizer. [A***] Show that $f(x_1,x_2)$ is globally concave (here you need to use $0<x_1<\frac{3}{2}$) and that the solution is a global maximizer.*

Answer: We have

$$\frac{\partial f(x_1,x_2)}{\partial x_1} = -\frac{1}{x_1}+\frac{x_2^2-x_2+1}{x_1^2}$$
$$\frac{\partial f(x_1,x_2)}{\partial x_2} = -\frac{2x_2-1}{x_1}$$

so the first-order conditions yield $x_1^*=\frac{3}{4}, x_2^*=\frac{1}{2}$ as

$$-\frac{1}{x_1^*}+\frac{(x_2^*)^2-x_2^*+1}{x_1^{*2}} = 0$$
$$\implies x_1^*=(x_2^*)^2-x_2^*+1$$
$$-\frac{2x_2^*-1}{x_1^*} = 0$$
$$\implies x_2^*=\frac{1}{2}\implies x_1^*=\left(\frac{1}{2}\right)^2-\frac{1}{2}+1=\frac{3}{4}.$$

The second derivatives are

$$\frac{\partial^2 f(x_1,x_2)}{\partial x_1^2}=\frac{1}{x_1^2}-2\frac{x_2^2-x_2+1}{x_1^3},\ \frac{\partial^2 f(x_1,x_2)}{\partial x_1\partial x_2}=\frac{2x_2-1}{x_1^2},\ \frac{\partial^2 f(x_1,x_2)}{\partial x_2^2}=-\frac{2}{x_1}$$

so the Hessian is

$$H(x_1,x_2)=\begin{bmatrix}\frac{1}{x_1^2}-2\frac{x_2^2-x_2+1}{x_1^3} & \frac{2x_2-1}{x_1^2}\\ \frac{2x_2-1}{x_1^2} & -\frac{2}{x_1}\end{bmatrix}.$$

At $x_1^*=\frac{3}{4}, x_2^*=\frac{1}{2}$ we have

$$H\left(\frac{3}{4},\frac{1}{2}\right)=\begin{bmatrix}-\frac{16}{9} & 0\\ 0 & -\frac{8}{3}\end{bmatrix}$$

which is negative definite so $x_1^*=\frac{3}{4}, x_2^*=\frac{1}{2}$ is a local maximizer. In Problem 4.5 on page 176 the function is shown to be globally concave and so $x_1^*=\frac{3}{4}, x_2^*=\frac{1}{2}$ is a global maximizer.

4.2.1 Profit Maximization

Problem 4.24 *[D***] Suppose a competitive firm has a production function*

$$Q=F(L,K)=4L^{\frac{1}{4}}K^{\frac{1}{4}}$$

with $P=36, W=9, R=4$. Find the first-order conditions for profit maximization and solve. Show that your solution is a global profit maximizer.

Answer: Using $Q = 4L^{\frac{1}{4}}K^{\frac{1}{4}}$ in the accounting definition of profits

$$\pi \equiv PQ - WL - RK$$

gives profits as a function of L, K as

$$\pi(L,K) = P \times 4L^{\frac{1}{4}}K^{\frac{1}{4}} - WL - RK = 36 \times 4L^{\frac{1}{4}}K^{\frac{1}{4}} - 9L - 4K.$$

We have

$$\begin{aligned}\frac{\partial \pi(L,K)}{\partial L} &= 36L^{-\frac{3}{4}}K^{\frac{1}{4}} - 9 \\ \frac{\partial \pi(L,K)}{\partial K} &= 36L^{\frac{1}{4}}K^{-\frac{3}{4}} - 4\end{aligned}$$

so that the first-order conditions yield

$$\begin{aligned}36(L^*)^{-\frac{3}{4}}(K^*)^{\frac{1}{4}} - 9 &= 0 \\ &\implies (L^*)^{-\frac{3}{4}}(K^*)^{\frac{1}{4}} = \frac{1}{4} \\ &\implies -\frac{3}{4}\ln(L^*) + \frac{1}{4}\ln(K^*) = -2\ln(2) \\ 36(L^*)^{\frac{1}{4}}(K^*)^{-\frac{3}{4}} - 4 &= 0 \\ &\implies (L^*)^{\frac{1}{4}}(K^*)^{-\frac{3}{4}} = \frac{1}{9} \\ &\implies \frac{1}{4}\ln(L^*) - \frac{3}{4}\ln(K^*) = -2\ln(3).\end{aligned}$$

Combining these two results and using matrix notation we find that

$$\begin{aligned}\begin{bmatrix} -\frac{3}{4} & \frac{1}{4} \\ \frac{1}{4} & -\frac{3}{4}\end{bmatrix}\begin{bmatrix}\ln(L^*) \\ \ln(K^*)\end{bmatrix} &= \begin{bmatrix}-2\ln(2) \\ -2\ln(3)\end{bmatrix} \\ \det\begin{bmatrix} -\frac{3}{4} & \frac{1}{4} \\ \frac{1}{4} & -\frac{3}{4}\end{bmatrix} &= \frac{1}{2}\end{aligned}$$

so that Cramer's rule yields

$$\begin{aligned}\ln(L^*) &= \frac{\det\begin{bmatrix}-2\ln(2) & \frac{1}{4} \\ -2\ln(3) & -\frac{3}{4}\end{bmatrix}}{\frac{1}{2}} = 3\ln(2) + \ln(3) \\ &\implies L^* = e^{\ln(L^*)} = e^{3\ln(2)+\ln(3)} = 24 \\ \ln(K^*) &= \frac{\det\begin{bmatrix}-\frac{3}{4} & -2\ln(2) \\ \frac{1}{4} & -2\ln(3)\end{bmatrix}}{\frac{1}{2}} = 3\ln(3) + \ln(2) \\ &\implies K^* = e^{\ln(K^*)} = e^{3\ln(3)+\ln(2)} = 54.\end{aligned}$$

Here is another method for solving for L^*, K^*. Taking the ratio of both sides of

$$(L^*)^{-\frac{3}{4}}(K^*)^{\frac{1}{4}} = \frac{1}{4} \text{ and } (L^*)^{\frac{1}{4}}(K^*)^{-\frac{3}{4}} = \frac{1}{9}$$

we have

$$\begin{aligned} \frac{(L^*)^{-\frac{3}{4}}(K^*)^{\frac{1}{4}}}{(L^*)^{\frac{1}{4}}(K^*)^{-\frac{3}{4}}} &= \frac{K^*}{L^*} = \frac{\frac{1}{4}}{\frac{1}{9}} = 9 \times 4^{-1} \\ &\Longrightarrow K^* = 9 \times 4^{-1} \times L^*. \end{aligned}$$

Now

$$\begin{aligned} (L^*)^{\frac{1}{4}}(K^*)^{-\frac{3}{4}} &= \frac{1}{9} \\ &\Longrightarrow (L^*)^{\frac{1}{4}}\left(9 \times 4^{-1} \times L^*\right)^{-\frac{3}{4}} = 9^{-1} \\ &\Longrightarrow (L^*)^{-\frac{1}{2}} = 9^{-1}\left(9 \times 4^{-1}\right)^{\frac{3}{4}} \\ &\Longrightarrow L^* = \left(9^{-1}\left(9 \times 4^{-1}\right)^{\frac{3}{4}}\right)^{-2} = 24 \\ &\Longrightarrow K^* = 9 \times 4^{-1} L^* = 9 \times 4^{-1} \times 24 = 54. \end{aligned}$$

From

$$\frac{\partial \pi(L,K)}{\partial L} = 36L^{-\frac{3}{4}}K^{\frac{1}{4}} - 9, \quad \frac{\partial \pi(L,K)}{\partial K} = 36L^{\frac{1}{4}}K^{-\frac{3}{4}} - 4$$

we have

$$\frac{\partial^2 \pi(L,K)}{\partial L^2} = -27L^{-\frac{7}{4}}K^{\frac{1}{4}}, \frac{\partial^2 \pi(L,K)}{\partial L \partial K} = 9L^{-\frac{3}{4}}K^{-\frac{3}{4}}, \frac{\partial^2 \pi(L,K)}{\partial K^2} = -27L^{\frac{1}{4}}K^{-\frac{7}{4}}$$

so that the Hessian is

$$H(L,K) = \begin{bmatrix} -27L^{-\frac{7}{4}}K^{\frac{1}{4}} & 9L^{-\frac{3}{4}}K^{-\frac{3}{4}} \\ 9L^{-\frac{3}{4}}K^{-\frac{3}{4}} & -27L^{\frac{1}{4}}K^{-\frac{7}{4}} \end{bmatrix}.$$

This is negative definite for all L, K since

$$\begin{aligned} M_1 &= -27L^{-\frac{7}{4}}K^{\frac{1}{4}} < 0 \\ M_2 &= \det \begin{bmatrix} -27L^{-\frac{7}{4}}K^{\frac{1}{4}} & 9L^{-\frac{3}{4}}K^{-\frac{3}{4}} \\ 9L^{-\frac{3}{4}}K^{-\frac{3}{4}} & -27L^{\frac{1}{4}}K^{-\frac{7}{4}} \end{bmatrix} \\ &= 729L^{-\frac{6}{4}}K^{-\frac{6}{4}} - 81L^{-\frac{6}{4}}K^{-\frac{6}{4}} = 648L^{-\frac{6}{4}}K^{-\frac{6}{4}} > 0. \end{aligned}$$

Therefore $L^* = 24, K^* = 54$ is a global maximizer.

Problem 4.25 *[C***] Suppose that a competitive firm has a production function given by* $Q = 5L^{\frac{2}{5}}K^{\frac{1}{5}}$ *with nominal prices* P, W, R *and real prices* $w \equiv \frac{W}{P}, r = \frac{R}{P}$. *Find the profit maximizing* L^*, K^*. *Show that* $\frac{\partial L^*}{\partial w} < 0, \frac{\partial K^*}{\partial r} < 0$, *and that* $\frac{\partial L^*}{\partial r} = \frac{\partial K^*}{\partial w} < 0$.

Answer: From

$$\pi(L,K) = P \times 5L^{\frac{2}{5}}K^{\frac{1}{5}} - WL - RK$$

the first-order conditions yield

$$\begin{aligned}
2P(L^*)^{-\frac{3}{5}}(K^*)^{\frac{1}{5}} - W &= 0 \\
&\implies (L^*)^{-\frac{3}{5}}(K^*)^{\frac{1}{5}} = \frac{W}{2P} = \frac{w}{2} \\
&\implies -\frac{3}{5}\ln(L^*) + \frac{1}{5}\ln(K^*) = \ln\left(\frac{w}{2}\right) = \ln(w) - \ln(2) \\
P(L^*)^{\frac{2}{5}}(K^*)^{-\frac{4}{5}} - R &= 0 \\
&\implies (L^*)^{\frac{2}{5}}(K^*)^{-\frac{4}{5}} = \frac{R}{P} = r \\
&\implies \frac{2}{5}\ln(L^*) - \frac{4}{5}\ln(K^*) = \ln(r).
\end{aligned}$$

In matrix notation we have

$$\begin{aligned}
\begin{bmatrix} -\frac{3}{5} & \frac{1}{5} \\ \frac{2}{5} & -\frac{4}{5} \end{bmatrix}\begin{bmatrix} \ln(L^*) \\ \ln(K^*) \end{bmatrix} &= \begin{bmatrix} \ln(w) - \ln(2) \\ \ln(r) \end{bmatrix} \\
\det\begin{bmatrix} -\frac{3}{5} & \frac{1}{5} \\ \frac{2}{5} & -\frac{4}{5} \end{bmatrix} &= \frac{2}{5}
\end{aligned}$$

so that Cramer's rule yields

$$\begin{aligned}
\ln(L^*) &= \frac{\det\begin{bmatrix} \ln(w) - \ln(2) & \frac{1}{5} \\ \ln(r) & -\frac{4}{5} \end{bmatrix}}{\frac{2}{5}} \\
&= -2\ln(w) - \frac{1}{2}\ln(r) + 2\ln(2) \\
\ln(K^*) &= \frac{\det\begin{bmatrix} -\frac{3}{5} & \ln(w) - \ln(2) \\ \frac{2}{5} & \ln(r) \end{bmatrix}}{\frac{2}{5}} \\
&= -\ln(w) - \frac{3}{2}\ln(r) + \ln(2).
\end{aligned}$$

Thus

$$\begin{aligned}
L^* &= e^{\ln(L^*)} = e^{-2\ln(w) - \frac{1}{2}\ln(r) + 2\ln(2)} = 4w^{-2}r^{-\frac{1}{2}} \\
K^* &= e^{\ln(K^*)} = e^{-\ln(w) - \frac{3}{2}\ln(r) + \ln(2)} = 2w^{-1}r^{-\frac{3}{2}}.
\end{aligned}$$

Here is another method for solving for L^*, K^*. Taking the ratio of both sides of

$$(L^*)^{-\frac{3}{5}}(K^*)^{\frac{1}{5}} = \frac{1}{2}w \text{ and } (L^*)^{\frac{2}{5}}(K^*)^{-\frac{4}{5}} = r$$

we find that

$$\begin{aligned}
\frac{(L^*)^{-\frac{3}{5}}(K^*)^{\frac{1}{5}}}{(L^*)^{\frac{2}{5}}(K^*)^{-\frac{4}{5}}} &= \frac{K^*}{L^*} = \frac{\frac{1}{2}w}{r} = 2^{-1}wr^{-1} \\
&\implies K^* = 2^{-1}wr^{-1}L^*.
\end{aligned}$$

Now

$$\begin{aligned}
(L^*)^{\frac{2}{5}}(K^*)^{-\frac{4}{5}} &= r \\
&\implies (L^*)^{\frac{2}{5}}\left(2^{-1}wr^{-1}L^*\right)^{-\frac{4}{5}} = r \\
&\implies 2^{\frac{4}{5}}(L^*)^{-\frac{2}{5}}w^{-\frac{4}{5}}r^{\frac{4}{5}} = r \\
&\implies (L^*)^{-\frac{2}{5}} = 2^{-\frac{4}{5}}w^{\frac{4}{5}}r^{\frac{1}{5}} \\
&\implies L^* = 4w^{-2}r^{-\frac{1}{2}} \\
&\implies K^* = 2^{-1}wr^{-1}L^* = 2^{-1}wr^{-1}4w^{-2}r^{-\frac{1}{2}} = 2w^{-1}r^{-\frac{3}{2}}.
\end{aligned}$$

The Hessian of the profit function

$$\pi(L,K) = P \times 5L^{\frac{2}{5}}K^{\frac{1}{5}} - WL - RK$$

is

$$H(L,K) = \begin{bmatrix} -\frac{6P}{5}L^{-\frac{8}{5}}K^{\frac{1}{5}} & \frac{2P}{5}L^{-\frac{3}{5}}K^{-\frac{4}{5}} \\ \frac{2P}{5}L^{-\frac{3}{5}}K^{-\frac{4}{5}} & -\frac{4P}{5}L^{\frac{2}{5}}K^{-\frac{9}{5}} \end{bmatrix}.$$

Using principal minors we find that

$$\begin{aligned}
M_1 &= -\frac{6P}{5}L^{-\frac{8}{5}}K^{\frac{1}{5}} < 0 \\
M_2 &= \left(-\frac{6P}{5}L^{-\frac{8}{5}}K^{\frac{1}{5}}\right)\left(-\frac{4P}{5}L^{\frac{2}{5}}K^{-\frac{9}{5}}\right) - \left(\frac{2P}{5}L^{-\frac{3}{5}}K^{-\frac{4}{5}}\right)^2 = \frac{20P^2}{25}L^{-\frac{6}{5}}K^{-\frac{8}{5}} > 0
\end{aligned}$$

so $H(L,K)$ is negative definite for all L, K, and so L^*, K^* is a global profit maximizer.

The demand curves slope downwards as

$$\frac{\partial L^*}{\partial w} = -8w^{-3}r^{-\frac{1}{2}} < 0 \text{ and } \frac{\partial K^*}{\partial r} = -3w^{-1}r^{-\frac{5}{2}} < 0.$$

Also

$$\frac{\partial L^*}{\partial r} = \frac{\partial K^*}{\partial w} = -2w^{-2}r^{-\frac{3}{2}} < 0.$$

Problem 4.26 *[B***] Suppose the production function takes the form* $Q = 4L^{\frac{1}{4}}K^{\frac{1}{2}}$. *Find the first-order conditions for profit maximization and solve. Show that you have a global maximum. Show that* $\frac{\partial L^*}{\partial w} < 0$, $\frac{\partial K^*}{\partial r} < 0$, *and that* $\frac{\partial L^*}{\partial r} = \frac{\partial K^*}{\partial w} < 0$. *Calculate the elasticity of the labour demand curve.*

Answer: From

$$\pi(L,K) = P \times 4L^{\frac{1}{4}}K^{\frac{1}{2}} - WL - RK$$

the first-order conditions yield

$$\begin{aligned}
P\left(L^{*}\right)^{-\frac{3}{4}}\left(K^{*}\right)^{\frac{1}{2}}-W &= 0 \\
&\Longrightarrow \left(L^{*}\right)^{-\frac{3}{4}}\left(K^{*}\right)^{\frac{1}{2}}=w \\
&\Longrightarrow -\frac{3}{4}\ln\left(L^{*}\right)+\frac{1}{2}\ln\left(K^{*}\right)=\ln\left(w\right) \\
2P\left(L^{*}\right)^{\frac{1}{4}}\left(K^{*}\right)^{-\frac{1}{2}}-R &= 0 \\
&\Longrightarrow \left(L^{*}\right)^{\frac{1}{4}}\left(K^{*}\right)^{-\frac{1}{2}}=2^{-1}r \\
&\Longrightarrow \frac{1}{4}\ln\left(L^{*}\right)-\frac{1}{2}\ln\left(K^{*}\right)=\ln\left(r\right)-\ln\left(2\right).
\end{aligned}$$

In matrix notation we have

$$\begin{aligned}
\begin{bmatrix} -\frac{3}{4} & \frac{1}{2} \\ \frac{1}{4} & -\frac{1}{2} \end{bmatrix}\begin{bmatrix} \ln\left(L^{*}\right) \\ \ln\left(K^{*}\right) \end{bmatrix} &= \begin{bmatrix} \ln\left(w\right) \\ \ln\left(r\right)-\ln\left(2\right) \end{bmatrix} \\
\det\begin{bmatrix} -\frac{3}{4} & \frac{1}{2} \\ \frac{1}{4} & -\frac{1}{2} \end{bmatrix} &= \frac{1}{4}
\end{aligned}$$

so that Cramer's rule yields

$$\begin{aligned}
\ln\left(L^{*}\right) &= \frac{\det\begin{bmatrix} \ln\left(w\right) & \frac{1}{2} \\ \ln\left(r\right)-\ln\left(2\right) & -\frac{1}{2} \end{bmatrix}}{\frac{1}{4}} \\
&= -2\ln\left(w\right)-2\ln\left(r\right)+2\ln\left(2\right) \\
\ln\left(K^{*}\right) &= \frac{\det\begin{bmatrix} -\frac{3}{4} & \ln\left(w\right) \\ \frac{1}{4} & \ln\left(r\right)-\ln\left(2\right) \end{bmatrix}}{\frac{1}{4}} \\
&= -3\ln\left(r\right)-\ln\left(w\right)+3\ln\left(2\right).
\end{aligned}$$

Thus

$$\begin{aligned}
L^{*} &= e^{\ln\left(L^{*}\right)}=e^{-2\ln(w)-2\ln(r)+2\ln(2)}=4w^{-2}r^{-2} \\
K^{*} &= e^{\ln\left(K^{*}\right)}=e^{-3\ln(r)-\ln(w)+3\ln(2)}=8w^{-1}r^{-3}.
\end{aligned}$$

The elasticity of labour demand is $-2<0$ and so the demand for labour is downward sloping.

Here is another method for solving for L^{*},K^{*}. Taking the ratio of both sides of

$$\left(L^{*}\right)^{-\frac{3}{4}}\left(K^{*}\right)^{\frac{1}{2}}=w \text{ and } \left(L^{*}\right)^{\frac{1}{4}}\left(K^{*}\right)^{-\frac{1}{2}}=2^{-1}r$$

we find that

$$\begin{aligned}
\frac{\left(L^{*}\right)^{-\frac{3}{4}}\left(K^{*}\right)^{\frac{1}{2}}}{\left(L^{*}\right)^{\frac{1}{4}}\left(K^{*}\right)^{-\frac{1}{2}}} &= \frac{K^{*}}{L^{*}}=\frac{w}{2^{-1}r}=2wr^{-1} \\
&\Longrightarrow K^{*}=2wr^{-1}L^{*}.
\end{aligned}$$

We have

$$\begin{aligned}
(L^*)^{-\frac{3}{4}}(K^*)^{\frac{1}{2}} &= w \\
&\Longrightarrow (L^*)^{-\frac{3}{4}}\left(2wr^{-1}L^*\right)^{\frac{1}{2}} = w \\
&\Longrightarrow (L^*)^{-\frac{1}{4}}2^{\frac{1}{2}}w^{\frac{1}{2}}r^{-\frac{1}{2}} = w \\
&\Longrightarrow (L^*)^{-\frac{1}{4}} = 2^{-\frac{1}{2}}w^{\frac{1}{2}}r^{\frac{1}{2}} \\
&\Longrightarrow L^* = 4w^{-2}r^{-2}
\end{aligned}$$

and

$$K^* = 2wr^{-1}L^* = 2wr^{-1}4w^{-2}r^{-2} = 8w^{-1}r^{-3}.$$

The Hessian of the profit function

$$\pi(L,K) = P \times 4L^{\frac{1}{4}}K^{\frac{1}{2}} - WL - RK$$

is

$$H(L,K) = \begin{bmatrix} -\frac{3P}{4}L^{-\frac{7}{4}}K^{\frac{1}{2}} & \frac{P}{2}L^{-\frac{3}{4}}K^{-\frac{1}{2}} \\ \frac{P}{2}L^{-\frac{3}{4}}K^{-\frac{1}{2}} & -PL^{\frac{1}{4}}K^{-\frac{3}{2}} \end{bmatrix}.$$

Using principal minors we find that

$$\begin{aligned}
M_1 &= -\frac{3P}{4}L^{-\frac{7}{4}}K^{\frac{1}{2}} < 0 \\
M_2 &= \left(-\frac{3P}{4}L^{-\frac{7}{4}}K^{\frac{1}{2}}\right)\left(-PL^{\frac{1}{4}}K^{-\frac{3}{2}}\right) - \left(\frac{P}{2}L^{-\frac{3}{4}}K^{-\frac{1}{2}}\right)^2 = \frac{2P^2}{4}L^{-\frac{3}{2}}K^{-1} > 0
\end{aligned}$$

so $H(L,K)$ is negative definite for all L, K, and L^*, K^* is a global profit maximizer.

The demand curves slope downwards as

$$\frac{\partial L^*}{\partial w} = -8w^{-3}r^{-2} < 0 \text{ and } \frac{\partial K^*}{\partial r} = -24w^{-1}r^{-4} < 0.$$

As well

$$\frac{\partial L^*}{\partial r} = \frac{\partial K^*}{\partial w} = -8w^{-2}r^{-3} < 0.$$

The elasticity of the demand for labour is -2, the exponent on w in $L^* = 4w^{-2}r^{-2}$.

Problem 4.27 *[A ***] Suppose that a perfectly competitive firm has a production function given by*

$$Q = L^{\alpha}K^{\beta}$$

where $0 < \alpha < 1$, $0 < \beta < 1$, and $\alpha + \beta < 1$. Find the first-order conditions for profit maximization and solve. Show that this is a profit maximizer. Show that $\frac{\partial L^*}{\partial w} < 0$, $\frac{\partial K^*}{\partial r} < 0$. Calculate the elasticity of the labour demand curve.

Answer: The profit function is

$$\pi(L,K) = PL^{\alpha}K^{\beta} - WL - RK$$

so that

$$\begin{aligned}
\frac{\partial \pi(L,K)}{\partial L} &= \alpha PL^{\alpha-1}K^{\beta} - W \\
\frac{\partial \pi(L,K)}{\partial K} &= \beta PL^{\alpha}K^{\beta-1} - R
\end{aligned}$$

and the first-order conditions yield

$$\begin{aligned}
\alpha P(L^*)^{\alpha-1}(K^*)^{\beta} - W &= 0 \\
&\implies (L^*)^{\alpha-1}(K^*)^{\beta} = \frac{W}{\alpha P} = \frac{w}{\alpha} \\
&\implies (\alpha-1)\ln(L^*) + \beta\ln(K^*) = \ln\left(\frac{w}{\alpha}\right) \\
\beta P(L^*)^{\alpha}(K^*)^{\beta-1} - R &= 0 \\
&\implies (L^*)^{\alpha}(K^*)^{\beta-1} = \frac{R}{\beta P} = \frac{r}{\beta} \\
&\implies \alpha\ln(L^*) + (\beta-1)\ln(K^*) = \ln\left(\frac{r}{\beta}\right).
\end{aligned}$$

In matrix notation we have

$$\begin{bmatrix} \alpha-1 & \beta \\ \alpha & \beta-1 \end{bmatrix}\begin{bmatrix} \ln(L^*) \\ \ln(K^*) \end{bmatrix} = \begin{bmatrix} \ln\left(\frac{w}{\alpha}\right) \\ \ln\left(\frac{r}{\beta}\right) \end{bmatrix}$$

so that using

$$\det\begin{bmatrix} \alpha-1 & \beta \\ \alpha & \beta-1 \end{bmatrix} = (\alpha-1)(\beta-1) - \beta\alpha = 1-\alpha-\beta$$

and Cramer's rule we have

$$\begin{aligned}
\ln(L^*) &= \frac{\det\begin{bmatrix} \ln\left(\frac{w}{\alpha}\right) & \beta \\ \ln\left(\frac{r}{\beta}\right) & (\beta-1) \end{bmatrix}}{1-\alpha-\beta} = \frac{(\beta-1)\ln\left(\frac{w}{\alpha}\right) - \beta\ln\left(\frac{r}{\beta}\right)}{1-\alpha-\beta} \\
\ln(K^*) &= \frac{\det\begin{bmatrix} (\alpha-1) & \ln\left(\frac{w}{\alpha}\right) \\ \alpha & \ln\left(\frac{r}{\beta}\right) \end{bmatrix}}{1-\alpha-\beta} = \frac{(\alpha-1)\ln\left(\frac{r}{\beta}\right) - \alpha\ln\left(\frac{w}{\alpha}\right)}{1-\alpha-\beta}.
\end{aligned}$$

Thus

$$\begin{aligned}
L^* &= e^{\ln(L^*)} = \left(\frac{w}{\alpha}\right)^{\frac{(\beta-1)}{1-\alpha-\beta}}\left(\frac{r}{\beta}\right)^{\frac{-\beta}{1-\alpha-\beta}} = \alpha^{\frac{1-\beta}{1-\alpha-\beta}}\beta^{\frac{\beta}{1-\alpha-\beta}}w^{\frac{\beta-1}{1-\alpha-\beta}}r^{\frac{-\beta}{1-\alpha-\beta}} \\
K^* &= e^{\ln(K^*)} = \left(\frac{w}{\alpha}\right)^{\frac{-\alpha}{1-\alpha-\beta}}\left(\frac{r}{\beta}\right)^{\frac{(\alpha-1)}{1-\alpha-\beta}} = \alpha^{\frac{\alpha}{1-\alpha-\beta}}\beta^{\frac{1-\alpha}{1-\alpha-\beta}}w^{\frac{-\alpha}{1-\alpha-\beta}}r^{\frac{\alpha-1}{1-\alpha-\beta}}.
\end{aligned}$$

The Hessian of the profit function is

$$H(L,K) = \begin{bmatrix} \alpha(\alpha-1) PL^{\alpha-2}K^{\beta} & \alpha\beta PL^{\alpha-1}K^{\beta-1} \\ \alpha\beta PL^{\alpha-1}K^{\beta-1} & \beta(\beta-1) PL^{\alpha}K^{\beta-2} \end{bmatrix}.$$

Using principal minors we find that

$$\begin{aligned} M_1 &= \alpha(\alpha-1) PL^{\alpha-2}K^{\beta} < 0 \\ M_2 &= \left(\alpha(\alpha-1) PL^{\alpha-2}K^{\beta}\right)\left(\beta(\beta-1) PL^{\alpha}K^{\beta-2}\right) - \left(\alpha\beta PL^{\alpha-1}K^{\beta-1}\right)^2 \\ &= \alpha\beta P^2 L^{2\alpha-2}K^{2\beta-2}\left(1-(\alpha+\beta)\right) > 0 \end{aligned}$$

so $H(L,K)$ is negative definite for all L, K, and L^*, K^* is a global profit maximizer.

The demand curve for labour slopes downwards since

$$\frac{\partial L^*}{\partial w} = \frac{\beta-1}{1-\alpha-\beta}\alpha^{\frac{1-\beta}{1-\alpha-\beta}}\beta^{\frac{\beta}{1-\alpha-\beta}}w^{\frac{\beta-1}{1-\alpha-\beta}-1}r^{\frac{-\beta}{1-\alpha-\beta}} < 0$$

since $\beta < 1, \alpha+\beta < 1$ implies that $\frac{\beta-1}{1-\alpha-\beta} < 0$. Here $\frac{\beta-1}{1-\alpha-\beta} < 0$ is the elasticity of the labour demand curve. The demand curve for capital slopes downwards since

$$\frac{\partial K^*}{\partial r} = \frac{\alpha-1}{1-\alpha-\beta}\alpha^{\frac{\alpha}{1-\alpha-\beta}}\beta^{\frac{1-\alpha}{1-\alpha-\beta}}w^{\frac{-\alpha}{1-\alpha-\beta}}r^{\frac{\alpha-1}{1-\alpha-\beta}-1} < 0$$

since $\alpha < 1, \alpha+\beta < 1$ implies that $\frac{\alpha-1}{1-\alpha-\beta} < 0$.

Problem 4.28 *[C***] Suppose that the firm's production function is given by*

$$Q = F(L,K) = 3L^{\frac{1}{3}} + 6K^{\frac{1}{6}}.$$

*Find the first-order conditions for profit maximization and solve. Find the firm's supply curve Q^* as a function of P, W, R. Show that the demand curve for labour is downward sloping and the supply curve is upward sloping. [B***] If w increases what happens to K^* and Q^*?*

Answer: The profit function is

$$\pi(L,K) = P\left(3L^{\frac{1}{3}} + 6K^{\frac{1}{6}}\right) - WL - RK$$

so that

$$\begin{aligned} \frac{\partial \pi(L,K)}{\partial L} &= PL^{-\frac{2}{3}} - W \\ \frac{\partial \pi(L,K)}{\partial K} &= PK^{-\frac{5}{6}} - R \end{aligned}$$

and the first-order conditions yield the labour and capital demand curves $L^* = w^{-\frac{3}{2}}, K^* = r^{-\frac{6}{5}}$ as

$$\begin{aligned} P(L^*)^{-\frac{2}{3}} - W &= 0 \\ &\implies (L^*)^{-\frac{2}{3}} = \frac{W}{P} = w \\ &\implies L^* = w^{-\frac{3}{2}} \\ P(K^*)^{-\frac{5}{6}} - R &= 0 \\ &\implies (K^*)^{-\frac{5}{6}} = \frac{R}{P} = r \\ &\implies K^* = r^{-\frac{6}{5}}. \end{aligned}$$

The Hessian of the profit function is

$$H(L,K) = \begin{bmatrix} -\frac{2P}{3}L^{-\frac{5}{3}} & 0 \\ 0 & -\frac{5P}{6}K^{-\frac{11}{6}} \end{bmatrix}.$$

Since this is a diagonal matrix with negative diagonal elements, it follows that $H(L,K)$ is negative definite for all L, K, and so $L^* = w^{-\frac{3}{2}}, K^* = r^{-\frac{6}{5}}$ is a global profit maximizer.

Labour demand slope downwards as

$$L^* = w^{-\frac{3}{2}} \implies \frac{\partial L^*}{\partial w} = -\frac{3}{2}w^{-\frac{5}{2}} < 0.$$

Note that w does not appear in K^* so that

$$\begin{aligned} K^* &= r^{-\frac{6}{5}} \\ &\implies \frac{\partial K^*}{\partial w} = 0. \end{aligned}$$

The supply curve is given by

$$\begin{aligned} Q^* &= 3(L^*)^{\frac{1}{3}} + 6(K^*)^{\frac{1}{6}} = 3\left(w^{-\frac{3}{2}}\right)^{\frac{1}{3}} + 6\left(r^{-\frac{6}{5}}\right)^{\frac{1}{6}} = 3w^{-\frac{1}{2}} + 6r^{-\frac{1}{5}} \\ \implies \frac{\partial Q^*}{\partial w} &= -\frac{3}{2}w^{-\frac{3}{2}} < 0. \end{aligned}$$

An increase in w causes Q^* to fall. Writing Q^* as a function of P, W, R yields

$$Q^* = 3w^{-\frac{1}{2}} + 6r^{-\frac{1}{5}} = 3\left(\frac{W}{P}\right)^{-\frac{1}{2}} + 6\left(\frac{R}{P}\right)^{-\frac{1}{5}} = 3P^{\frac{1}{2}}W^{-\frac{1}{2}} + 6P^{\frac{1}{5}}R^{-\frac{1}{5}}$$

so that the supply curve slopes upwards as

$$\frac{\partial Q^*}{\partial P} = \frac{3}{2}P^{-\frac{1}{2}}W^{-\frac{1}{2}} + \frac{6}{5}P^{-\frac{4}{5}}R^{-\frac{1}{5}} > 0.$$

Problem 4.29 *Consider a special case of the CES (constant elasticity of substitution) production function*

$$Q = F(L, K) = 2\left(L^{\frac{3}{4}} + K^{\frac{3}{4}}\right)^{\frac{2}{3}}.$$

*[B***] Find the first-order conditions and solve. Show that $\frac{\partial L^*}{\partial w} < 0$. Find the supply curve Q^* as a function of P, W, R and calculate the elasticity of supply.*

Answer: From

$$\pi(L, K) = P \times 2\left(L^{\frac{3}{4}} + K^{\frac{3}{4}}\right)^{\frac{2}{3}} - WL - RK$$

the first-order conditions yield

$$\begin{aligned}
P\left((L^*)^{\frac{3}{4}} + (K^*)^{\frac{3}{4}}\right)^{-\frac{1}{3}} (L^*)^{-\frac{1}{4}} - W &= 0 \\
&\implies \left((L^*)^{\frac{3}{4}} + (K^*)^{\frac{3}{4}}\right)^{-\frac{1}{3}} (L^*)^{-\frac{1}{4}} = w \\
P\left((L^*)^{\frac{3}{4}} + (K^*)^{\frac{3}{4}}\right)^{-\frac{1}{3}} (K^*)^{-\frac{1}{4}} - R &= 0 \\
&\implies \left((L^*)^{\frac{3}{4}} + (K^*)^{\frac{3}{4}}\right)^{-\frac{1}{3}} (K^*)^{-\frac{1}{4}} = r.
\end{aligned}$$

Taking the ratio of both sides we find that

$$\begin{aligned}
\frac{\left((L^*)^{\frac{3}{4}} + (K^*)^{\frac{3}{4}}\right)^{-\frac{1}{3}} (K^*)^{-\frac{1}{4}}}{\left((L^*)^{\frac{3}{4}} + (K^*)^{\frac{3}{4}}\right)^{-\frac{1}{3}} (L^*)^{-\frac{1}{4}}} &= \left(\frac{K^*}{L^*}\right)^{-\frac{1}{4}} = w^{-1}r \\
&\implies K^* = w^4 r^{-4} L^*.
\end{aligned}$$

We have

$$\begin{aligned}
\left((L^*)^{\frac{3}{4}} + (K^*)^{\frac{3}{4}}\right)^{-\frac{1}{3}} (L^*)^{-\frac{1}{4}} &= w \\
&\implies \left((L^*)^{\frac{3}{4}} + \left(w^4 r^{-4} L^*\right)^{\frac{3}{4}}\right)^{-\frac{1}{3}} (L^*)^{-\frac{1}{4}} = w
\end{aligned}$$

so that

$$\begin{aligned}
(L^*)^{-\frac{1}{4}} w^{-1} \left(w^{-3} + r^{-3}\right)^{-\frac{1}{3}} (L^*)^{-\frac{1}{4}} &= w \\
&\implies (L^*)^{-\frac{1}{2}} = w^2 \left(w^{-3} + r^{-3}\right)^{\frac{1}{3}} \\
&\implies L^* = \left(w^{-3} + r^{-3}\right)^{-\frac{2}{3}} w^{-4}
\end{aligned}$$

and

$$K^* = w^4 r^{-4} L^* = w^4 r^{-4} \left(w^{-3} + r^{-3}\right)^{-\frac{2}{3}} w^{-4} = \left(w^{-3} + r^{-3}\right)^{-\frac{2}{3}} r^{-4}.$$

To show that the labour demand curve slopes downwards we use

$$\begin{aligned} L^* &= \left(w^{-3}+r^{-3}\right)^{-\frac{2}{3}} w^{-4} = \left(w^{-3}+r^{-3}\right)^{-\frac{2}{3}} \left(w^{6}\right)^{-\frac{2}{3}} = \left(w^{6}\left(w^{-3}+r^{-3}\right)\right)^{-\frac{2}{3}} \\ &= \left(w^{3}+w^{6}r^{-3}\right)^{-\frac{2}{3}} \end{aligned}$$

so that

$$\frac{\partial L^*}{\partial w} = \frac{\partial}{\partial w}\left(w^{3}+w^{6}r^{-3}\right)^{-\frac{2}{3}} = -\frac{2}{3}\left(w^{3}+w^{6}r^{-3}\right)^{-\frac{5}{3}}\left(3w^{2}+6w^{5}r^{-3}\right) < 0.$$

The supply curve is found as

$$\begin{aligned} Q^* &= 2\left(\left(L^*\right)^{\frac{3}{4}}+\left(K^*\right)^{\frac{3}{4}}\right)^{\frac{2}{3}} \\ &= 2\left(\left(\left(w^{-3}+r^{-3}\right)^{-\frac{2}{3}} w^{-4}\right)^{\frac{3}{4}}+\left(\left(w^{-3}+r^{-3}\right)^{-\frac{2}{3}} r^{-4}\right)^{\frac{3}{4}}\right)^{\frac{2}{3}} \\ &= 2\left(\left(w^{-3}+r^{-3}\right)^{-\frac{1}{2}} w^{-3}+\left(w^{-3}+r^{-3}\right)^{-\frac{1}{2}} r^{-3}\right)^{\frac{2}{3}} \\ &= 2\left(\left(w^{-3}+r^{-3}\right)^{-\frac{1}{2}}\left(w^{-3}+r^{-3}\right)\right)^{\frac{2}{3}} \\ &= 2\left(\left(w^{-3}+r^{-3}\right)^{\frac{1}{2}}\right)^{\frac{2}{3}} = 2\left(w^{-3}+r^{-3}\right)^{\frac{1}{3}} \\ &= 2\left(\left(\frac{W}{P}\right)^{-3}+\left(\frac{R}{P}\right)^{-3}\right)^{\frac{1}{3}} = 2P\left(W^{-3}+R^{-3}\right)^{\frac{1}{3}} \end{aligned}$$

so that the elasticity of supply is the exponent on P or 1.

Problem 4.30 *Consider a special case of the CES (constant elasticity of substitution production function*

$$Q = F(L,K) = 3\left(L^{-1}+K^{-1}\right)^{-\frac{1}{3}}.$$

*[B***] Find the first-order conditions for profit maximization and solve. Find the supply curve Q^* as a function of P, W, R and calculate the elasticity of supply.*

Answer: From

$$\pi(L,K) = P \times 3\left(L^{-1}+K^{-1}\right)^{-\frac{1}{3}} - WL - RK$$

the first-order conditions yield

$$\begin{aligned} P\left(\left(L^*\right)^{-1}+\left(K^*\right)^{-1}\right)^{-\frac{4}{3}}\left(L^*\right)^{-2} - W &= 0 \\ &\implies \left(\left(L^*\right)^{-1}+\left(K^*\right)^{-1}\right)^{-\frac{4}{3}}\left(L^*\right)^{-2} = w \\ P\left(\left(L^*\right)^{-1}+\left(K^*\right)^{-1}\right)^{-\frac{4}{3}}\left(K^*\right)^{-2} - R &= 0 \\ &\implies \left(\left(L^*\right)^{-1}+\left(K^*\right)^{-1}\right)^{-\frac{4}{3}}\left(K^*\right)^{-2} = r. \end{aligned}$$

Taking the ratio of both sides we find that

$$\frac{\left((L^*)^{-1}+(K^*)^{-1}\right)^{-\frac{4}{3}}(K^*)^{-2}}{\left((L^*)^{-1}+(K^*)^{-1}\right)^{-\frac{4}{3}}(L^*)^{-2}} = \left(\frac{K^*}{L^*}\right)^{-2} = \frac{r}{w}$$

$$\Longrightarrow K^* = w^{\frac{1}{2}}r^{-\frac{1}{2}}L^*.$$

We have

$$\left((L^*)^{-1}+(K^*)^{-1}\right)^{-\frac{4}{3}}(L^*)^{-2} = w$$

$$\Longrightarrow \left((L^*)^{-1}+\left(w^{\frac{1}{2}}r^{-\frac{1}{2}}L^*\right)^{-1}\right)^{-\frac{4}{3}}(L^*)^{-2} = w$$

$$\Longrightarrow L^* = \left(w^{\frac{1}{2}}+r^{\frac{1}{2}}\right)^{-2}w^{-\frac{1}{2}}$$

and

$$K^* = w^{\frac{1}{2}}r^{-\frac{1}{2}}L^* = w^{\frac{1}{2}}r^{-\frac{1}{2}}\left(w^{\frac{1}{2}}+r^{\frac{1}{2}}\right)^{-2}w^{-\frac{1}{2}} = \left(w^{\frac{1}{2}}+r^{\frac{1}{2}}\right)^{-2}r^{-\frac{1}{2}}.$$

To find the supply curve use

$$\begin{aligned}
Q^* &= 3\left((L^*)^{-1}+(K^*)^{-1}\right)^{-\frac{1}{3}} \\
&= 3\left(\left(\left(w^{\frac{1}{2}}+r^{\frac{1}{2}}\right)^{-2}w^{-\frac{1}{2}}\right)^{-1}+\left(\left(w^{\frac{1}{2}}+r^{\frac{1}{2}}\right)^{-2}r^{-\frac{1}{2}}\right)^{-1}\right)^{-\frac{1}{3}} \\
&= 3\left(\left(w^{\frac{1}{2}}+r^{\frac{1}{2}}\right)^{2}\left(w^{\frac{1}{2}}+r^{\frac{1}{2}}\right)\right)^{-\frac{1}{3}} \\
&= 3\left(w^{\frac{1}{2}}+r^{\frac{1}{2}}\right)^{-1} = 3\left(\left(\frac{W}{P}\right)^{\frac{1}{2}}+\left(\frac{R}{P}\right)^{\frac{1}{2}}\right)^{-1} \\
&= 3P^{\frac{1}{2}}\left(W^{\frac{1}{2}}+R^{\frac{1}{2}}\right)^{-1}
\end{aligned}$$

so that the elasticity of supply is the exponent on P or $\frac{1}{2}$.

Problem 4.31 *Consider the CES production function*

$$Q = \frac{1}{\gamma}\left(L^{\rho}+K^{\rho}\right)^{\frac{\gamma}{\rho}}.$$

*[A ***] Find the first-order conditions for profit maximization and solve. Find the supply curve Q^* as a function of P, W, R, and show that the elasticity of supply is $\frac{\gamma}{1-\gamma}$.*

Answer: From

$$\pi(L,K) = P \times \frac{1}{\gamma}(L^{\rho} + K^{\rho})^{\frac{\gamma}{\rho}} - WL - RK$$

the first-order conditions yield

$$\begin{aligned} P\left((L^*)^{\rho} + (K^*)^{\rho}\right)^{\frac{\gamma-\rho}{\rho}}(L^*)^{\rho-1} - W &= 0 \\ &\Longrightarrow \left((L^*)^{\rho} + (K^*)^{\rho}\right)^{\frac{\gamma-\rho}{\rho}}(L^*)^{\rho-1} = w \\ P\left((L^*)^{\rho} + (K^*)^{\rho}\right)^{\frac{\gamma-\rho}{\rho}}(K^*)^{\rho-1} - R &= 0 \\ &\Longrightarrow \left((L^*)^{\rho} + (K^*)^{\rho}\right)^{\frac{\gamma-\rho}{\rho}}(K^*)^{\rho-1} = r. \end{aligned}$$

Taking the ratio of both sides we find that

$$\begin{aligned} \frac{\left((L^*)^{\rho} + (K^*)^{\rho}\right)^{\frac{\gamma-\rho}{\rho}}(K^*)^{\rho-1}}{\left((L^*)^{\rho} + (K^*)^{\rho}\right)^{\frac{\gamma-\rho}{\rho}}(L^*)^{\rho-1}} &= \left(\frac{K^*}{L^*}\right)^{\rho-1} = rw^{-1} \\ &\Longrightarrow K^* = w^{\frac{1}{1-\rho}} r^{-\frac{1}{1-\rho}} L^*. \end{aligned}$$

We then have

$$\begin{aligned} \left((L^*)^{\rho} + (K^*)^{\rho}\right)^{\frac{\gamma-\rho}{\rho}}(L^*)^{\rho-1} &= w \\ &\Longrightarrow \left((L^*)^{\rho} + \left(w^{\frac{1}{1-\rho}} r^{-\frac{1}{1-\rho}} L^*\right)^{\rho}\right)^{\frac{\gamma-\rho}{\rho}}(L^*)^{\rho-1} = w \end{aligned}$$

so that

$$\begin{aligned} \left((L^*)^{\rho}\left(1 + w^{\frac{\rho}{1-\rho}} r^{-\frac{\rho}{1-\rho}}\right)\right)^{\frac{\gamma-\rho}{\rho}}(L^*)^{\rho-1} &= w \\ &\Longrightarrow w^{\frac{\gamma-\rho}{1-\rho}}\left(w^{-\frac{\rho}{1-\rho}} + r^{-\frac{\rho}{1-\rho}}\right)^{\frac{\gamma-\rho}{\rho}}(L^*)^{\gamma-1} = w \\ &\Longrightarrow (L^*)^{1-\gamma} = w^{\frac{\gamma-\rho}{1-\rho}} w^{-1}\left(w^{-\frac{\rho}{1-\rho}} + r^{-\frac{\rho}{1-\rho}}\right)^{\frac{\gamma-\rho}{\rho}} \\ &\Longrightarrow L^* = \left(w^{-\frac{\rho}{1-\rho}} + r^{-\frac{\rho}{1-\rho}}\right)^{\frac{\gamma-\rho}{\rho(1-\gamma)}} w^{-\frac{1}{1-\rho}} \end{aligned}$$

and

$$\begin{aligned} K^* &= w^{\frac{1}{1-\rho}} r^{-\frac{1}{1-\rho}} L^* = w^{\frac{1}{1-\rho}} r^{-\frac{1}{1-\rho}}\left(w^{-\frac{\rho}{1-\rho}} + r^{-\frac{\rho}{1-\rho}}\right)^{\frac{\gamma-\rho}{\rho(1-\gamma)}} w^{-\frac{1}{1-\rho}} \\ &= \left(w^{-\frac{\rho}{1-\rho}} + r^{-\frac{\rho}{1-\rho}}\right)^{\frac{\gamma-\rho}{\rho(1-\gamma)}} r^{-\frac{1}{1-\rho}}. \end{aligned}$$

To find the supply curve use

$$\begin{aligned}
Q^* &= \frac{1}{\gamma}\left((L^*)^\rho + (K^*)^\rho\right)^{\frac{\gamma}{\rho}} \\
&= \frac{1}{\gamma}\left(\left(\left(w^{-\frac{\rho}{1-\rho}} + r^{-\frac{\rho}{1-\rho}}\right)^{\frac{\gamma-\rho}{\rho(1-\gamma)}} w^{-\frac{1}{1-\rho}}\right)^\rho + \left(\left(w^{-\frac{\rho}{1-\rho}} + r^{-\frac{\rho}{1-\rho}}\right)^{\frac{\gamma-\rho}{\rho(1-\gamma)}} r^{-\frac{1}{1-\rho}}\right)^\rho\right)^{\frac{\gamma}{\rho}} \\
&= \frac{1}{\gamma}\left(\left(w^{-\frac{\rho}{1-\rho}} + r^{-\frac{\rho}{1-\rho}}\right)^{\frac{\gamma-\rho}{1-\gamma}} w^{-\frac{\rho}{1-\rho}} + \left(w^{-\frac{\rho}{1-\rho}} + r^{-\frac{\rho}{1-\rho}}\right)^{\frac{\gamma-\rho}{1-\gamma}} r^{-\frac{\rho}{1-\rho}}\right)^{\frac{\gamma}{\rho}} \\
&= \frac{1}{\gamma}\left(\left(w^{-\frac{\rho}{1-\rho}} + r^{-\frac{\rho}{1-\rho}}\right)^{\frac{\gamma-\rho}{1-\gamma}} \left(w^{-\frac{\rho}{1-\rho}} + r^{-\frac{\rho}{1-\rho}}\right)\right)^{\frac{\gamma}{\rho}} = \frac{1}{\gamma}\left(\left(w^{-\frac{\rho}{1-\rho}} + r^{-\frac{\rho}{1-\rho}}\right)^{\frac{1-\rho}{1-\gamma}}\right)^{\frac{\gamma}{\rho}} \\
&= \frac{1}{\gamma}\left(w^{-\frac{\rho}{1-\rho}} + r^{-\frac{\rho}{1-\rho}}\right)^{\frac{1-\rho}{1-\gamma}\frac{\gamma}{\rho}} = \frac{1}{\gamma}\left(\left(\frac{W}{P}\right)^{-\frac{\rho}{1-\rho}} + \left(\frac{R}{P}\right)^{-\frac{\rho}{1-\rho}}\right)^{\frac{1-\rho}{1-\gamma}\frac{\gamma}{\rho}} \\
&= \frac{1}{\gamma}P^{\frac{\gamma}{1-\gamma}}\left(W^{-\frac{\rho}{1-\rho}} + R^{-\frac{\rho}{1-\rho}}\right)^{\frac{1-\rho}{1-\gamma}\frac{\gamma}{\rho}}
\end{aligned}$$

so that the elasticity of supply is the exponent on P or $\frac{\gamma}{1-\gamma}$.

4.3 Constrained Optimization

Problem 4.32 *[D***] Construct the Lagrangian for the problem of minimizing*

$$f(x_1, x_2) = x_1^2 + \frac{1}{2}x_2^2 - x_1 x_2$$

subject to the constraint $x_1 + x_2 = 1$. *[C***] Find the first-order conditions and solve. [B***] Verify that second-order conditions for a constrained global minimum are satisfied.*

Answer: We rewrite the constraint as

$$x_1 + x_2 = 1 \Longrightarrow 1 - x_1 - x_2 = 0$$

so that the Lagrangian is

$$\mathcal{L}(\lambda, x_1, x_2) = x_1^2 + \frac{1}{2}x_2^2 - x_1 x_2 + \lambda(1 - x_1 - x_2).$$

Differentiating the Lagrangian with respect to λ, x_1, x_2 results in

$$\begin{aligned}
\frac{\partial \mathcal{L}}{\partial \lambda} &= \frac{\partial}{\partial \lambda}\left(x_1^2 + \frac{1}{2}x_2^2 - x_1 x_2 + \lambda(1 - x_1 - x_2)\right) = 1 - x_1 - x_2 \\
\frac{\partial \mathcal{L}}{\partial x_1} &= \frac{\partial}{\partial x_1}\left(x_1^2 + \frac{1}{2}x_2^2 - x_1 x_2 + \lambda(1 - x_1 - x_2)\right) = 2x_1 - x_2 - \lambda \\
\frac{\partial \mathcal{L}}{\partial x_2} &= \frac{\partial}{\partial x_2}\left(x_1^2 + \frac{1}{2}x_2^2 - x_1 x_2 + \lambda(1 - x_1 - x_2)\right) = x_2 - x_1 - \lambda.
\end{aligned}$$

The first-order conditions yield

$$\begin{aligned} 1 - x_1^* - x_2^* &= 0 \\ &\Longrightarrow x_1^* + x_2^* = 1 \\ 2x_1^* - x_2^* - \lambda^* &= 0 \\ &\Longrightarrow 2x_1^* - x_2^* = \lambda^* \\ x_2^* - x_1^* - \lambda^* &= 0 \\ &\Longrightarrow x_2^* - x_1^* = \lambda^*. \end{aligned}$$

Combining the last two results yields

$$x_2^* - x_1^* = \lambda^* = 2x_1^* - x_2^* \Longrightarrow 3x_1^* = 2x_2^* \Longrightarrow x_1^* = \frac{2}{3}x_2^*.$$

Putting this in the constraint $x_1^* + x_2^* = 1$ yields

$$x_1^* + x_2^* = 1 \Longrightarrow \frac{2}{3}x_2^* + x_2^* = 1 \Longrightarrow \frac{5}{3}x_2^* = 1 \Longrightarrow x_2^* = \frac{3}{5}.$$

We have $x_1^* = \frac{2}{3}x_2^* = \frac{2}{3} \times \frac{3}{5} = \frac{2}{5}$. Finally

$$x_2^* - x_1^* = \lambda^* \Longrightarrow \frac{3}{5} - \frac{2}{5} = \lambda^* \Longrightarrow \lambda^* = \frac{1}{5}$$

so the solution to the first-order conditions is $x_1^* = \frac{2}{5}, x_2^* = \frac{3}{5}, \lambda^* = \frac{1}{5}$. Notice that $x_1^* = \frac{2}{5}, x_2^* = \frac{3}{5}$ satisfy the constraint as $x_1^* + x_2^* = \frac{1}{3} + \frac{2}{3} = 1$.

Now to verify that we have a constrained minimum. Since the constraint $1 - x_1 - x_2$ is linear, we calculate the Hessian of the objective function from

$$\begin{aligned} f(x_1, x_2) &= x_1^2 + \frac{1}{2}x_2^2 - x_1 x_2 \\ &\Longrightarrow \frac{\partial f(x_1, x_2)}{\partial x_1} = 2x_1 - x_2, \frac{\partial f(x_1, x_2)}{\partial x_2} = x_2 - x_1 \\ &\Longrightarrow \frac{\partial^2 f(x_1, x_2)}{\partial x_1^2} = 2, \frac{\partial^2 f(x_1, x_2)}{\partial x_2 \partial x_1} = -1, \frac{\partial^2 f(x_1, x_2)}{\partial x_2^2} = 1 \end{aligned}$$

as

$$\begin{aligned} H_f(x_1, x_2) &= \begin{bmatrix} 2 & -1 \\ -1 & 1 \end{bmatrix} \\ &\Longrightarrow M_1 = 2 > 0, M_2 = \det \begin{bmatrix} 2 & -1 \\ -1 & 1 \end{bmatrix} = 1 > 0. \end{aligned}$$

Thus $f(x_1, x_2)$ is globally convex. Since 1) $f(x_1, x_2)$ is globally convex and 2) the constraint is linear, it follows that $x_1^* = \frac{2}{5}, x_2^* = \frac{3}{5}, \lambda^* = \frac{1}{5}$ is a constrained global minimizer.

Problem 4.33 *[D***] Construct the Lagrangian for the problem of minimizing*

$$f(x_1, x_2) = 3x_1 + 2x_2, \ x_1 > 0, x_2 > 0$$

subject to the constraint $\ln(x_1) + \ln(x_2) = 1$. *[C***] Find the first-order conditions and solve. [B***] Verify that second-order conditions for a constrained global constrained minimum are satisfied.*

Answer: The Lagrangian is

$$\mathcal{L}(\lambda, x_1, x_2) = 3x_1 + 2x_2 + \lambda(1 - \ln(x_1) - \ln(x_2)).$$

We have

$$\begin{aligned}
\frac{\partial \mathcal{L}(\lambda, x_1, x_2)}{\partial \lambda} &= \frac{\partial}{\partial \lambda}(3x_1 + 2x_2 + \lambda(1 - \ln(x_1) - \ln(x_2))) = 1 - \ln(x_1) - \ln(x_2) \\
\frac{\partial \mathcal{L}(\lambda, x_1, x_2)}{\partial x_1} &= \frac{\partial}{\partial x_1}(3x_1 + 2x_2 + \lambda(1 - \ln(x_1) - \ln(x_2))) = 3 - \frac{\lambda}{x_1} \\
\frac{\partial \mathcal{L}(\lambda, x_1, x_2)}{\partial x_2} &= \frac{\partial}{\partial x_2}(x_1 + 2x_2 + \lambda(1 - \ln(x_1) - \ln(x_2))) = 2 - \frac{\lambda}{x_2}
\end{aligned}$$

and so the first-order conditions yield

$$\begin{aligned}
1 - \ln(x_1^*) - \ln(x_2^*) &= 0 \\
&\implies \ln(x_1^*) + \ln(x_2^*) = 1 \\
3 - \frac{\lambda^*}{x_1^*} &= 0 \\
&\implies x_1^* = \frac{\lambda^*}{3} \\
2 - \frac{\lambda^*}{x_2^*} &= 0 \\
&\implies x_2^* = \frac{\lambda^*}{2}.
\end{aligned}$$

To find λ^* substitute $x_1^* = \frac{\lambda^*}{3}, x_2^* = \frac{\lambda^*}{2}$ into the constraint as

$$\begin{aligned}
\ln(x_1^*) + \ln(x_2^*) &= 1 \\
&\implies \ln\left(\frac{\lambda^*}{3}\right) + \ln\left(\frac{\lambda^*}{2}\right) = 1 \\
&\implies \ln(\lambda^*) + \ln(\lambda^*) - \ln(6) = 1 \\
&\implies 2\ln(\lambda^*) = 1 + \ln(6) \\
&\implies \ln(\lambda^*) = \frac{1}{2} + \frac{1}{2}\ln(6) \\
&\implies \lambda^* = e^{\frac{1}{2} + \frac{1}{2}\ln(6)} = 6^{\frac{1}{2}} e^{\frac{1}{2}} \approx 4.04.
\end{aligned}$$

Now solve for x_1^*, x_2^* as

$$x_1^* = \frac{\lambda^*}{3} = \frac{6^{\frac{1}{2}} e^{\frac{1}{2}}}{3} \approx 1.35, \ x_2^* = \frac{\lambda^*}{2} = \frac{6^{\frac{1}{2}} e^{\frac{1}{2}}}{2} \approx 2.02.$$

To verify that this is a constrained global minimizer we note that

$$f(x_1, x_2) = 3x_1 + 2x_2$$

is linear and so we calculate the Hessian of the constraint from

$$\begin{aligned}
g(x_1, x_2) &= 1 - \ln(x_1) - \ln(x_2) \\
\frac{\partial g(x_1, x_2)}{\partial x_1} &= -\frac{1}{x_1} \\
&\Longrightarrow \frac{\partial^2 g(x_1, x_2)}{\partial x_1^2} = \frac{1}{x_1^2}, \frac{\partial^2 g(x_1, x_2)}{\partial x_2 \partial x_1} = 0 \\
\frac{\partial g(x_1, x_2)}{\partial x_2} &= -\frac{1}{x_2} \\
&\Longrightarrow \frac{\partial^2 g(x_1, x_2)}{\partial x_2 \partial x_1} = 0, \frac{\partial^2 g(x_1, x_2)}{\partial x_2^2} = \frac{1}{x_2^2}
\end{aligned}$$

as

$$H_g(x_1, x_2) = \begin{bmatrix} \frac{1}{x_1^2} & 0 \\ 0 & \frac{1}{x_2^2} \end{bmatrix}.$$

Since $H_g(x_1, x_2)$ is a diagonal matrix with positive diagonal elements, $H_g(x_1, x_2)$ is positive definite for all x_1, x_2 and so the constraint is globally convex. Since 1) the objective function is linear, 2) the constraint is globally convex, and 3) the Lagrange multiplier is positive as $\lambda^* = 6^{\frac{1}{2}} e^{\frac{1}{2}} > 0$, it follows that $\lambda^* = 2^{\frac{1}{2}} e^{\frac{1}{2}}$, $x_1^* = 2^{\frac{1}{2}} e^{\frac{1}{2}}$, $x_2^* = 2^{-\frac{1}{2}} e^{\frac{1}{2}}$ is a constrained global minimizer.

Problem 4.34 *[D***] Construct the Lagrangian for the problem of maximizing*

$$f(x_1, x_2) = 3\ln(x_1) + 2\ln(x_2) \ \ for\ x_1 > 0, x_2 > 0$$

subject to the constraint $x_1^2 + x_2^2 = 2$. *[C***] Find the first-order conditions and solve. [B***] Verify that second-order conditions for a constrained global maximum are satisfied.*

Answer: Here we write the constraint as

$$x_1^2 + x_2^2 = 2 \Longrightarrow \frac{1}{2}x_1^2 + \frac{1}{2}x_2^2 = 1 \Longrightarrow 1 - \frac{1}{2}x_1^2 - \frac{1}{2}x_2^2 = 0$$

which makes the mathematics slightly easier than writing it as $2 - x_1^2 - x_2^2 = 0$ which would still be correct. The Lagrangian then is

$$\mathcal{L}(\lambda, x_1, x_2) = 3\ln(x_1) + 2\ln(x_2) + \lambda\left(1 - \frac{1}{2}x_1^2 - \frac{1}{2}x_2^2\right).$$

From

$$\begin{aligned}
\frac{\partial \mathcal{L}(\lambda, x_1, x_2)}{\partial \lambda} &= 1 - \frac{1}{2}x_1^2 - \frac{1}{2}x_2^2 \\
\frac{\partial \mathcal{L}(\lambda, x_1, x_2)}{\partial x_1} &= \frac{3}{x_1} - \lambda x_1 \\
\frac{\partial \mathcal{L}(\lambda, x_1, x_2)}{\partial x_2} &= \frac{2}{x_2} - \lambda x_2
\end{aligned}$$

the first-order conditions yield

$$\begin{aligned}
1-\frac{1}{2}\left(x_1^*\right)^2-\frac{1}{2}\left(x_2^*\right)^2 &= 0 \\
&\Longrightarrow \frac{1}{2}\left(x_1^*\right)^2+\frac{1}{2}\left(x_2^*\right)^2=1 \\
\frac{3}{x_1^*}-\lambda^* x_1^* &= 0 \\
&\Longrightarrow \left(x_1^*\right)^2=\frac{3}{\lambda^*} \Longrightarrow x_1^*=\left(\frac{3}{\lambda^*}\right)^{\frac{1}{2}} \\
\frac{2}{x_2^*}-\lambda^* x_2^* &= 0 \\
&\Longrightarrow \left(x_2^*\right)^2=\frac{2}{\lambda^*} \Longrightarrow x_2^*=\left(\frac{2}{\lambda^*}\right)^{\frac{1}{2}}.
\end{aligned}$$

To find λ^* substitute $\left(x_1^*\right)^2=\frac{3}{\lambda^*},\left(x_2^*\right)^2=\frac{2}{\lambda^*}$ into the constraint as

$$\begin{aligned}
\frac{1}{2}\left(x_1^*\right)^2+\frac{1}{2}\left(x_2^*\right)^2 &= \frac{1}{2}\frac{3}{\lambda^*}+\frac{1}{2}\frac{2}{\lambda^*}=1 \\
&\Longrightarrow \frac{5}{2}\frac{1}{\lambda^*}=1 \Longrightarrow \lambda^*=\frac{5}{2}.
\end{aligned}$$

We can now solve for x_1^*, x_2^* as

$$\begin{aligned}
x_1^* &= \left(\frac{3}{\lambda^*}\right)^{\frac{1}{2}}=\left(\frac{3}{\frac{5}{2}}\right)^{\frac{1}{2}}=\left(\frac{6}{5}\right)^{\frac{1}{2}} \approx 1.10 \\
x_2^* &= \left(\frac{2}{\lambda^*}\right)^{\frac{1}{2}}=\left(\frac{2}{\frac{5}{2}}\right)^{\frac{1}{2}}=\left(\frac{4}{5}\right)^{\frac{1}{2}} \approx 0.9.
\end{aligned}$$

The solution is $\lambda^*=\frac{5}{2}, x_1^*=\left(\frac{6}{5}\right)^{\frac{1}{2}}, x_2^*=\left(\frac{4}{5}\right)^{\frac{1}{2}}$.

To show this is a maximum we first calculate the Hessian of $f\left(x_1, x_2\right)$ from

$$\begin{aligned}
f\left(x_1, x_2\right) &= 3\ln\left(x_1\right)+2\ln\left(x_2\right) \\
&\Longrightarrow \frac{\partial f\left(x_1, x_2\right)}{\partial x_1}=\frac{3}{x_1} \\
&\Longrightarrow \frac{\partial^2 f\left(x_1, x_2\right)}{\partial x_1^2}=-\frac{3}{x_1^2}, \frac{\partial^2 f\left(x_1, x_2\right)}{\partial x_2 \partial x_1}=0 \\
&\Longrightarrow \frac{\partial f\left(x_1, x_2\right)}{\partial x_2}=\frac{2}{x_2} \\
&\Longrightarrow \frac{\partial^2 f\left(x_1, x_2\right)}{\partial x_2 \partial x_1}=0, \frac{\partial^2 f\left(x_1, x_2\right)}{\partial x_2^2}=-\frac{2}{x_2^2}
\end{aligned}$$

as

$$H_f\left(x_1, x_2\right)=\begin{bmatrix} -\frac{3}{x_1^2} & 0 \\ 0 & -\frac{2}{x_2^2} \end{bmatrix}.$$

Since this is a diagonal matrix with negative diagonal elements, it follows that $H_f(x_1, x_2)$ is negative definite for all x_1, x_2, and so the objective function is globally concave. Now we calculate the Hessian of the constraint from

$$\begin{aligned} g(x_1,x_2) &= 1-\frac{1}{2}x_1^2-\frac{1}{2}x_2^2 \\ &\Longrightarrow \frac{\partial g(x_1,x_2)}{\partial x_1}=-x_1 \\ &\Longrightarrow \frac{\partial^2 g(x_1,x_2)}{\partial x_1^2}=-1, \ \frac{\partial^2 g(x_1,x_2)}{\partial x_2 \partial x_1}=0 \\ &\Longrightarrow \frac{\partial g(x_1,x_2)}{\partial x_2}=-x_2 \\ &\Longrightarrow \frac{\partial^2 g(x_1,x_2)}{\partial x_2 \partial x_1}=0, \ \frac{\partial^2 g(x_1,x_2)}{\partial x_2^2}=-1 \end{aligned}$$

as

$$H_g(x_1,x_2)=\begin{bmatrix} -1 & 0 \\ 0 & -1 \end{bmatrix}.$$

Since this is a diagonal matrix with negative diagonal elements, $H_g(x_1, x_2)$ is negative definite for all x_1, x_2, and so the constraint is globally concave. Since 1) the objective function is globally concave, 2) the constraint function is globally concave, and 3) the Lagrange multiplier is positive as $\lambda^* = \frac{5}{2} > 0$, it follows that we have a constrained global maximum.

Problem 4.35 *[C***] Let A be a symmetric 2×2 matrix and let x be a 2×1 vector as*

$$A=\begin{bmatrix} 9 & -2 \\ -2 & 6 \end{bmatrix}, \ x=\begin{bmatrix} x_1 \\ x_2 \end{bmatrix}.$$

Consider the problem of maximizing $f(x) = x^T Ax$ or

$$f(x_1,x_2)=9x_1^2-4x_1x_2+6x_2^2$$

subject to the constraint that $\|x\|^2 = x^T x = 1$ or $1 = x_1^2 + x_2^2$. Find x^* and $f(x^*)$.

Answer: The Lagrangian for this problem is

$$\mathcal{L}(\lambda,x_1,x_2)=9x_1^2-4x_1x_2+6x_2^2+\lambda\left(1-x_1^2-x_2^2\right)$$

with

$$\begin{aligned} \frac{\partial \mathcal{L}(\lambda,x_1,x_2)}{\partial \lambda} &= 1-x_1^2-x_2^2 \\ \frac{\partial \mathcal{L}(\lambda,x_1,x_2)}{\partial x_1} &= 18x_1-4x_2-2\lambda x_1 \\ \frac{\partial \mathcal{L}(\lambda,x_1,x_2)}{\partial x_2} &= -4x_1+12x_2-2\lambda x_2 \end{aligned}$$

which leads to the first-order conditions

$$\begin{aligned}
1-(x_1^*)^2-(x_2^*)^2 &= 0\\
18x_1^*-4x_2^*-2\lambda^* x_1^* &= 0\\
-4x_1^*+12x_2^*-2\lambda^* x_2^* &= 0.
\end{aligned}$$

Writing the last two conditions in matrix notation after cancelling the $2's$ we find that

$$\begin{bmatrix} 9 & -2\\ -2 & 6\end{bmatrix}\begin{bmatrix} x_1^*\\ x_2^*\end{bmatrix}=\lambda^*\begin{bmatrix} x_1^*\\ x_2^*\end{bmatrix}\Longrightarrow Ax^*=\lambda^* x^*.$$

It follows that x^* is an eigenvector of A and λ^* is an eigenvalue. Since A has two eigenvalues from the roots of

$$\lambda^2-15\lambda+50=0\Longrightarrow\lambda=\frac{-(-15)\pm\sqrt{(-15)^2-4\times 50}}{2}=\frac{15\pm 5}{2}\Longrightarrow\lambda_1=10,\ \lambda_2=5.$$

Since $Ax^*=\lambda^* x^*$ and the constraint requires that $(x^*)^T x^*=1$ we have

$$f(x^*)=(x^*)^T Ax^*=(x^*)^T x^*\lambda^*=\lambda^*.$$

Thus we choose x^* corresponding to the largest eigenvalue or $\lambda^*=10$. Now an eigenvector x^* corresponding to λ^* is any non-zero column of

$$\begin{aligned}
adj\,[A-\lambda^* I] &= adj\,[A-\lambda^* I]=adj\left[\begin{bmatrix} 9 & -2\\ -2 & 6\end{bmatrix}-10\begin{bmatrix} 1 & 0\\ 0 & 1\end{bmatrix}\right]\\
&= adj\left[\begin{bmatrix} -1 & -2\\ -2 & -4\end{bmatrix}\right]=\begin{bmatrix} -4 & 2\\ 2 & -1\end{bmatrix}
\end{aligned}$$

so we can take the second column with $(x^*)^T x^*=5$ and normalize so that $(x^*)^T x^*=1$ as

$$x^*=\begin{bmatrix} x_1^*\\ x_2^*\end{bmatrix}=\frac{1}{\sqrt{5}}\begin{bmatrix} 2\\ -1\end{bmatrix}=\begin{bmatrix} \frac{2}{\sqrt{5}}\\ -\frac{1}{\sqrt{5}}\end{bmatrix}.$$

Problem 4.36 *[B***] Let A be a symmetric 2×2 matrix and let x be a 2×1 vector as*

$$A=\begin{bmatrix} a_{11} & a_{12}\\ a_{12} & a_{22}\end{bmatrix},\ x=\begin{bmatrix} x_1\\ x_2\end{bmatrix}.$$

Consider the problem of maximizing $f(x)=x^T Ax$ or

$$f(x_1,x_2)=a_{11}x_1^2+2a_{12}x_1x_2+a_{22}x_2^2$$

subject to the constraint that $||x||^2=x^T x=1$ or $1=x_1^2+x_2^2$. Find x^* and $f(x^*)$.

Answer: The Lagrangian for this problem is

$$\mathcal{L}(\lambda, x_1, x_2) = a_{11}x_1^2 + 2a_{12}x_1x_2 + a_{22}x_2^2 + \lambda\left(1 - x_1^2 - x_2^2\right)$$

which leads to the first-order conditions

$$\begin{aligned} 1 - (x_1^*)^2 - (x_2^*)^2 &= 0 \\ 2a_{11}x_1^* + 2a_{12}x_2^* - 2\lambda^* x_1^* &= 0 \\ 2a_{12}x_1^* + 2a_{22}x_2^* - 2\lambda^* x_2^* &= 0. \end{aligned}$$

Writing the last two conditions in matrix notation after cancelling the $2's$ we find that

$$\begin{bmatrix} a_{11} & a_{12} \\ a_{12} & a_{22} \end{bmatrix} \begin{bmatrix} x_1^* \\ x_2^* \end{bmatrix} = \lambda^* \begin{bmatrix} x_1^* \\ x_2^* \end{bmatrix}$$

or $Ax^* = \lambda^* x^*$. It follows that x^* is an eigenvector of A and λ^* is an eigenvalue. The matrix A has two eigenvalues, the roots of

$$\lambda^2 - (a_{11} + a_{22})\lambda + \left(a_{11}a_{22} - a_{12}^2\right) = 0.$$

Since we are maximizing $f(x_1, x_2)$ we choose the largest eigenvalue. This is because

$$f(x^*) = (x^*)^T Ax^* = \lambda^* (x^*)^T x^* = \lambda^*$$

since $Ax^* = \lambda^* x^*$ and $(x^*)^T x^* = 1$. To maximize we choose x^* as the eigenvector corresponding to the largest eigenvalue λ^*.

Problem 4.37 *[A ***] Suppose that A is a symmetric matrix and x is an $n \times 1$ vector. Consider the problem of maximizing (minimizing) $x^T Ax$ subject to the constraint that $x^T x = 1$. Show that x^* is the eigenvector of A associated with the maximum (minimum) eigenvalue of A, normalized so that $\|x^*\| = 1$.*

Answer: (See also Problem 3.58 on page 160 where this is solved without calculus.) The Lagrangian for this problem is

$$\mathcal{L}(\lambda, x) = x^T Ax + \lambda\left(1 - x^T x\right)$$

from which the first-order conditions yield

$$\begin{aligned} \frac{\partial \mathcal{L}(\lambda^*, x^*)}{\partial \lambda} &= 1 - (x^*)^T x^* = 0 \\ \frac{\partial \mathcal{L}(\lambda^*, x^*)}{\partial x} &= 2Ax^* - 2\lambda^* x^* = 0 \\ &\Longrightarrow Ax^* = \lambda^* x^*. \end{aligned}$$

From the second first-order condition it follows that x^* is an eigenvector of A and λ^* is the associated eigenvalue. If x^* is any eigenvector of A, we can always satisfy the constraint by using $x^* / \|x^*\|$. Furthermore

$$Ax^* = \lambda^* x^* \Longrightarrow x^{*T} Ax^* = Ax^* = \lambda^* (x^*)^T x^* = \lambda^*$$

so to maximize (minimize) $x^T Ax$ we pick the maximum (minimum) eigenvalue λ^* and its associated eigenvector x^* normalized so that $(x^*)^T x^* = 1$.

4.3.1 Utility Maximization

Problem 4.38 *Consider a Cobb-Douglas utility function*

$$U(Q_1, Q_2) = \frac{1}{4}\ln(Q_1) + \frac{3}{4}\ln(Q_2)$$

with budget constraint $Y = P_1Q_1 + P_2Q_2$. *[D***] Construct the Lagrangian, [C***] derive the first-order conditions, and [B***] solve. [A***] Verify that you have a constrained global maximum. [A**] Suppose instead the utility function were*

$$\tilde{U}(Q_1, Q_2) = \ln(Q_1) + 3\ln(Q_2).$$

What results change and what results stay the same? Explain why.

Answer: The Lagrangian is

$$\mathcal{L}(\lambda, Q_1, Q_2) = \frac{1}{4}\ln(Q_1) + \frac{3}{4}\ln(Q_2) + \lambda(Y - P_1Q_1 - P_2Q_2)$$

so that

$$\begin{aligned}
\frac{\partial \mathcal{L}}{\partial \lambda} &= Y - P_1Q_1 - P_2Q_2 \\
\frac{\partial \mathcal{L}}{\partial Q_1} &= \frac{1}{4Q_1} - \lambda P_1 \\
\frac{\partial \mathcal{L}}{\partial Q_2} &= \frac{3}{4Q_2} - \lambda P_2.
\end{aligned}$$

The first-order conditions yield

$$\begin{aligned}
Y - P_1Q_1^* - P_2Q_2^* &= 0 \\
&\implies Y = P_1Q_1^* + P_2Q_2^* \\
\frac{1}{4Q_1^*} - \lambda^* P_1 &= 0 \\
&\implies \frac{1}{4} = \lambda^* P_1 Q_1^* \\
\frac{3}{4Q_2^*} - \lambda^* P_2 &= 0 \\
&\implies \frac{3}{4} = \lambda^* P_2 Q_2^*.
\end{aligned}$$

Adding up the last two results we have

$$\begin{aligned}
\frac{1}{4} + \frac{3}{4} &= 1 = \lambda^* P_1Q_1^* + \lambda^* P_2Q_2^* = \lambda^* \underbrace{(P_1Q_1^* + P_2Q_2^*)}_{Y} = \lambda^* Y \\
&\implies \lambda^* = \frac{1}{Y}.
\end{aligned}$$

It then follows that

$$\begin{aligned}
\frac{1}{4} &= \lambda^* P_1 Q_1^* = \frac{1}{Y} P_1 Q_1^* \\
&\Longrightarrow Q_1^* = \frac{Y}{4P_1} \\
\frac{3}{4} &= \lambda^* P_2 Q_2^* = \frac{1}{Y} P_2 Q_2^* \\
&\Longrightarrow Q_2^* = \frac{3Y}{4P_2}.
\end{aligned}$$

To show this corresponds to a constrained global maximum we need to show that: 1) the objective function is globally concave and 2) the constraint is linear. First the objective function, the utility function, is globally concave since it has a Hessian

$$H(Q_1, Q_2) = \begin{bmatrix} -\frac{1}{4Q_1^2} & 0 \\ 0 & -\frac{3}{4Q_2^2} \end{bmatrix}$$

which is a diagonal matrix with negative diagonal elements, and hence is negative definite for all Q_1, Q_2. Second the constraint $Y - P_1Q_1 - P_2Q_2 = 0$ is a linear function of Q_1, Q_2. Thus the solution corresponds to a constrained global maximum.

With $\tilde{U}(Q_1, Q_2) = \ln(Q_1) + 3\ln(Q_2)$ the Lagrangian is

$$\mathcal{L}(\lambda, Q_1, Q_2) = \ln(Q_1) + 3\ln(Q_2) + \lambda(Y - P_1Q_1 - P_2Q_2)$$

so that

$$\begin{aligned}
\frac{\partial \mathcal{L}}{\partial \lambda} &= Y - P_1Q_1 - P_2Q_2 \\
\frac{\partial \mathcal{L}}{\partial Q_1} &= \frac{1}{Q_1} - \lambda P_1 \\
\frac{\partial \mathcal{L}}{\partial Q_2} &= \frac{3}{Q_2} - \lambda P_2
\end{aligned}$$

and the first-order conditions yield

$$\begin{aligned}
Y - P_1Q_1^* - P_2Q_2^* &= 0 \\
&\Longrightarrow Y = P_1Q_1^* + P_2Q_2^* \\
\frac{1}{Q_1^*} - \lambda^* P_1 &= 0 \\
&\Longrightarrow 1 = \lambda^* P_1 Q_1^* \\
\frac{3}{Q_2^*} - \lambda^* P_2 &= 0 \\
&\Longrightarrow 3 = \lambda^* P_2 Q_2^*.
\end{aligned}$$

Adding up the last two first-order conditions we have

$$\begin{aligned} 1+3 &= 4 = \lambda^* P_1 Q_1^* + \lambda^* P_2 Q_2^* = \lambda^* \underbrace{(P_1 Q_1^* + P_2 Q_2^*)}_{Y} = \lambda^* Y \\ &\implies \lambda^* = \frac{4}{Y}. \end{aligned}$$

It then follows that

$$\begin{aligned} 1 &= \lambda^* P_1 Q_1^* = \frac{4}{Y} P_1 Q_1^* \\ &\implies Q_1^* = \frac{Y}{4P_1} \\ 3 &= \lambda^* P_2 Q_2^* = \frac{4}{Y} P_2 Q_2^* \\ &\implies Q_2^* = \frac{3Y}{4P_2}. \end{aligned}$$

Notice that the demand curves $Q_1^* = \frac{Y}{4P_1}, Q_2^* = \frac{3Y}{4P_2}$ are identical for both utility functions. It is the Lagrange multipliers: $\lambda^* = \frac{1}{Y}$ and $\lambda^* = \frac{4}{Y}$ that differ. The demand curves are the same because $\tilde{U}(Q_1, Q_2)$ is a monotonic transformation of $U(Q_1, Q_2)$ as $\tilde{U}(Q_1, Q_2) = 4U(Q_1, Q_2)$ since

$$\tilde{U}(Q_1, Q_2) = \ln(Q_1) + 3\ln(Q_2) = 4U(Q_1, Q_2) = 4\left(\frac{1}{4}\ln(Q_1) + \frac{3}{4}\ln(Q_2)\right).$$

Problem 4.39 *Consider a Cobb-Douglas utility function*

$$U(Q_1, Q_2) = \alpha \ln(Q_1) + (1-\alpha)\ln(Q_2)$$

*where $0 < \alpha < 1$. [C***] Find the utility maximizing values Q_1^*, Q_2^*. [B***] Verify the second-order conditions for a constrained global maximum. Show how α determines the budget share of each good. [A***] The indirect utility function is defined as $U^*(P_1, P_2, Y) \equiv U(Q_1^*, Q_2^*)$. Find the indirect utility function for this problem. Verify that the Lagrange multiplier is the marginal utility of income and verify Roy's identity as*

$$\frac{\partial U^*(P_1, P_2, Y)}{\partial Y} = \lambda^* \text{ and } Q_1^* = -\frac{\frac{\partial U^*(P_1, P_2, Y)}{\partial P_1}}{\frac{\partial U^*(P_1, P_2, Y)}{\partial Y}}.$$

Answer: The Lagrangian is

$$\mathcal{L}(\lambda, Q_1, Q_2) = \alpha \ln(Q_1) + (1-\alpha)\ln(Q_2) + \lambda(Y - P_1 Q_1 - P_2 Q_2)$$

so that

$$\begin{aligned} \frac{\partial \mathcal{L}}{\partial \lambda} &= Y - P_1 Q_1 - P_2 Q_2 \\ \frac{\partial \mathcal{L}}{\partial Q_1} &= \frac{\alpha}{Q_1} - \lambda P_1 \\ \frac{\partial \mathcal{L}}{\partial Q_2} &= \frac{1-\alpha}{Q_2} - \lambda P_2 \end{aligned}$$

and the first-order conditions yield

$$\begin{aligned}
Y - P_1 Q_1^* - P_2 Q_2^* &= 0 \\
&\implies Y = P_1 Q_1^* + P_2 Q_2^* \\
\frac{\alpha}{Q_1^*} - \lambda^* P_1 &= 0 \\
&\implies \alpha = \lambda^* P_1 Q_1^* \\
\frac{1-\alpha}{Q_2^*} - \lambda^* P_2 &= 0 \\
&\implies 1-\alpha = \lambda^* P_2 Q_2^*.
\end{aligned}$$

Adding up the last two results we have

$$\begin{aligned}
\alpha + (1-\alpha) &= 1 = \lambda^* P_1 Q_1^* + \lambda^* P_2 Q_2^* = \lambda^* \underbrace{(P_1 Q_1^* + P_2 Q_2^*)}_{Y} = \lambda^* Y \\
&\implies \lambda^* = \frac{1}{Y}.
\end{aligned}$$

It then follows that

$$\begin{aligned}
\alpha &= \lambda^* P_1 Q_1^* \implies Q_1^* = \frac{\alpha Y}{P_1} \\
1-\alpha &= \lambda^* P_2 Q_2^* \implies Q_2^* = \frac{(1-\alpha) Y}{P_2}.
\end{aligned}$$

The budget shares are

$$\begin{aligned}
\frac{P_1 Q_1^*}{Y} &= \alpha \\
\frac{P_2 Q_2^*}{Y} &= 1-\alpha.
\end{aligned}$$

To show this corresponds to a constrained global maximum we need to show that: 1) the objective function is globally concave and 2) the constraint is linear. First the objective function, the utility function, is globally concave since it has a Hessian

$$H(Q_1, Q_2) = \begin{bmatrix} -\frac{\alpha}{Q_1^2} & 0 \\ 0 & -\frac{1-\alpha}{Q_2^2} \end{bmatrix}$$

which is a diagonal matrix with negative diagonal elements, and hence is negative definite for all Q_1, Q_2. Second the constraint $Y - P_1 Q_1 - P_2 Q_2 = 0$ is a linear function of Q_1, Q_2. Thus the solution corresponds to a constrained global maximum.

The indirect utility function is

$$\begin{aligned}
U^*(P_1, P_2, Y) &= \alpha \ln(Q_1^*) + (1-\alpha)\ln(Q_2^*) = \alpha \ln\left(\frac{\alpha Y}{P_1}\right) + (1-\alpha)\ln\left(\frac{(1-\alpha)Y}{P_2}\right) \\
&= c + \ln(Y) - \alpha \ln(P_1) - (1-\alpha)\ln(P_2)
\end{aligned}$$

where $c = \alpha \ln(\alpha) + (1-\alpha)\ln(1-\alpha)$.

We have

$$\begin{aligned}\frac{\partial U^*(P_1,P_2,Y)}{\partial Y} &= \frac{\partial}{\partial Y}\left(c+\ln(Y)-\alpha\ln(P_1)-(1-\alpha)\ln(P_2)\right) = \frac{1}{Y} = \lambda^* \\ \frac{\partial U^*(P_1,P_2,Y)}{\partial P_1} &= \frac{\partial}{\partial P_1}\left(c+\ln(Y)-\alpha\ln(P_1)-(1-\alpha)\ln(P_2)\right) = -\frac{\alpha}{P_1}\end{aligned}$$

so that

$$-\frac{\frac{\partial U^*(P_1,P_2,Y)}{\partial P_1}}{\frac{\partial U^*(P_1,P_2,Y)}{\partial Y}} = -\frac{-\frac{\alpha}{P_1}}{\frac{1}{Y}} = \frac{\alpha Y}{P_1} = Q_1^*.$$

Problem 4.40 *[A ***] Consider a Cobb-Douglas utility function with n goods*

$$U(Q_1,Q_2,\ldots,Q_n) = \alpha_1\ln(Q_1)+\alpha_2\ln(Q_2)+\cdots+\alpha_n\ln(Q_n)$$

where $\alpha_i \geq 0$ and $\alpha_1+\alpha_2+\cdots+\alpha_n = 1$. The budget constraint is

$$Y = P_1Q_1+P_2Q_2+\cdots+P_nQ_n.$$

Find the utility maximizing values of Q_i^*, $i=1,2,\ldots,n$. Verify the second-order conditions for a constrained global maximum. Find the indirect utility function

$$U^*(P_1,P_2,\ldots,P_n,Y) = U(Q_1^*,Q_2^*,\ldots,Q_n^*).$$

Show that $\frac{\partial U^*}{\partial Y} = \lambda^*$ so that the Lagrange multiplier is the marginal utility of income. Verify Roy's identity $Q_i^* = -\frac{\partial U^*}{\partial P_i}/\frac{\partial U^*}{\partial Y}$.

Answer: The Lagrangian is

$$\begin{aligned}\mathcal{L}(\lambda,Q_1,Q_2,\ldots,Q_n) &= \alpha_1\ln(Q_1)+\alpha_2\ln(Q_2)+\cdots+\alpha_n\ln(Q_n) \\ &\quad +\lambda(Y-P_1Q_1-P_2Q_2-\cdots-P_nQ_n)\end{aligned}$$

so that

$$\begin{aligned}\frac{\partial\mathcal{L}}{\partial\lambda} &= Y-P_1Q_1-P_2Q_2-\cdots-P_nQ_n \\ \frac{\partial\mathcal{L}}{\partial Q_i} &= \frac{\alpha_i}{Q_i}-\lambda P_i \text{ for } i=1,2,\ldots,n.\end{aligned}$$

The first-order conditions yield

$$\begin{aligned}Y-P_1Q_1^*-P_2Q_2^*-\cdots-P_nQ_n^* &= 0 \\ &\implies Y = P_1Q_1^*+P_2Q_2^*+\cdots+P_nQ_n^* \\ \frac{\alpha_i}{Q_i^*}-\lambda^*P_i &= 0 \\ &\implies \alpha_i = \lambda^*P_iQ_i^* \text{ for } i=1,2,\ldots,n.\end{aligned}$$

So we have

$$\begin{aligned} 1 &= \alpha_1 + \alpha_2 + \cdots + \alpha_n = \lambda^* P_1 Q_1^* + \lambda^* P_2 Q_2^* + \cdots + \lambda^* P_n Q_n^* \\ &= \lambda^* \underbrace{(P_1 Q_1^* + P_2 Q_2^* + \cdots + P_n Q_n^*)}_{Y} = \lambda^* Y \\ \Longrightarrow\ & \lambda^* = \frac{1}{Y}. \end{aligned}$$

It then follows that

$$\begin{aligned} \alpha_i &= \lambda^* P_i Q_i^* = \frac{P_i Q_i^*}{Y} \\ \Longrightarrow\ & Q_i^* = \frac{\alpha_i Y}{P_i} \text{ for } i = 1, 2, \ldots, n. \end{aligned}$$

To show this corresponds to a constrained global maximum we need to show that: 1) the objective function is globally concave and 2) the constraint is linear. First the objective function, the utility function, is globally concave since it has a Hessian

$$H(Q_1, Q_2, \ldots, Q_n) = \begin{bmatrix} -\frac{\alpha_1}{Q_1^2} & 0 & \cdots & 0 \\ 0 & -\frac{\alpha_2}{Q_2^2} & \ddots & \vdots \\ \vdots & \ddots & \ddots & 0 \\ 0 & \cdots & 0 & -\frac{\alpha_n}{Q_n^2} \end{bmatrix}$$

which is a diagonal matrix with negative diagonal elements, and hence is negative definite for all $Q_1, Q_2, \ldots, Q_n$. Second the budget constraint is a linear function. Thus the solution corresponds to a constrained global maximum.

We have

$$\begin{aligned} U^*(P_1, P_2, \ldots, P_n, Y) &= \alpha_1 \ln(Q_1^*) + \alpha_2 \ln(Q_2^*) + \cdots + \alpha_n \ln(Q_n^*) \\ &= \alpha_1 \ln\left(\frac{\alpha_1 Y}{P_1}\right) + \alpha_2 \ln\left(\frac{\alpha_2 Y}{P_2}\right) + \cdots + \alpha_n \ln\left(\frac{\alpha_n Y}{P_n}\right) \\ &= c + \ln(Y) - (\alpha_1 \ln(P_1) + \alpha_2 \ln(P_2) + \cdots + \alpha_n \ln(P_n)) \end{aligned}$$

where

$$c = \alpha_1 \ln(\alpha_1) + \alpha_2 \ln(\alpha_2) + \cdots + \alpha_n \ln(\alpha_n).$$

We therefore have

$$\begin{aligned} \frac{\partial U^*}{\partial Y} &= \frac{\partial}{\partial Y}(c + \ln(Y) - (\alpha_1 \ln(P_1) + \alpha_2 \ln(P_2) + \cdots + \alpha_n \ln(P_n))) = \frac{1}{Y} = \lambda^* \\ \frac{\partial U^*}{\partial P_i} &= \frac{\partial}{\partial P_i}(c + \ln(Y) - (\alpha_1 \ln(P_1) + \alpha_2 \ln(P_2) + \cdots + \alpha_n \ln(P_n))) = -\frac{\alpha_i}{P_i} \end{aligned}$$

so that

$$-\frac{\frac{\partial U^*(P_1, P_2, \ldots, P_n, Y)}{\partial P_i}}{\frac{\partial U^*(P_1, P_2, \ldots, P_n, Y)}{\partial Y}} = -\frac{-\frac{\alpha_i}{P_i}}{\frac{1}{Y}} = \frac{\alpha_i Y}{P_i} = Q_i^*.$$

Problem 4.41 *Consider a household with the utility function*

$$U(Q_1, Q_2) = 3Q_1^{\frac{1}{3}} + 3Q_2^{\frac{1}{3}}.$$

*[D***] What is the Lagrangian for the utility maximization problem? [C***] Find the first-order conditions and solve. [B***] Show that the solution corresponds to a constrained global maximum.*

Answer: We have

$$\mathcal{L}(\lambda, Q_1, Q_2) = 3Q_1^{\frac{1}{3}} + 3Q_2^{\frac{1}{3}} + \lambda(Y - P_1Q_1 - P_2Q_2)$$

so that

$$\begin{aligned} \frac{\partial \mathcal{L}(\lambda, Q_1, Q_2)}{\partial \lambda} &= Y - P_1Q_1 - P_2Q_2 \\ \frac{\partial \mathcal{L}(\lambda, Q_1, Q_2)}{\partial Q_1} &= Q_1^{-\frac{2}{3}} - \lambda P_1 \\ \frac{\partial \mathcal{L}(\lambda, Q_1, Q_2)}{\partial Q_2} &= Q_2^{-\frac{2}{3}} - \lambda P_2. \end{aligned}$$

The first-order conditions yield

$$\begin{aligned} Y - P_1Q_1^* - P_2Q_2^* &= 0 \\ &\implies Y = P_1Q_1^* + P_2Q_2^* \\ (Q_1^*)^{-\frac{2}{3}} - \lambda^* P_1 &= 0 \\ &\implies Q_1^* = (\lambda^* P_1)^{-\frac{3}{2}} \implies P_1Q_1^* = (\lambda^*)^{-\frac{3}{2}} P_1^{-\frac{1}{2}} \\ (Q_2^*)^{-\frac{2}{3}} - \lambda^* P_2 &= 0 \\ &\implies Q_2^* = (\lambda^* P_2)^{-\frac{3}{2}} \implies P_2Q_2^* = (\lambda^*)^{-\frac{3}{2}} P_2^{-\frac{1}{2}} \end{aligned}$$

so that

$$\begin{aligned} Y &= P_1Q_1^* + P_2Q_2^* = (\lambda^*)^{-\frac{3}{2}} P_1^{-\frac{1}{2}} + (\lambda^*)^{-\frac{3}{2}} P_2^{-\frac{1}{2}} = (\lambda^*)^{-\frac{3}{2}} \left(P_1^{-\frac{1}{2}} + P_2^{-\frac{1}{2}}\right) \\ &\implies (\lambda^*)^{-\frac{3}{2}} = \frac{Y}{P_1^{-\frac{1}{2}} + P_2^{-\frac{1}{2}}} \\ &\implies \lambda^* = \left(\frac{Y}{P_1^{-\frac{1}{2}} + P_2^{-\frac{1}{2}}}\right)^{-\frac{2}{3}} = Y^{-\frac{2}{3}} \left(P_1^{-\frac{1}{2}} + P_2^{-\frac{1}{2}}\right)^{\frac{2}{3}}. \end{aligned}$$

Therefore

$$\begin{aligned} Q_1^* &= (\lambda^*)^{-\frac{3}{2}} P_1^{-\frac{3}{2}} = \frac{YP_1^{-\frac{3}{2}}}{P_1^{-\frac{1}{2}} + P_2^{-\frac{1}{2}}} \\ Q_2^* &= (\lambda^*)^{-\frac{3}{2}} P_2^{-\frac{3}{2}} = \frac{YP_2^{-\frac{3}{2}}}{P_1^{-\frac{1}{2}} + P_2^{-\frac{1}{2}}}. \end{aligned}$$

To show this corresponds to a constrained global maximum we need to show that: 1) the objective function is globally concave and 2) the constraint is linear. First the objective function, the utility function, is globally concave since it has a Hessian

$$H(Q_1, Q_2) = \begin{bmatrix} -\frac{2}{3}Q_1^{-\frac{5}{3}} & 0 \\ 0 & -\frac{2}{3}Q_2^{-\frac{5}{3}} \end{bmatrix}$$

which is a diagonal matrix with negative diagonal elements, and hence is negative definite for all Q_1, Q_2. Second the constraint $Y - P_1Q_1 - P_2Q_2 = 0$ is a linear function of Q_1, Q_2. Thus the solution is a constrained global maximizer.

Problem 4.42 *Consider a household with utility function*

$$U(Q_1, Q_2) = 4Q_1^{\frac{1}{4}} + 4Q_2^{\frac{1}{4}}.$$

*[D***] Find the Lagrangian for the utility maximization problem, the first-order conditions, and [C***] solve. [B***] Show that the demand curve for Q_1 is downward sloping and that Q_1 is a normal good. Show that the solution is a constrained global maximizer.*

Answer: We have a Lagrangian

$$\mathcal{L}(\lambda, Q_1, Q_2) = 4Q_1^{\frac{1}{4}} + 4Q_2^{\frac{1}{4}} + \lambda(Y - P_1Q_1 - P_2Q_2)$$

so that

$$\begin{aligned} \frac{\partial \mathcal{L}}{\partial \lambda} &= Y - P_1Q_1 - P_2Q_2 \\ \frac{\partial \mathcal{L}}{\partial Q_1} &= Q_1^{-\frac{3}{4}} - \lambda P_1 \\ \frac{\partial \mathcal{L}}{\partial Q_2} &= Q_2^{-\frac{3}{4}} - \lambda P_2. \end{aligned}$$

The first-order conditions yield

$$\begin{aligned} Y - P_1Q_1^* - P_2Q_2^* &= 0 \\ &\Longrightarrow Y = P_1Q_1^* + P_2Q_2^* \\ (Q_1^*)^{-\frac{3}{4}} - \lambda^* P_1 &= 0 \\ &\Longrightarrow Q_1^* = (\lambda^* P_1)^{-\frac{4}{3}} \Longrightarrow P_1Q_1^* = (\lambda^*)^{-\frac{4}{3}} P_1^{-\frac{1}{3}} \\ (Q_2^*)^{-\frac{3}{4}} - \lambda^* P_2 &= 0 \\ &\Longrightarrow Q_2^* = (\lambda^* P_2)^{-\frac{4}{3}} \Longrightarrow P_2Q_2^* = (\lambda^*)^{-\frac{4}{3}} P_2^{-\frac{1}{3}} \end{aligned}$$

so that

$$\begin{aligned} Y &= P_1Q_1^* + P_2Q_2^* = (\lambda^*)^{-\frac{4}{3}} P_1^{-\frac{1}{3}} + (\lambda^*)^{-\frac{4}{3}} P_2^{-\frac{1}{3}} \\ &\Longrightarrow (\lambda^*)^{-\frac{4}{3}} \left(P_1^{-\frac{1}{3}} + P_2^{-\frac{1}{3}}\right) = Y \\ &\Longrightarrow (\lambda^*)^{-\frac{4}{3}} = \frac{Y}{\left(P_1^{-\frac{1}{3}} + P_2^{-\frac{1}{3}}\right)} \\ &\Longrightarrow \lambda^* = \left(P_1^{-\frac{1}{3}} + P_2^{-\frac{1}{3}}\right)^{\frac{3}{4}} Y^{-\frac{3}{4}}. \end{aligned}$$

Therefore

$$\begin{aligned} Q_1^* &= (\lambda^*)^{-\frac{4}{3}} P_1^{-\frac{4}{3}} = \frac{YP_1^{-\frac{4}{3}}}{P_1^{-\frac{1}{3}} + P_2^{-\frac{1}{3}}} \\ Q_2^* &= (\lambda^*)^{-\frac{4}{3}} P_2^{-\frac{4}{3}} = \frac{YP_2^{-\frac{4}{3}}}{P_1^{-\frac{1}{3}} + P_2^{-\frac{1}{3}}}. \end{aligned}$$

To show this correspond to a constrained global maximum we need to show that: 1) the objective function is globally concave, 2) the constraint is linear. First (as we showed above) the objective function, the utility function $U(Q_1, Q_2)$ is globally concave. Second the constraint $Y - P_1Q_1 - P_2Q_2 = 0$ is a linear function of Q_1, Q_2. Thus the solution is a constrained global maximizer.

The demand curve for Q_1 is downward sloping as

$$Q_1^* = \frac{YP_1^{-\frac{4}{3}}}{P_1^{-\frac{1}{3}} + P_2^{-\frac{1}{3}}} = \frac{Y}{P_1^{\frac{4}{3}}\left(P_1^{-\frac{1}{3}} + P_2^{-\frac{1}{3}}\right)} = \frac{Y}{P_1 + P_1^{\frac{4}{3}} P_2^{-\frac{1}{3}}}$$

so that

$$\frac{\partial Q_1^*}{\partial P_1} = \frac{\partial}{\partial P_1}\left(\frac{Y}{P_1 + P_1^{\frac{4}{3}} P_2^{-\frac{1}{3}}}\right) = -\frac{Y}{\left(P_1 + P_1^{\frac{4}{3}} P_2^{-\frac{1}{3}}\right)^2}\left(1 + \frac{4}{3}P_1^{\frac{1}{3}} P_2^{-\frac{1}{3}}\right) < 0.$$

Also Q_1 is normal

$$\frac{\partial Q_1^*}{\partial Y} = \frac{\partial}{\partial Y}\left(\frac{YP_1^{-\frac{4}{3}}}{P_1^{-\frac{1}{3}} + P_2^{-\frac{1}{3}}}\right) = \frac{P_1^{-\frac{4}{3}}}{P_1^{-\frac{1}{3}} + P_2^{-\frac{1}{3}}} > 0.$$

Problem 4.43 *Consider a household with the utility function*

$$U(Q_1, Q_2) = -Q_1^{-1} - Q_2^{-1}.$$

*[C***] Find the first-order conditions and solve. [B***] Show that the solution is a constrained global maximizer. Show that Q_1 and Q_2 are complements.*

Answer: We have

$$\mathcal{L}(\lambda, Q_1, Q_2) = -Q_1^{-1} - Q_2^{-1} + \lambda(Y - P_1Q_1 - P_2Q_2)$$

so that

$$\begin{aligned}\frac{\partial \mathcal{L}(\lambda, Q_1, Q_2)}{\partial \lambda} &= Y - P_1Q_1 - P_2Q_2 \\ \frac{\partial \mathcal{L}(\lambda, Q_1, Q_2)}{\partial Q_1} &= Q_1^{-2} - \lambda P_1 \\ \frac{\partial \mathcal{L}(\lambda, Q_1, Q_2)}{\partial Q_2} &= Q_2^{-2} - \lambda P_2.\end{aligned}$$

The first-order conditions yield

$$\begin{aligned}Y - P_1Q_1^* - P_2Q_2^* &= 0 \\ &\implies Y = P_1Q_1^* + P_2Q_2^* \\ (Q_1^*)^{-2} - \lambda^* P_1 &= 0 \\ &\implies Q_1^* = (\lambda^*)^{-\frac{1}{2}} P_1^{-\frac{1}{2}} \implies P_1Q_1^* = (\lambda^*)^{-\frac{1}{2}} P_1^{\frac{1}{2}} \\ (Q_2^*)^{-2} - \lambda^* P_2 &= 0 \\ &\implies Q_2^* = (\lambda^*)^{-\frac{1}{2}} P_2^{-\frac{1}{2}} \implies P_2Q_2^* = (\lambda^*)^{-\frac{1}{2}} P_2^{\frac{1}{2}}\end{aligned}$$

so that

$$\begin{aligned}Y &= P_1Q_1^* + P_2Q_2^* = (\lambda^*)^{-\frac{1}{2}} P_1^{\frac{1}{2}} + (\lambda^*)^{-\frac{1}{2}} P_2^{\frac{1}{2}} = (\lambda^*)^{-\frac{1}{2}} \left(P_1^{\frac{1}{2}} + P_2^{\frac{1}{2}}\right) \\ &\implies (\lambda^*)^{-\frac{1}{2}} = \frac{Y}{P_1^{\frac{1}{2}} + P_2^{\frac{1}{2}}} \\ &\implies \lambda^* = Y^{-2}\left(P_1^{\frac{1}{2}} + P_2^{\frac{1}{2}}\right)^2.\end{aligned}$$

Therefore

$$\begin{aligned}Q_1^* &= (\lambda^*)^{-\frac{1}{2}} P_1^{-\frac{1}{2}} = \frac{YP_1^{-\frac{1}{2}}}{P_1^{\frac{1}{2}} + P_2^{\frac{1}{2}}} \\ Q_2^* &= (\lambda^*)^{-\frac{1}{2}} P_2^{-\frac{1}{2}} = \frac{YP_2^{-\frac{1}{2}}}{P_1^{\frac{1}{2}} + P_2^{\frac{1}{2}}}.\end{aligned}$$

The two goods are complements since

$$\begin{aligned}Q_1^* &= \frac{YP_1^{-\frac{1}{2}}}{P_1^{\frac{1}{2}} + P_2^{\frac{1}{2}}} \\ \implies \frac{\partial Q_1^*}{\partial P_2} &= -\frac{YP_1^{-\frac{1}{2}}}{\left(P_1^{\frac{1}{2}} + P_2^{\frac{1}{2}}\right)^2}\frac{1}{2}P_2^{-\frac{1}{2}} < 0.\end{aligned}$$

To show this corresponds to a constrained global maximum we need to show that: 1) the objective function is globally concave and 2) the constraint is linear. First the objective function, the utility function, is globally concave since it has a Hessian

$$\begin{aligned} U(Q_1, Q_2) &= -Q_1^{-1} - Q_2^{-1} \\ &\Longrightarrow H(Q_1, Q_2) = \begin{bmatrix} -2Q_1^{-3} & 0 \\ 0 & -2Q_2^{-3} \end{bmatrix} \end{aligned}$$

which is a diagonal matrix with negative diagonal elements, and hence is negative definite for all Q_1, Q_2. The constraint $Y - P_1 Q_1 - P_2 Q_2 = 0$ is a linear function of Q_1, Q_2. Thus the solution is a constrained global maximizer.

Problem 4.44 *The CES utility function (constant elasticity of substitution) is given by*

$$U(Q_1, Q_2) = \alpha_1^{\rho} \frac{Q_1^{1-\rho} - 1}{1-\rho} + \alpha_2^{\rho} \frac{Q_2^{1-\rho} - 1}{1-\rho}$$

*where $\rho > 0$, $\alpha_1 > 0$, $\alpha_1 > 0$, and $\alpha_1 + \alpha_2 = 1$. This includes the Cobb-Douglas as a special case where $\rho = 1$ (from L'Hôpital's rule). [A ***] Calculate the demand functions Q_i^* for the CES utility function. Verify the second-order conditions for a constrained global maximum.*

Answer: The Lagrangian is

$$\mathcal{L}(\lambda, Q_1, Q_2) = \alpha_1^{\rho} \frac{Q_1^{1-\rho} - 1}{1-\rho} + \alpha_1^{\rho} \frac{Q_2^{1-\rho} - 1}{1-\rho} + \lambda (Y - P_1 Q_1 - P_2 Q_2)$$

so that the first-order conditions yield

$$\begin{aligned} Y - P_1 Q_1^* - P_2 Q_2^* &= 0 \\ &\Longrightarrow Y = P_1 Q_1^* + P_2 Q_2^* \\ \alpha_1^{\rho} (Q_1^*)^{-\rho} - \lambda^* P_1 &= 0 \\ &\Longrightarrow P_1 Q_1^* = \alpha_1 (\lambda^*)^{-\frac{1}{\rho}} P_1^{1-\frac{1}{\rho}} = \alpha_1 (\lambda^*)^{-\frac{1}{\rho}} P_1^{-\frac{1-\rho}{\rho}} \\ \alpha_1^{\rho} (Q_2^*)^{-\rho} - \lambda^* P_2 &= 0 \\ &\Longrightarrow P_2 Q_2^* = \alpha_2 (\lambda^*)^{-\frac{1}{\rho}} P_2^{1-\frac{1}{\rho}} = \alpha_2 (\lambda^*)^{-\frac{1}{\rho}} P_2^{-\frac{1-\rho}{\rho}}. \end{aligned}$$

Putting the last two results in the budget constraint yields

$$\begin{aligned} Y &= P_1 Q_1^* + P_2 Q_2^* = \alpha_1 (\lambda^*)^{-\frac{1}{\rho}} P_1^{-\frac{1-\rho}{\rho}} + \alpha_2 (\lambda^*)^{-\frac{1}{\rho}} P_2^{-\frac{1-\rho}{\rho}} \\ &= (\lambda^*)^{-\frac{1}{\rho}} \left(\alpha_1 P_1^{-\frac{1-\rho}{\rho}} + \alpha_2 P_2^{-\frac{1-\rho}{\rho}} \right) \\ \Longrightarrow (\lambda^*)^{-\frac{1}{\rho}} &= \frac{Y}{\alpha_1 P_1^{-\frac{1-\rho}{\rho}} + \alpha_2 P_2^{-\frac{1-\rho}{\rho}}} \\ \Longrightarrow \lambda^* &= \left(\frac{Y}{\alpha_1 P_1^{-\frac{1-\rho}{\rho}} + \alpha_2 P_2^{-\frac{1-\rho}{\rho}}} \right)^{-\rho}. \end{aligned}$$

Once we have the Lagrange multiplier we can calculate the demand curves using

$$\begin{aligned}
\alpha_1^\rho (Q_1^*)^{-\rho} - \lambda^* P_1 &= 0 \\
&\Longrightarrow Q_1^* = \alpha_1 (\lambda^*)^{-\frac{1}{\rho}} P_1^{-\frac{1}{\rho}} \\
&\Longrightarrow Q_1^* = \frac{\alpha_1 P_1^{-\frac{1}{\rho}} Y}{\alpha_1 P_1^{-\frac{1-\rho}{\rho}} + \alpha_2 P_2^{-\frac{1-\rho}{\rho}}} \\
\alpha_2^\rho (Q_2^*)^{-\rho} - \lambda^* P_2 &= 0 \\
&\Longrightarrow Q_2^* = \alpha_2 (\lambda^*)^{-\frac{1}{\rho}} P_2^{-\frac{1}{\rho}} \\
&\Longrightarrow Q_2^* = \frac{\alpha_2 P_2^{-\frac{1}{\rho}} Y}{\alpha_1 P_1^{-\frac{1-\rho}{\rho}} + \alpha_2 P_2^{-\frac{1-\rho}{\rho}}}.
\end{aligned}$$

To show this corresponds to a constrained global maximum we need to show that: 1) the objective function is globally concave, 2) the constraint is linear. First the objective function, the utility function, is globally concave since it has a Hessian

$$H(Q_1, Q_2) = \begin{bmatrix} -\rho\alpha_1^\rho Q_1^{\rho-1} & 0 \\ 0 & -\rho\alpha_2^\rho Q_2^{\rho-1} \end{bmatrix}$$

which is a diagonal matrix with negative diagonal elements since $\rho > 0$, and hence is negative definite for all Q_1, Q_2. Second the constraint is linear function. Thus the solution Q_1^*, Q_2^* corresponds to a constrained global maximum.

4.3.2 Expenditure Minimization

Problem 4.45 *[A ***] Suppose that an individual with utility function*

$$U(Q_1, Q_2) = 4Q_1^{\frac{1}{4}} + 4Q_2^{\frac{1}{4}}$$

is thrown in jail, and that the warden of the jail wishes to minimize expenditure $E \equiv P_1Q_1 + P_2Q_2$ on the individual. The law states however that prisoners must obtain 4 units of utility, so that the law constrains the warden to chose Q_1, Q_2 so that $U(Q_1, Q_2) = 4$. Construct the Lagrangian for this problem, find the first-order conditions and solve. Verify that you have a constrained global minimum.

Answer: The constraint is

$$4 = 4Q_1^{\frac{1}{4}} + 4Q_2^{\frac{1}{4}} \Longrightarrow 1 - Q_1^{\frac{1}{4}} - Q_2^{\frac{1}{4}} = 0$$

and the objective function is expenditure or $P_1Q_1 + P_2Q_2$. The Lagrangian then is

$$\mathcal{L}(\lambda, Q_1, Q_2) = P_1Q_1 + P_2Q_2 + \lambda\left(1 - Q_1^{\frac{1}{4}} - Q_2^{\frac{1}{4}}\right)$$

so that

$$\begin{aligned}
\frac{\partial \mathcal{L}}{\partial \lambda} &= 1 - Q_1^{\frac{1}{4}} - Q_2^{\frac{1}{4}} \\
\frac{\partial \mathcal{L}}{\partial Q_1} &= P_1 - \frac{\lambda}{4} Q_1^{-\frac{3}{4}} \\
\frac{\partial \mathcal{L}}{\partial Q_2} &= P_2 - \frac{\lambda}{4} Q_2^{-\frac{3}{4}}.
\end{aligned}$$

The first-order conditions yield

$$\begin{aligned}
1 - (Q_1^*)^{\frac{1}{4}} - (Q_2^*)^{\frac{1}{4}} &= 0 \\
&\implies 1 = (Q_1^*)^{\frac{1}{4}} + (Q_2^*)^{\frac{1}{4}} \\
P_1 - \frac{\lambda^*}{4} (Q_1^*)^{-\frac{3}{4}} &= 0 \\
&\implies P_1 = \frac{\lambda^*}{4} (Q_1^*)^{-\frac{3}{4}} \implies Q_1^* = \left(\frac{\lambda^*}{4}\right)^{\frac{4}{3}} P_1^{-\frac{4}{3}} \\
P_2 - \frac{\lambda^*}{4} (Q_2^*)^{-\frac{3}{4}} &= 0 \\
&\implies P_2 = \frac{\lambda^*}{4} (Q_2^*)^{-\frac{3}{4}} \implies Q_2^* = \left(\frac{\lambda^*}{4}\right)^{\frac{4}{3}} P_2^{-\frac{4}{3}}.
\end{aligned}$$

Substituting the last two results into the first result yields

$$\begin{aligned}
1 &= (Q_1^*)^{\frac{1}{4}} + (Q_2^*)^{\frac{1}{4}} = \left(\left(\frac{\lambda^*}{4}\right)^{\frac{4}{3}} P_1^{-\frac{4}{3}}\right)^{\frac{1}{4}} + \left(\left(\frac{\lambda^*}{4}\right)^{\frac{4}{3}} P_1^{-\frac{4}{3}}\right)^{\frac{1}{4}} \\
&\implies \left(\frac{\lambda^*}{4}\right)^{\frac{1}{3}} \left(P_1^{-\frac{1}{3}} + P_2^{-\frac{1}{3}}\right) = 1 \\
&\implies \left(\frac{\lambda^*}{4}\right)^{\frac{1}{3}} = \frac{1}{P_1^{-\frac{1}{3}} + P_2^{-\frac{1}{3}}} \\
&\implies \lambda^* = 4\left(P_1^{-\frac{1}{3}} + P_2^{-\frac{1}{3}}\right)^{-3}.
\end{aligned}$$

Once λ^* is found we can derive the optimal Q_1^*, Q_2^* as

$$\begin{aligned}
Q_1^* &= \left(\frac{\lambda^*}{4}\right)^{\frac{4}{3}} P_1^{-\frac{4}{3}} \\
&\implies Q_1^* = \left(\left(P_1^{-\frac{1}{3}} + P_2^{-\frac{1}{3}}\right)^{-3}\right)^{\frac{4}{3}} P_1^{-\frac{4}{3}} \\
&\implies Q_1^* = \left(P_1^{-\frac{1}{3}} + P_2^{-\frac{1}{3}}\right)^{-4} P_1^{-\frac{4}{3}}
\end{aligned}$$

and

$$
\begin{aligned}
Q_2^* &= \left(\frac{\lambda^*}{4}\right)^{\frac{4}{3}} P_2^{-\frac{4}{3}} \\
&\implies Q_2^* = \left(\left(P_1^{-\frac{1}{3}} + P_2^{-\frac{1}{3}}\right)^{-3}\right)^{\frac{4}{3}} P_2^{-\frac{4}{3}} \\
&\implies Q_2^* = \left(P_1^{-\frac{1}{3}} + P_2^{-\frac{1}{3}}\right)^{-4} P_2^{-\frac{4}{3}}.
\end{aligned}
$$

Here Q_1^*, Q_2^* are known as Hicksian demand curves.

To show this corresponds to a constrained global minimum we need to show that: 1) the objective function is linear, 2) the constraint function is globally convex, 3) the Lagrange multiplier is positive.

First the objective function, expenditure $P_1Q_1 + P_2Q_2$ is linear. The Hessian of the constraint $g(Q_1, Q_2) = 1 - Q_1^{\frac{1}{4}} - Q_2^{\frac{1}{4}} = 0$ is

$$
H(Q_1, Q_2) = \begin{bmatrix} \frac{3}{16} Q_1^{-\frac{7}{4}} & 0 \\ 0 & \frac{3}{16} Q_2^{-\frac{7}{4}} \end{bmatrix}
$$

which is a diagonal matrix with positive diagonal elements, and hence is positive definite for all Q_1, Q_2, and so the constraint is globally convex. Finally the Lagrange multiplier is positive as $\lambda^* = 4\left(P_1^{-\frac{1}{3}} + P_2^{-\frac{1}{3}}\right)^{-3} > 0$. Thus the solution corresponds to a constrained global minimum.

Problem 4.46 *[A***] Suppose an individual with utility*

$$
U(Q_1, Q_2) = 3Q_1^{\frac{1}{3}} + 3Q_2^{\frac{1}{3}}
$$

is thrown in prison. By law the prison warden must insure that the individual obtains enough Q_1, Q_2 to obtain $U_{\min}$ units of utility. Expenditure is then $E \equiv P_1Q_1 + P_2Q_2$ which the warden attempts to minimize. Write down the Lagrangian, the first-order conditions, and solve. Verify that you have a constrained global minimum.

Answer: The constraint is

$$
U_{\min} = U(Q_1, Q_2) = 3Q_1^{\frac{1}{3}} + 3Q_2^{\frac{1}{3}} \implies U_{\min} - 3Q_1^{\frac{1}{3}} - 3Q_2^{\frac{1}{3}} = 0
$$

and the objective function is $E \equiv P_1Q_1 + P_2Q_2$, which the warden attempts to minimize subject to the constraint. The Lagrangian is therefore

$$
\mathcal{L}(\lambda, Q_1, Q_2) = P_1Q_1 + P_2Q_2 + \lambda\left(U_{\min} - 3Q_1^{\frac{1}{3}} - 3Q_2^{\frac{1}{3}}\right)
$$

so that

$$\begin{aligned}\frac{\partial \mathcal{L}(\lambda, Q_1, Q_2)}{\partial \lambda} &= U_{\min} - 3Q_1^{\frac{1}{3}} - 3Q_2^{\frac{1}{3}} \\ \frac{\partial \mathcal{L}(\lambda, Q_1, Q_2)}{\partial Q_1} &= P_1 - \lambda Q_1^{-2/3} \\ \frac{\partial \mathcal{L}(\lambda, Q_1, Q_2)}{\partial Q_2} &= P_2 - \lambda Q_2^{-2/3}.\end{aligned}$$

The first-order conditions yield

$$\begin{aligned}U_{\min} - 3(Q_1^*)^{\frac{1}{3}} - 3(Q_2^*)^{\frac{1}{3}} &= 0 \\ &\implies U_{\min} = 3(Q_1^*)^{\frac{1}{3}} + 3(Q_2^*)^{\frac{1}{3}} \\ P_1 - \lambda^*(Q_1^*)^{-\frac{2}{3}} &= 0 \\ &\implies P_1 = \lambda^*(Q_1^*)^{-\frac{2}{3}} \implies Q_1^* = (\lambda^*)^{\frac{3}{2}} P_1^{-\frac{3}{2}} \\ P_2 - \lambda^*(Q_2^*)^{-\frac{2}{3}} &= 0 \\ &\implies P_2 = \lambda^*(Q_2^*)^{-\frac{2}{3}} \implies Q_2^* = (\lambda^*)^{\frac{3}{2}} P_2^{-\frac{3}{2}}\end{aligned}$$

so that

$$\begin{aligned}U_{\min} &= 3(Q_1^*)^{\frac{1}{3}} + 3(Q_2^*)^{\frac{1}{3}} = 3\left((\lambda^*)^{\frac{3}{2}} P_1^{-\frac{3}{2}}\right)^{\frac{1}{3}} + 3\left((\lambda^*)^{\frac{3}{2}} P_2^{-\frac{3}{2}}\right)^{\frac{1}{3}} \\ &\implies 3(\lambda^*)^{\frac{1}{2}}\left(P_1^{-\frac{1}{2}} + P_2^{-\frac{1}{2}}\right) = U_{\min} \\ &\implies \lambda^* = \frac{U_{\min}^2}{9}\left(P_1^{-\frac{1}{2}} + P_2^{-\frac{1}{2}}\right)^{-2}.\end{aligned}$$

Once λ^* is found we can derive the optimal Q_1^*, Q_2^* as

$$\begin{aligned}Q_1^* &= (\lambda^*)^{\frac{3}{2}} P_1^{-\frac{3}{2}} = \left(\frac{U_{\min}^2}{9}\left(P_1^{-\frac{1}{2}} + P_2^{-\frac{1}{2}}\right)^{-2}\right)^{\frac{3}{2}} P_1^{-\frac{3}{2}} \\ &= \frac{(U_{\min})^3}{27}\left(P_1^{-\frac{1}{2}} + P_2^{-\frac{1}{2}}\right)^{-3} P_1^{-\frac{3}{2}}\end{aligned}$$

and

$$\begin{aligned}Q_2^* &= (\lambda^*)^{\frac{3}{2}} P_2^{-\frac{3}{2}} = \left(\frac{U_{\min}^2}{9}\left(P_1^{-\frac{1}{2}} + P_2^{-\frac{1}{2}}\right)^{-2}\right)^{\frac{3}{2}} P_2^{-\frac{3}{2}} \\ &= \frac{(U_{\min})^3}{27}\left(P_1^{-\frac{1}{2}} + P_2^{-\frac{1}{2}}\right)^{-3} P_2^{-\frac{3}{2}}.\end{aligned}$$

To show this corresponds to a constrained global minimum we need to show that: 1) the objective function is linear, 2) the constraint function is globally convex, 3) the Lagrange multiplier is positive. First the objective function, expenditure $P_1Q_1 + P_2Q_2$ is linear. The Hessian of the constraint

$$g(Q_1, Q_2) = U_{\min} - 3Q_1^{\frac{1}{3}} - 3Q_2^{\frac{1}{3}} = 0$$

is

$$H(Q_1, Q_2) = \begin{bmatrix} \frac{2}{3}Q_1^{-\frac{5}{3}} & 0 \\ 0 & \frac{2}{3}Q_2^{-\frac{5}{3}} \end{bmatrix}$$

which is a diagonal matrix with positive diagonal elements, and hence is positive definite for all Q_1, Q_2, and so the constraint is globally convex. Finally the Lagrange multiplier is positive as

$$\lambda^* = \frac{U_{\min}^2}{9}\left(P_1^{-\frac{1}{2}} + P_2^{-\frac{1}{2}}\right)^{-2} > 0.$$

Thus the solution corresponds to a constrained global minimum.

4.3.3 Cost Minimization

Problem 4.47 *[B***] Consider a firm with production function*

$$Q = F(L, K) = 3L^{\frac{1}{3}}K^{\frac{1}{3}}.$$

*From the Lagrangian for the problem of minimizing the cost $C \equiv WL + RK$ of producing Q units of output. Find the first-order conditions and solve. Show that the solution corresponds to a constrained global minimum. [A***] Find the cost function $C^*(Q, W, R) = WL^* + RK^*$ and verify Shephard's lemma $\frac{\partial C^*}{\partial W} = L^*$.*

Answer: The Lagrangian is

$$\mathcal{L}(\lambda, L, K) = WL + RK + \lambda\left(Q - 3L^{\frac{1}{3}}K^{\frac{1}{3}}\right)$$

so that

$$\begin{aligned} \frac{\partial \mathcal{L}(\lambda, L, K)}{\partial \lambda} &= Q - 3L^{\frac{1}{3}}K^{\frac{1}{3}} \\ \frac{\partial \mathcal{L}(\lambda, L, K)}{\partial L} &= W - \lambda L^{-\frac{2}{3}}K^{\frac{1}{3}} \\ \frac{\partial \mathcal{L}(\lambda, L, K)}{\partial K} &= R - \lambda L^{\frac{1}{3}}K^{-\frac{2}{3}}. \end{aligned}$$

The first-order conditions yield

$$\begin{aligned} Q - 3(L^*)^{\frac{1}{3}}(K^*)^{\frac{1}{3}} &= 0 \\ &\implies Q = 3(L^*)^{\frac{1}{3}}(K^*)^{\frac{1}{3}} \\ W - \lambda^*(L^*)^{-\frac{2}{3}}(K^*)^{\frac{1}{3}} &= 0 \\ &\implies W = \lambda^*(L^*)^{-\frac{2}{3}}(K^*)^{\frac{1}{3}} \\ R - \lambda^*(L^*)^{\frac{1}{3}}(K^*)^{-\frac{2}{3}} &= 0 \\ &\implies R = \lambda^*(L^*)^{\frac{1}{3}}(K^*)^{-\frac{2}{3}}. \end{aligned}$$

We have

$$\begin{aligned}\frac{W}{R} &= \frac{\lambda^*(L^*)^{-\frac{2}{3}}(K^*)^{\frac{1}{3}}}{\lambda^*(L^*)^{\frac{1}{3}}(K^*)^{-\frac{2}{3}}} = \frac{K^*}{L^*}\\ &\implies K^* = WR^{-1}L^*\end{aligned}$$

so that

$$\begin{aligned}Q &= 3(L^*)^{\frac{1}{3}}(K^*)^{\frac{1}{3}} = 3(L^*)^{\frac{1}{3}}\left(WR^{-1}L^*\right)^{\frac{1}{3}} = 3(L^*)^{\frac{2}{3}}W^{\frac{1}{3}}R^{-\frac{1}{3}}\\ &\implies L^* = 3^{-\frac{3}{2}}Q^{\frac{3}{2}}W^{-\frac{1}{2}}R^{\frac{1}{2}}\end{aligned}$$

and

$$\begin{aligned}K^* &= WR^{-1}L^* = WR^{-1}\times 3^{-\frac{3}{2}}Q^{\frac{3}{2}}W^{-\frac{1}{2}}R^{\frac{1}{2}} = 3^{-\frac{3}{2}}Q^{\frac{3}{2}}W^{\frac{1}{2}}R^{-\frac{1}{2}}\\ &\implies K^* = 3^{-\frac{3}{2}}Q^{\frac{3}{2}}W^{\frac{1}{2}}R^{-\frac{1}{2}}.\end{aligned}$$

Now we find the Lagrange multiplier from $W = \lambda^*(L^*)^{-\frac{2}{3}}(K^*)^{\frac{1}{3}}$ as

$$\begin{aligned}\lambda^* &= (L^*)^{\frac{2}{3}}(K^*)^{-\frac{1}{3}}W = \left(3^{-\frac{3}{2}}Q^{\frac{3}{2}}W^{-\frac{1}{2}}R^{\frac{1}{2}}\right)^{\frac{2}{3}}\left(3^{-\frac{3}{2}}Q^{\frac{3}{2}}W^{\frac{1}{2}}R^{-\frac{1}{2}}\right)^{-\frac{1}{3}}W\\ &= 3^{-\frac{1}{2}}Q^{\frac{1}{2}}W^{\frac{1}{2}}R^{\frac{1}{2}}.\end{aligned}$$

To show that our solution corresponds to a constrained global minimum we have: 1) the objective function $WL+RK$ is linear. 2) The constraint is globally convex as

$$\begin{aligned}g(L,K) &= Q - 3L^{\frac{1}{3}}K^{\frac{1}{3}}\\ &\implies H(L,K) = \begin{bmatrix}\frac{2}{3}L^{-\frac{5}{3}}K^{\frac{1}{3}} & -\frac{1}{3}L^{-\frac{2}{3}}K^{-\frac{2}{3}}\\ -\frac{1}{3}L^{-\frac{2}{3}}K^{-\frac{2}{3}} & \frac{2}{3}L^{\frac{1}{3}}K^{-\frac{5}{3}}\end{bmatrix}\end{aligned}$$

with

$$\begin{aligned}M_1 &= \frac{2}{3}L^{-\frac{5}{3}}K^{\frac{1}{3}} > 0\\ M_2 &= \left(\frac{2}{3}L^{-\frac{5}{3}}K^{\frac{1}{3}}\right)\left(\frac{2}{3}L^{\frac{1}{3}}K^{-\frac{5}{3}}\right) - \left(-\frac{1}{3}L^{-\frac{2}{3}}K^{-\frac{2}{3}}\right)^2\\ &= \frac{4}{9}L^{-\frac{4}{3}}K^{-\frac{4}{3}} - \frac{1}{9}L^{-\frac{4}{3}}K^{-\frac{4}{3}} = \frac{3}{9}L^{-\frac{4}{3}}K^{-\frac{4}{3}} > 0.\end{aligned}$$

Since $M_1 > 0$ and $M_2 > 0$ for all L, K it follows that $H(L,K)$ is positive definite for all L, K and so $g(L,K) = Q - 3L^{\frac{1}{3}}K^{\frac{1}{3}}$ is globally convex. 3) The Lagrange multiplier is positive as $\lambda^* = \frac{1}{4}QW^{\frac{1}{2}}R^{\frac{1}{2}} > 0$.

The cost function is then

$$\begin{aligned}C^*(Q,W,R) &= WL^* + RK^* = W\times 3^{-\frac{3}{2}}Q^{\frac{3}{2}}W^{-\frac{1}{2}}R^{\frac{1}{2}} + R\times 3^{-\frac{3}{2}}Q^{\frac{3}{2}}W^{\frac{1}{2}}R^{-\frac{1}{2}}\\ &= 2\times 3^{-\frac{3}{2}}Q^{\frac{3}{2}}W^{\frac{1}{2}}R^{\frac{1}{2}}\end{aligned}$$

so that

$$\frac{\partial C^*(Q,W,R)}{\partial W} = \frac{\partial}{\partial W}\left(2\times 3^{-\frac{3}{2}}Q^{\frac{3}{2}}W^{\frac{1}{2}}R^{\frac{1}{2}}\right) = 3^{-\frac{3}{2}}Q^{\frac{3}{2}}W^{-\frac{1}{2}}R^{\frac{1}{2}} = L^*.$$

Problem 4.48 *Consider a firm with production function* $Q = \frac{1}{\alpha}L^\alpha K^\alpha$ *for* $\alpha < \frac{1}{2}$. *[B***] Find the first-order conditions for cost minimization and solve. [A***] Show that the solution corresponds to a constrained global minimum. Find the cost function* $C^* = WL^* + RK^*$ *and show that there are decreasing returns to scale. Verify that the Lagrange multiplier is marginal cost or* $\frac{\partial C^*}{\partial Q} = \lambda^*$, *and Shephard's lemma that*

$$\frac{\partial C^*}{\partial W} = L^*, \frac{\partial C^*}{\partial R} = K^*.$$

Answer: The Lagrangian is

$$\mathcal{L}(\lambda, L, K) = WL + RK + \lambda\left(Q - \frac{1}{\alpha}L^\alpha K^\alpha\right)$$

so that

$$\begin{aligned}
\frac{\partial \mathcal{L}(\lambda,L,K)}{\partial \lambda} &= Q - \frac{1}{\alpha}L^\alpha K^\alpha \\
\frac{\partial \mathcal{L}(\lambda,L,K)}{\partial L} &= W - \lambda L^{\alpha-1}K^\alpha \\
\frac{\partial \mathcal{L}(\lambda,L,K)}{\partial K} &= R - \lambda L^\alpha K^{\alpha-1}.
\end{aligned}$$

The first-order conditions yield

$$\begin{aligned}
Q - \frac{1}{\alpha}(L^*)^\alpha (K^*)^\alpha &= 0 \\
&\implies Q = \frac{1}{\alpha}(L^*)^\alpha (K^*)^\alpha \\
W - \lambda^*(L^*)^{\alpha-1}(K^*)^\alpha &= 0 \\
&\implies W = \lambda^*(L^*)^{\alpha-1}(K^*)^\alpha \\
R - \lambda^*(L^*)^\alpha (K^*)^{\alpha-1} &= 0 \\
&\implies R = \lambda^*(L^*)^\alpha (K^*)^{\alpha-1}.
\end{aligned}$$

Solving we have

$$\begin{aligned}
\frac{W}{R} &= \frac{\lambda^*(L^*)^{\alpha-1}(K^*)^\alpha}{\lambda^*(L^*)^\alpha (K^*)^{\alpha-1}} = \frac{K^*}{L^*} \\
&\implies K^* = WR^{-1}L^*
\end{aligned}$$

so that

$$\begin{aligned}
Q &= \frac{1}{\alpha}(L^*)^\alpha (K^*)^\alpha = \frac{1}{\alpha}(L^*)^\alpha \left(WR^{-1}L^*\right)^\alpha = \frac{1}{\alpha}(L^*)^{2\alpha}W^\alpha R^{-\alpha} \\
&\implies (L^*)^{2\alpha} = \alpha Q W^{-\alpha}R^\alpha \\
&\implies L^* = \alpha^{\frac{1}{2\alpha}}Q^{\frac{1}{2\alpha}}W^{-\frac{1}{2}}R^{\frac{1}{2}}
\end{aligned}$$

and

$$K^* = WR^{-1}L^* = WR^{-1}\alpha^{\frac{1}{2\alpha}}Q^{\frac{1}{2\alpha}}W^{-\frac{1}{2}}R^{\frac{1}{2}} = \alpha^{\frac{1}{2\alpha}}Q^{\frac{1}{2\alpha}}W^{\frac{1}{2}}R^{-\frac{1}{2}}.$$

Now we find the Lagrange multiplier from

$$W = \lambda^* (L^*)^{\alpha-1} (K^*)^{\alpha}$$

as

$$\begin{aligned} W &= \lambda^* (L^*)^{\alpha-1} (K^*)^{\alpha} = \lambda^* \left(\alpha^{\frac{1}{2\alpha}}Q^{\frac{1}{2\alpha}}W^{-\frac{1}{2}}R^{\frac{1}{2}}\right)^{\alpha-1} \left(\alpha^{\frac{1}{2\alpha}}Q^{\frac{1}{2\alpha}}W^{\frac{1}{2}}R^{-\frac{1}{2}}\right)^{\alpha} \\ &\implies \lambda^* = \alpha^{\frac{1}{2\alpha}-1}Q^{\frac{1}{2\alpha}-1}W^{\frac{1}{2}}R^{\frac{1}{2}} \end{aligned}$$

To show that our solution corresponds to a constrained global minimum we have: 1) the objective function $WL + RK$ is linear. 2) The constraint is convex as

$$\begin{aligned} g(L,K) &= Q - \frac{1}{\alpha}L^{\alpha}K^{\alpha} \\ &\implies H(L,K) = \begin{bmatrix} (1-\alpha) L^{\alpha-2}K^{\alpha} & -\alpha L^{\alpha-1}K^{\alpha-1} \\ -\alpha L^{\alpha-1}K^{\alpha-1} & (1-\alpha) L^{\alpha}K^{\alpha-2} \end{bmatrix} \end{aligned}$$

and so using $\alpha < \frac{1}{2}$ we have

$$\begin{aligned} M_1 &= (1-\alpha) L^{\alpha-2}K^{\alpha} > 0 \\ M_2 &= \left((1-\alpha) L^{\alpha-2}K^{\alpha}\right)\left((1-\alpha) L^{\alpha}K^{\alpha-2}\right) - \left(-\alpha L^{\alpha-1}K^{\alpha-1}\right)^2 \\ &= (1-\alpha)^2 L^{2\alpha-2}K^{2\alpha-2} - \alpha^2 L^{2\alpha-2}K^{2\alpha-2} \\ &= \left((1-\alpha)^2 - \alpha^2\right) L^{2\alpha-2}K^{2\alpha-2} = (1-2\alpha) L^{2\alpha-2}K^{2\alpha-2} > 0. \end{aligned}$$

Since $M_1 > 0$ and $M_2 > 0$ for all L, K it follows that $H(L,K)$ is positive definite for all L, K and so $g(L,K) = Q - \frac{1}{\alpha}L^{\alpha}K^{\alpha}$ is globally convex. 3) The Lagrange multiplier is positive as $\lambda^* = \alpha^{\frac{1}{2\alpha}-1}Q^{\frac{1}{2\alpha}-1}W^{\frac{1}{2}}R^{\frac{1}{2}} > 0$.

The cost function is then

$$\begin{aligned} C^*(Q,W,R) &= WL^* + RK^* = W \times \alpha^{\frac{1}{2\alpha}}Q^{\frac{1}{2\alpha}}W^{-\frac{1}{2}}R^{\frac{1}{2}} + R \times \alpha^{\frac{1}{2\alpha}}Q^{\frac{1}{2\alpha}}W^{\frac{1}{2}}R^{-\frac{1}{2}} \\ &= \alpha^{\frac{1}{2\alpha}}Q^{\frac{1}{2\alpha}}W^{\frac{1}{2}}R^{\frac{1}{2}} + \alpha^{\frac{1}{2\alpha}}Q^{\frac{1}{2\alpha}}W^{\frac{1}{2}}R^{\frac{1}{2}} \\ &= 2\alpha^{\frac{1}{2\alpha}}Q^{\frac{1}{2\alpha}}W^{\frac{1}{2}}R^{\frac{1}{2}}. \end{aligned}$$

The Lagrange multiplier is marginal cost since

$$\frac{\partial C^*(Q,W,R)}{\partial Q} = \frac{\partial}{\partial Q}\left(2\alpha^{\frac{1}{2\alpha}}Q^{\frac{1}{2\alpha}}W^{\frac{1}{2}}R^{\frac{1}{2}}\right) = \alpha^{\frac{1}{2\alpha}-1}Q^{\frac{1}{2\alpha}-1}W^{\frac{1}{2}}R^{\frac{1}{2}} = \lambda^*$$

and Shephard's lemma holds as

$$\begin{aligned} \frac{\partial C^*(Q,W,R)}{\partial W} &= \frac{\partial}{\partial W}\left(2\alpha^{\frac{1}{2\alpha}}Q^{\frac{1}{2\alpha}}W^{\frac{1}{2}}R^{\frac{1}{2}}\right) = \alpha^{\frac{1}{2\alpha}}Q^{\frac{1}{2\alpha}}W^{-\frac{1}{2}}R^{\frac{1}{2}} = L^* \\ \frac{\partial C^*(Q,W,R)}{\partial R} &= \frac{\partial}{\partial R}\left(2\alpha^{\frac{1}{2\alpha}}Q^{\frac{1}{2\alpha}}W^{\frac{1}{2}}R^{\frac{1}{2}}\right) = \alpha^{\frac{1}{2\alpha}}Q^{\frac{1}{2\alpha}}W^{\frac{1}{2}}R^{-\frac{1}{2}} = K^*. \end{aligned}$$

There are decreasing returns to scale since average costs are

$$\begin{aligned}AC^*(Q,W,R) &\equiv \frac{C^*(Q,W,R)}{Q}\\ &= \frac{2\alpha^{\frac{1}{2\alpha}}Q^{\frac{1}{2\alpha}}W^{\frac{1}{2}}R^{\frac{1}{2}}}{Q} = 2\alpha^{\frac{1}{2\alpha}}Q^{\frac{1}{2\alpha}-1}W^{\frac{1}{2}}R^{\frac{1}{2}}\end{aligned}$$

with

$$\frac{\partial AC^*(Q,W,R)}{\partial Q} = \left(\frac{1}{2\alpha}-1\right)2\alpha^{\frac{1}{2\alpha}}Q^{\frac{1}{2\alpha}-2}W^{\frac{1}{2}}R^{\frac{1}{2}} > 0$$

since

$$\alpha < \frac{1}{2} \Longrightarrow \frac{1}{2\alpha} - 1 = \frac{1-2\alpha}{2\alpha} = \frac{\overbrace{2\left(\frac{1}{2}-\alpha\right)}^{+}}{2\alpha} > 0.$$

Problem 4.49 *Consider the problem of minimizing the cost: $C \equiv WL + RK$ of producing Q units of output where*

$$Q = F(L,K) = 4L^{\frac{1}{4}} + 4K^{\frac{1}{4}}.$$

*[B***] Write down the Lagrangian for this problem, find the first-order conditions and solve. Show that the solution is a constrained global minimum. [A***] Verify that the Lagrange multiplier is marginal cost or $\frac{\partial C^*}{\partial Q} = \lambda^*$, and Shephard's lemma that*

$$\frac{\partial C^*}{\partial W} = L^*, \quad \frac{\partial C^*}{\partial R} = K^*.$$

Answer: We have

$$\mathcal{L}(\lambda,L,K) = WL + RK + \lambda\left(Q - 4L^{\frac{1}{4}} - 4K^{\frac{1}{4}}\right)$$

with

$$\begin{aligned}\frac{\partial \mathcal{L}(\lambda,L,K)}{\partial \lambda} &= Q - 4L^{\frac{1}{4}} - 4K^{\frac{1}{4}}\\ \frac{\partial \mathcal{L}(\lambda,L,K)}{\partial L} &= W - \lambda L^{-\frac{3}{4}}\\ \frac{\partial \mathcal{L}(\lambda,L,K)}{\partial K} &= R - \lambda K^{-\frac{3}{4}}.\end{aligned}$$

The first-order conditions yield

$$\begin{aligned}Q - 4(L^*)^{\frac{1}{4}} - 4(K^*)^{\frac{1}{4}} &= 0\\ &\Longrightarrow Q = 4(L^*)^{\frac{1}{4}} + 4(K^*)^{\frac{1}{4}}\\ W - \lambda^*(L^*)^{-\frac{3}{4}} &= 0\\ &\Longrightarrow L^* = (\lambda^*)^{\frac{4}{3}}W^{-\frac{4}{3}}\\ R - \lambda^*(K^*)^{-\frac{3}{4}} &= 0\\ &\Longrightarrow K^* = (\lambda^*)^{\frac{4}{3}}R^{-\frac{4}{3}}.\end{aligned}$$

Dividing the second into the third result yields

$$\frac{K^*}{L^*} = \left(\frac{R}{W}\right)^{-\frac{4}{3}} \Longrightarrow K^* = R^{-\frac{4}{3}} W^{\frac{4}{3}} L^*.$$

We have

$$\begin{aligned}
Q &= 4\left(L^*\right)^{\frac{1}{4}} + 4\left(R^{-\frac{4}{3}} W^{\frac{4}{3}} L^*\right)^{\frac{1}{4}} \\
&= 4\left(L^*\right)^{\frac{1}{4}}\left(1 + R^{-\frac{1}{3}} W^{\frac{1}{3}}\right) = 4\left(L^*\right)^{\frac{1}{4}}\left(W^{-\frac{1}{3}} + R^{-\frac{1}{3}}\right) W^{\frac{1}{3}} \\
&\Longrightarrow \left(L^*\right)^{\frac{1}{4}} = \frac{Q}{4}\left(W^{-\frac{1}{3}} + R^{-\frac{1}{3}}\right)^{-1} W^{-\frac{1}{3}} \\
&\Longrightarrow L^* = \left(\frac{Q}{4}\right)^4 \left(W^{-\frac{1}{3}} + R^{-\frac{1}{3}}\right)^{-4} W^{-\frac{4}{3}}.
\end{aligned}$$

Now

$$\begin{aligned}
K^* &= R^{-\frac{4}{3}} W^{\frac{4}{3}} L^* = R^{-\frac{4}{3}} W^{\frac{4}{3}} \left(\frac{Q}{4}\right)^4 \left(W^{-\frac{1}{3}} + R^{-\frac{1}{3}}\right)^{-4} W^{-\frac{4}{3}} \\
&= \left(\frac{Q}{4}\right)^4 \left(W^{-\frac{1}{3}} + R^{-\frac{1}{3}}\right)^{-4} R^{-\frac{4}{3}}
\end{aligned}$$

and

$$\begin{aligned}
\lambda^* &= R\left(K^*\right)^{\frac{3}{4}} \\
&\Longrightarrow \lambda^* = R\left(\left(\frac{Q}{4}\right)^4 R^{-\frac{4}{3}} \left(W^{-\frac{1}{3}} + R^{-\frac{1}{3}}\right)^{-4}\right)^{\frac{3}{4}} \\
&\Longrightarrow \lambda^* = \left(\frac{Q}{4}\right)^3 \left(W^{-\frac{1}{3}} + R^{-\frac{1}{3}}\right)^{-3}.
\end{aligned}$$

Since 1) the objective function $WL + RK$ is linear, 2) the constraint

$$g(L, K) = Q - 4L^{\frac{1}{4}} - 4K^{\frac{1}{4}} = 0$$

has a Hessian

$$H(L, K) = \begin{bmatrix} \frac{3}{4}L^{-\frac{7}{4}} & 0 \\ 0 & \frac{3}{4}K^{-\frac{7}{4}} \end{bmatrix}$$

which is positive definite (diagonal matrix with positive diagonal elements) and so the constraint is globally convex and 3) the Lagrange multiplier is positive as

$$\lambda^* = \left(\frac{Q}{4}\right)^3 \left(W^{-\frac{1}{3}} + R^{-\frac{1}{3}}\right)^{-3} > 0$$

it follows that the solution is a constrained global minimizer.

The cost function here is

$$\begin{aligned} C^*(Q,W,R) &= WL^* + RK^* \\ &= W\left(\frac{Q}{4}\right)^4 \left(W^{-\frac{1}{3}} + R^{-\frac{1}{3}}\right)^{-4} W^{-\frac{4}{3}} \\ &\quad + R\left(\frac{Q}{4}\right)^4 \left(W^{-\frac{1}{3}} + R^{-\frac{1}{3}}\right)^{-4} R^{-\frac{4}{3}} \\ &= \left(\frac{Q}{4}\right)^4 \left(W^{-\frac{1}{3}} + R^{-\frac{1}{3}}\right)^{-4} \left(W^{-\frac{1}{3}} + R^{-\frac{1}{3}}\right) \\ &= \left(\frac{Q}{4}\right)^4 \left(W^{-\frac{1}{3}} + R^{-\frac{1}{3}}\right)^{-3}. \end{aligned}$$

The Lagrange multiplier is marginal cost since

$$\begin{aligned} \frac{\partial C^*(Q,W,R)}{\partial Q} &= \frac{\partial}{\partial Q}\left(\left(\frac{Q}{4}\right)^4 \left(W^{-\frac{1}{3}} + R^{-\frac{1}{3}}\right)^{-3}\right) \\ &= \left(\frac{Q}{4}\right)^3 \left(W^{-\frac{1}{3}} + R^{-\frac{1}{3}}\right)^{-3} = \lambda^* \end{aligned}$$

and Shephard's lemma holds as

$$\begin{aligned} \frac{\partial C^*(Q,W,R)}{\partial W} &= \frac{\partial}{\partial W}\left(\left(\frac{Q}{4}\right)^3 \left(W^{-\frac{1}{3}} + R^{-\frac{1}{3}}\right)^{-3}\right) \\ &= \left(\frac{Q}{4}\right)^4 \left(W^{-\frac{1}{3}} + R^{-\frac{1}{3}}\right)^{-4} W^{-\frac{4}{3}} = L^* \\ \frac{\partial C^*(Q,W,R)}{\partial R} &= \frac{\partial}{\partial R}\left(\left(\frac{Q}{4}\right)^3 \left(W^{-\frac{1}{3}} + R^{-\frac{1}{3}}\right)^{-3}\right) \\ &= \left(\frac{Q}{4}\right)^4 \left(W^{-\frac{1}{3}} + R^{-\frac{1}{3}}\right)^{-4} R^{-\frac{4}{3}} = K^*. \end{aligned}$$

Problem 4.50 *Consider the problem of minimizing the cost $C \equiv WL + RK$ of producing Q units of output where*

$$Q = F(L,K) = \frac{1}{\alpha}L^\alpha + \frac{1}{\alpha}K^\alpha \text{ for } \alpha < 1.$$

*[B***] Write down the Lagrangian for this problem, find the first-order conditions and solve. [A***] Show that the solution corresponds to a constrained global minimum.*

Answer: We have

$$\mathcal{L}(\lambda, L, K) = WL + RK + \lambda\left(Q - \frac{1}{\alpha}L^\alpha - \frac{1}{\alpha}K^\alpha\right)$$

with

$$\begin{aligned}
\frac{\partial \mathcal{L}(\lambda, L, K)}{\partial \lambda} &= Q - \frac{1}{\alpha}L^{\alpha} - \frac{1}{\alpha}K^{\alpha} \\
\frac{\partial \mathcal{L}(\lambda, L, K)}{\partial L} &= W - \lambda L^{\alpha-1} \\
\frac{\partial \mathcal{L}(\lambda, L, K)}{\partial K} &= R - \lambda K^{\alpha-1}.
\end{aligned}$$

The first-order conditions yield

$$\begin{aligned}
Q - \frac{1}{\alpha}(L^*)^{\alpha} - \frac{1}{\alpha}(K^*)^{\alpha} &= 0 \\
&\Longrightarrow Q = \frac{1}{\alpha}(L^*)^{\alpha} + \frac{1}{\alpha}(K^*)^{\alpha} \\
W - \lambda^*(L^*)^{\alpha-1} &= 0 \\
&\Longrightarrow L^* = (\lambda^*)^{\frac{1}{1-\alpha}} W^{-\frac{1}{1-\alpha}} \\
R - \lambda^*(K^*)^{\alpha-1} &= 0 \\
&\Longrightarrow K^* = (\lambda^*)^{\frac{1}{1-\alpha}} R^{-\frac{1}{1-\alpha}}.
\end{aligned}$$

Dividing the second into the third result yields

$$\begin{aligned}
\frac{K^*}{L^*} &= \frac{(\lambda^*)^{\frac{1}{1-\alpha}} R^{-\frac{1}{1-\alpha}}}{(\lambda^*)^{\frac{1}{1-\alpha}} W^{-\frac{1}{1-\alpha}}} = R^{-\frac{1}{1-\alpha}} W^{\frac{1}{1-\alpha}} \\
&\Longrightarrow K^* = R^{-\frac{1}{1-\alpha}} W^{\frac{1}{1-\alpha}} L^*
\end{aligned}$$

so that

$$\begin{aligned}
Q &= \frac{1}{\alpha}(L^*)^{\alpha} + \frac{1}{\alpha}\left(R^{-\frac{1}{1-\alpha}} W^{\frac{1}{1-\alpha}} L^*\right)^{\alpha} = \frac{1}{\alpha}(L^*)^{\alpha}\left(1 + R^{-\frac{\alpha}{1-\alpha}} W^{\frac{\alpha}{1-\alpha}}\right) \\
&= \frac{1}{\alpha}(L^*)^{\alpha} W^{\frac{\alpha}{1-\alpha}}\left(W^{-\frac{\alpha}{1-\alpha}} + R^{-\frac{\alpha}{1-\alpha}}\right) \\
&\Longrightarrow (L^*)^{\alpha} = \alpha Q W^{-\frac{\alpha}{1-\alpha}}\left(W^{-\frac{\alpha}{1-\alpha}} + R^{-\frac{\alpha}{1-\alpha}}\right)^{-1} \\
&\Longrightarrow L^* = \alpha^{\frac{1}{\alpha}} Q^{\frac{1}{\alpha}} W^{-\frac{1}{1-\alpha}}\left(W^{-\frac{\alpha}{1-\alpha}} + R^{-\frac{\alpha}{1-\alpha}}\right)^{-\frac{1}{\alpha}}
\end{aligned}$$

and

$$\begin{aligned}
K^* &= R^{-\frac{1}{1-\alpha}} W^{\frac{1}{1-\alpha}} L^* = R^{-\frac{1}{1-\alpha}} W^{\frac{1}{1-\alpha}} \times \alpha^{\frac{1}{\alpha}} Q^{\frac{1}{\alpha}} W^{-\frac{1}{1-\alpha}}\left(W^{-\frac{\alpha}{1-\alpha}} + R^{-\frac{\alpha}{1-\alpha}}\right)^{-\frac{1}{\alpha}} \\
&\Longrightarrow K^* = \alpha^{\frac{1}{\alpha}} Q^{\frac{1}{\alpha}} R^{-\frac{1}{1-\alpha}}\left(W^{-\frac{\alpha}{1-\alpha}} + R^{-\frac{\alpha}{1-\alpha}}\right)^{-\frac{1}{\alpha}}.
\end{aligned}$$

The cost function is

$$\begin{aligned}
C^*(Q,W,R) &= WL^* + RK^* \\
&= W\alpha^{\frac{1}{\alpha}}Q^{\frac{1}{\alpha}}W^{-\frac{1}{1-\alpha}}\left(W^{-\frac{\alpha}{1-\alpha}} + R^{-\frac{\alpha}{1-\alpha}}\right)^{-\frac{1}{\alpha}} \\
&\quad + R\alpha^{\frac{1}{\alpha}}Q^{\frac{1}{\alpha}}R^{-\frac{1}{1-\alpha}}\left(W^{-\frac{\alpha}{1-\alpha}} + R^{-\frac{\alpha}{1-\alpha}}\right)^{-\frac{1}{\alpha}} \\
&= \alpha^{\frac{1}{\alpha}}Q^{\frac{1}{\alpha}}W^{-\frac{\alpha}{1-\alpha}}\left(W^{-\frac{\alpha}{1-\alpha}} + R^{-\frac{\alpha}{1-\alpha}}\right)^{-\frac{1}{\alpha}} \\
&\quad + \alpha^{\frac{1}{\alpha}}Q^{\frac{1}{\alpha}}R^{-\frac{\alpha}{1-\alpha}}\left(W^{-\frac{\alpha}{1-\alpha}} + R^{-\frac{\alpha}{1-\alpha}}\right)^{-\frac{1}{\alpha}} \\
&= \alpha^{\frac{1}{\alpha}}Q^{\frac{1}{\alpha}}\left(W^{-\frac{\alpha}{1-\alpha}} + R^{-\frac{\alpha}{1-\alpha}}\right)^{-\frac{1}{\alpha}}\left(W^{-\frac{\alpha}{1-\alpha}} + R^{-\frac{\alpha}{1-\alpha}}\right) \\
&= \alpha^{\frac{1}{\alpha}}Q^{\frac{1}{\alpha}}\left(W^{-\frac{\alpha}{1-\alpha}} + R^{-\frac{\alpha}{1-\alpha}}\right)^{1-\frac{1}{\alpha}} = \alpha^{\frac{1}{\alpha}}Q^{\frac{1}{\alpha}}\left(W^{-\frac{\alpha}{1-\alpha}} + R^{-\frac{\alpha}{1-\alpha}}\right)^{-\frac{1-\alpha}{\alpha}}.
\end{aligned}$$

There are decreasing returns to scale since average costs are

$$\begin{aligned}
AC^*(Q,W,R) &\equiv \frac{C^*(Q,W,R)}{Q} = \frac{\alpha^{\frac{1}{\alpha}}Q^{\frac{1}{\alpha}}\left(W^{-\frac{\alpha}{1-\alpha}} + R^{-\frac{\alpha}{1-\alpha}}\right)^{-\frac{1-\alpha}{\alpha}}}{Q} \\
&= \alpha^{\frac{1}{\alpha}}Q^{\frac{1}{\alpha}-1}\left(W^{-\frac{\alpha}{1-\alpha}} + R^{-\frac{\alpha}{1-\alpha}}\right)^{-\frac{1-\alpha}{\alpha}}
\end{aligned}$$

so that

$$\frac{\partial AC^*(Q,W,R)}{\partial Q} = \left(\frac{1}{\alpha} - 1\right)\alpha^{\frac{1}{\alpha}}Q^{\frac{1}{\alpha}-2}\left(W^{-\frac{\alpha}{1-\alpha}} + R^{-\frac{\alpha}{1-\alpha}}\right)^{-\frac{1-\alpha}{\alpha}} > 0$$

since

$$\alpha < 1 \Longrightarrow \frac{1}{\alpha} - 1 = \frac{1-\alpha}{\alpha} > 0.$$

Since 1) the objective function $WL + RK$ is linear, 2) the constraint

$$g(L,K) = Q - \frac{1}{\alpha}L^{\alpha} - \frac{1}{\alpha}K^{\alpha} = 0$$

has a Hessian

$$H(L,K) = \begin{bmatrix} (1-\alpha)L^{\alpha-2} & 0 \\ 0 & (1-\alpha)K^{\alpha-2} \end{bmatrix}$$

which is positive definite (diagonal matrix with positive diagonal elements) since $\alpha < 1$, and so the constraint is convex and 3) the Lagrange multiplier is positive since from the first-order conditions

$$W - \lambda^*(L^*)^{\alpha-1} = 0 \Longrightarrow \lambda^* = \frac{W}{(L^*)^{\alpha-1}} > 0$$

it follows that the solution corresponds to a constrained global minimum.

Problem 4.51 *Suppose the firm's production function is given by*

$$Q = F(L, K) = \left(L^{\frac{1}{2}} + K^{\frac{1}{2}}\right)^2.$$

*[A ***] Set up the Lagrangian for the problem of finding that amount of labour L^* and that amount of capital K^* that minimizes the cost of producing Q units of output. Find the first-order conditions from this Lagrangian and solve. Show that the solution corresponds to a constrained global minimum. Find the cost function $C^* \equiv WL^* + RK^*$ and show that there are constant returns to scale.*

Answer: The math is easier if write the constraint as

$$\begin{aligned} Q &= \left(L^{\frac{1}{2}} + K^{\frac{1}{2}}\right)^2 \Longrightarrow Q^{\frac{1}{2}} = L^{\frac{1}{2}} + K^{\frac{1}{2}} \\ &\Longrightarrow Q^{\frac{1}{2}} - L^{\frac{1}{2}} - K^{\frac{1}{2}} = 0. \end{aligned}$$

The Lagrangian is then

$$\mathcal{L}(\lambda, L, K) = WL + RK + \lambda\left(Q^{\frac{1}{2}} - L^{\frac{1}{2}} - K^{\frac{1}{2}}\right)$$

so that

$$\begin{aligned} \frac{\partial \mathcal{L}(\lambda, L, K)}{\partial \lambda} &= Q^{\frac{1}{2}} - L^{\frac{1}{2}} - K^{\frac{1}{2}} \\ \frac{\partial \mathcal{L}(\lambda, L, K)}{\partial L} &= W - \frac{\lambda}{2} L^{-\frac{1}{2}} \\ \frac{\partial \mathcal{L}(\lambda, L, K)}{\partial K} &= R - \frac{\lambda}{2} K^{-\frac{1}{2}}. \end{aligned}$$

The first-order conditions yield

$$\begin{aligned} Q^{\frac{1}{2}} - (L^*)^{\frac{1}{2}} - (K^*)^{\frac{1}{2}} &= 0 \\ &\Longrightarrow Q^{\frac{1}{2}} = (L^*)^{\frac{1}{2}} + (K^*)^{\frac{1}{2}} \\ W - \frac{\lambda^*}{2}(L^*)^{-\frac{1}{2}} &= 0 \\ &\Longrightarrow W = \frac{\lambda^*}{2}(L^*)^{-\frac{1}{2}} \\ R - \frac{\lambda^*}{2}(K^*)^{-\frac{1}{2}} &= 0 \\ &\Longrightarrow R = \frac{\lambda^*}{2}(K^*)^{-\frac{1}{2}}. \end{aligned}$$

Dividing the two last we have

$$\frac{W}{R} = \frac{\frac{\lambda^*}{2}(L^*)^{-\frac{1}{2}}}{\frac{\lambda^*}{2}(K^*)^{-\frac{1}{2}}} = \left(\frac{K^*}{L^*}\right)^{\frac{1}{2}} \Longrightarrow K^* = W^2 R^{-2} L^*$$

and so putting this in the first we have

$$\begin{aligned} Q^{\frac{1}{2}} &= (L^*)^{\frac{1}{2}} + (K^*)^{\frac{1}{2}} = (L^*)^{\frac{1}{2}} + \left(W^2R^{-2}L^*\right)^{\frac{1}{2}} \\ &= (L^*)^{\frac{1}{2}}\left(1 + WR^{-1}\right) = (L^*)^{\frac{1}{2}} W\left(W^{-1}+R^{-1}\right) \\ &\Longrightarrow (L^*)^{\frac{1}{2}} = Q^{\frac{1}{2}}W^{-1}\left(W^{-1}+R^{-1}\right)^{-1} \\ &\Longrightarrow L^* = QW^{-2}\left(W^{-1}+R^{-1}\right)^{-2} \end{aligned}$$

and

$$K^* = W^2R^{-2}L^* = W^2R^{-2}QW^{-2}\left(W^{-1}+R^{-1}\right)^{-2} = QR^{-2}\left(W^{-1}+R^{-1}\right)^{-2}.$$

Finally to find λ^* use

$$\begin{aligned} W &= \frac{\lambda^*}{2}(L^*)^{-\frac{1}{2}} = \frac{\lambda^*}{2}\left(QW^{-2}\left(W^{-1}+R^{-1}\right)^{-2}\right)^{-\frac{1}{2}} = \frac{\lambda^*}{2}Q^{-\frac{1}{2}}W\left(W^{-1}+R^{-1}\right) \\ &\Longrightarrow \lambda^* = 2Q^{\frac{1}{2}}\left(W^{-1}+R^{-1}\right)^{-1}. \end{aligned}$$

To show this is a constrained global minimizer we need to show that: 1) the objective function is linear, 2) the constraint function is globally convex, 3) the Lagrange multiplier is positive. First the objective function $WL+RK$ is linear. The Hessian of the constraint $g(L,K)$ is

$$H(L,K) = \begin{bmatrix} \frac{1}{4}L^{-\frac{3}{2}} & 0 \\ 0 & \frac{1}{4}K^{-\frac{3}{2}} \end{bmatrix}$$

which is a diagonal matrix with positive diagonal elements, and hence is positive definite for all L, K, and so the constraint is globally convex. Finally the Lagrange multiplier is positive as $\lambda^* = 2Q^{\frac{1}{2}}W^{-2}\left(W^{-1}+R^{-1}\right)^{-1} > 0$. Thus the solution corresponds to a constrained global minimum.

The cost function is

$$\begin{aligned} C^*(Q,W,R) &= WL^* + RK^* \\ &= W \times QW^{-2}\left(W^{-1}+R^{-1}\right)^{-2} + R \times QR^{-2}\left(W^{-1}+R^{-1}\right)^{-2} \\ &= QW^{-1}\left(W^{-1}+R^{-1}\right)^{-2} + QR^{-1}\left(W^{-1}+R^{-1}\right)^{-2} \\ &= Q\left(W^{-1}+R^{-1}\right)^{-2}\left(W^{-1}+R^{-1}\right) \\ &= Q\left(W^{-1}+R^{-1}\right)^{-1}. \end{aligned}$$

There are constant returns to scale since average cost is

$$AC^*(Q,W,R) \equiv \frac{C^*(Q,W,R)}{Q} = \frac{Q\left(W^{-1}+R^{-1}\right)^{-1}}{Q} = \left(W^{-1}+R^{-1}\right)^{-1}$$

so that

$$\frac{\partial AC^*(Q,W,R)}{\partial Q} = \frac{\partial}{\partial Q}\left(\left(W^{-1}+R^{-1}\right)^{-1}\right) = 0.$$

Problem 4.52 *[A ***] Suppose the firm's production function is given by*

$$Q = F(L,K) = \left(\frac{L^\alpha}{\alpha} + \frac{K^\alpha}{\alpha}\right)^{\frac{1}{\alpha}}$$

where $\alpha < 1$. Set up the Lagrangian for the problem of finding that amount of labour L^* and that amount of capital K^* that minimizes the cost of producing Q units of output. Find the first-order conditions from this Lagrangian and solve. Show that the solution is a constrained global minimizer. Find the cost function $C^* = WL^* + RK^*$ and show that there are constant returns to scale.

Answer: The math is easier if we use

$$Q = \left(\frac{L^\alpha}{\alpha} + \frac{K^\alpha}{\alpha}\right)^{\frac{1}{\alpha}} \Longrightarrow Q^\alpha = \frac{L^\alpha}{\alpha} + \frac{K^\alpha}{\alpha}$$

and we write the constraint as

$$g(L,K) = Q^\alpha - \frac{L^\alpha}{\alpha} - \frac{K^\alpha}{\alpha} = 0.$$

The Lagrangian is then

$$\mathcal{L}(\lambda, L, K) = WL + RK + \lambda\left(Q^\alpha - \frac{L^\alpha}{\alpha} - \frac{K^\alpha}{\alpha}\right)$$

so that

$$\begin{aligned}
\frac{\partial \mathcal{L}(\lambda, L, K)}{\partial \lambda} &= Q^\alpha - \frac{L^\alpha}{\alpha} - \frac{K^\alpha}{\alpha} \\
\frac{\partial \mathcal{L}(\lambda, L, K)}{\partial L} &= W - \lambda L^{\alpha-1} \\
\frac{\partial \mathcal{L}(\lambda, L, K)}{\partial K} &= R - \lambda K^{\alpha-1}.
\end{aligned}$$

The first-order conditions yield

$$\begin{aligned}
Q^\alpha - \frac{(L^*)^\alpha}{\alpha} - \frac{(K^*)^\alpha}{\alpha} &= 0 \\
&\Longrightarrow Q^\alpha = \frac{(L^*)^\alpha}{\alpha} + \frac{(K^*)^\alpha}{\alpha} \\
W - \alpha\lambda^* (L^*)^{\alpha-1} &= 0 \\
&\Longrightarrow W = \lambda^* (L^*)^{\alpha-1} \\
R - \alpha\lambda^* (K^*)^{\alpha-1} &= 0 \\
&\Longrightarrow R = \lambda^* (K^*)^{\alpha-1}.
\end{aligned}$$

Dividing the last two we have

$$\frac{W}{R} = \frac{\lambda^* (L^*)^{\alpha-1}}{\lambda^* (K^*)^{\alpha-1}} = \left(\frac{K^*}{L^*}\right)^{1-\alpha} \Longrightarrow K^* = W^{\frac{1}{1-\alpha}} R^{-\frac{1}{1-\alpha}} L^*$$

and so putting this last result in the production constraint yields

$$\begin{aligned} Q^\alpha &= \frac{(L^*)^\alpha}{\alpha} + \frac{(K^*)^\alpha}{\alpha} = \frac{(L^*)^\alpha}{\alpha} + \frac{\left(W^{\frac{1}{1-\alpha}} R^{-\frac{1}{1-\alpha}} L^*\right)^\alpha}{\alpha} \\ &= \frac{(L^*)^\alpha}{\alpha}\left(1 + W^{\frac{\alpha}{1-\alpha}} R^{-\frac{\alpha}{1-\alpha}}\right) = \frac{(L^*)^\alpha}{\alpha} W^{\frac{\alpha}{1-\alpha}}\left(W^{-\frac{\alpha}{1-\alpha}} + R^{-\frac{\alpha}{1-\alpha}}\right) \\ &\implies (L^*)^\alpha = \alpha Q^\alpha W^{-\frac{\alpha}{1-\alpha}}\left(W^{-\frac{\alpha}{1-\alpha}} + R^{-\frac{\alpha}{1-\alpha}}\right)^{-1} \\ &\implies L^* = \alpha^{\frac{1}{\alpha}} Q W^{-\frac{1}{1-\alpha}}\left(W^{-\frac{\alpha}{1-\alpha}} + R^{-\frac{\alpha}{1-\alpha}}\right)^{-\frac{1}{\alpha}}. \end{aligned}$$

We have

$$\begin{aligned} K^* &= W^{\frac{1}{1-\alpha}} R^{-\frac{1}{1-\alpha}} L^* = W^{\frac{1}{1-\alpha}} R^{-\frac{1}{1-\alpha}} \times \alpha^{\frac{1}{\alpha}} Q W^{-\frac{1}{1-\alpha}}\left(W^{-\frac{\alpha}{1-\alpha}} + R^{-\frac{\alpha}{1-\alpha}}\right)^{-\frac{1}{\alpha}} \\ &= \alpha^{\frac{1}{\alpha}} Q R^{-\frac{1}{1-\alpha}}\left(W^{-\frac{\alpha}{1-\alpha}} + R^{-\frac{\alpha}{1-\alpha}}\right)^{-\frac{1}{\alpha}} \\ &\implies K^* = \alpha^{\frac{1}{\alpha}} Q R^{-\frac{1}{1-\alpha}}\left(W^{-\frac{\alpha}{1-\alpha}} + R^{-\frac{\alpha}{1-\alpha}}\right)^{-\frac{1}{\alpha}}. \end{aligned}$$

Finally to find λ^* use

$$W = \lambda^* (L^*)^{\alpha-1} = \lambda^*\left(\alpha^{\frac{1}{\alpha}} Q W^{-\frac{1}{1-\alpha}}\left(W^{-\frac{\alpha}{1-\alpha}} + R^{-\frac{\alpha}{1-\alpha}}\right)^{-\frac{1}{\alpha}}\right)^{\alpha-1}$$

so that

$$\begin{aligned} \lambda^* &= \alpha^{\frac{1}{\alpha}-1} W\left(Q W^{-\frac{1}{1-\alpha}}\left(W^{-\frac{\alpha}{1-\alpha}} + R^{-\frac{\alpha}{1-\alpha}}\right)^{-\frac{1}{\alpha}}\right)^{1-\alpha} \\ &= \alpha^{\frac{1}{\alpha}-1} Q^{1-\alpha}\left(W^{-\frac{\alpha}{1-\alpha}} + R^{-\frac{\alpha}{1-\alpha}}\right)^{-\frac{1-\alpha}{\alpha}}. \end{aligned}$$

To show this corresponds to a constrained global minimum we need to show that: 1) the objective function is linear, 2) the constraint function is globally convex, 3) the Lagrange multiplier is positive. First the objective function, costs $WL + RK$ is linear. The Hessian of the constraint $g(L, K)$ is

$$H(L, K) = \begin{bmatrix} -(\alpha-1) L^{\alpha-2} & 0 \\ 0 & -(\alpha-1) K^{\alpha-2} \end{bmatrix}$$

which is a diagonal matrix with positive diagonal elements since $0 < \alpha < 1$, and hence is positive definite for all L, K, and so the constraint is globally convex. Finally the Lagrange multiplier is positive as

$$\lambda^* = \alpha^{\frac{1}{\alpha}-1} Q^{1-\alpha}\left(W^{-\frac{\alpha}{1-\alpha}} + R^{-\frac{\alpha}{1-\alpha}}\right)^{-\frac{1-\alpha}{\alpha}} > 0.$$

Thus the solution is a constrained global minimizer.

The cost function is

$$\begin{aligned}
C^*(Q,W,R) &= WL^* + RK^* \\
&= W \times \alpha^{\frac{1}{\alpha}} Q W^{-\frac{1}{1-\alpha}} \left(W^{-\frac{\alpha}{1-\alpha}} + R^{-\frac{\alpha}{1-\alpha}}\right)^{-\frac{1}{\alpha}} \\
&\quad + R \times \alpha^{\frac{1}{\alpha}} Q R^{-\frac{1}{1-\alpha}} \left(W^{-\frac{\alpha}{1-\alpha}} + R^{-\frac{\alpha}{1-\alpha}}\right)^{-\frac{1}{\alpha}} \\
&= \alpha^{\frac{1}{\alpha}} Q W^{-\frac{\alpha}{1-\alpha}} \left(W^{-\frac{\alpha}{1-\alpha}} + R^{-\frac{\alpha}{1-\alpha}}\right)^{-\frac{1}{\alpha}} + \alpha^{\frac{1}{\alpha}} Q R^{-\frac{\alpha}{1-\alpha}} \left(W^{-\frac{\alpha}{1-\alpha}} + R^{-\frac{\alpha}{1-\alpha}}\right)^{-\frac{1}{\alpha}} \\
&= \alpha^{\frac{1}{\alpha}} Q \left(W^{-\frac{\alpha}{1-\alpha}} + R^{-\frac{\alpha}{1-\alpha}}\right)^{-\frac{1}{\alpha}} \left(W^{-\frac{\alpha}{1-\alpha}} + R^{-\frac{\alpha}{1-\alpha}}\right) \\
&= \alpha^{\frac{1}{\alpha}} Q \left(W^{-\frac{\alpha}{1-\alpha}} + R^{-\frac{\alpha}{1-\alpha}}\right)^{1-\frac{1}{\alpha}} = \alpha^{\frac{1}{\alpha}} Q \left(W^{-\frac{\alpha}{1-\alpha}} + R^{-\frac{\alpha}{1-\alpha}}\right)^{-\frac{1-\alpha}{\alpha}}.
\end{aligned}$$

There are constant returns to scale since average cost is

$$AC^*(Q,W,R) \equiv \frac{C^*(Q,W,R)}{Q} = \frac{Q\left(W^{-\frac{\alpha}{1-\alpha}} + R^{-\frac{\alpha}{1-\alpha}}\right)^{-\frac{1-\alpha}{\alpha}}}{Q} = \left(W^{-\frac{\alpha}{1-\alpha}} + R^{-\frac{\alpha}{1-\alpha}}\right)^{-\frac{1-\alpha}{\alpha}}$$

and so $\frac{\partial AC^*(Q,W,R)}{\partial Q} = 0$.

Problem 4.53 *[A***] Suppose the firm's production function is given by*

$$Q = F(L,K) = \left(\frac{L^\alpha}{\alpha} + \frac{K^\alpha}{\alpha}\right)^{\frac{\beta}{\alpha}}$$

where $\alpha < 1$ and $\beta > 0$. Find the cost function $C^* = WL^* + RK^*$. Show that there are decreasing returns to scale if $\beta < 1$, constant returns to scale if $\beta = 1$, and increasing returns to scale if $\beta > 1$.

Answer: The math is easier if we use

$$Q = \left(\frac{L^\alpha}{\alpha} + \frac{K^\alpha}{\alpha}\right)^{\frac{\beta}{\alpha}} \Longrightarrow Q^{\frac{\alpha}{\beta}} = \frac{L^\alpha}{\alpha} + \frac{K^\alpha}{\alpha} \Longrightarrow Q^{\frac{\alpha}{\beta}} - \frac{L^\alpha}{\alpha} - \frac{K^\alpha}{\alpha} = 0.$$

The Lagrangian is then

$$\mathcal{L}(\lambda,L,K) = WL + RK + \lambda\left(Q^{\frac{\alpha}{\beta}} - \frac{L^\alpha}{\alpha} - \frac{K^\alpha}{\alpha}\right)$$

so that

$$\begin{aligned}
\frac{\partial \mathcal{L}(\lambda,L,K)}{\partial \lambda} &= Q^{\frac{\alpha}{\beta}} - \frac{L^\alpha}{\alpha} - \frac{K^\alpha}{\alpha} \\
\frac{\partial \mathcal{L}(\lambda,L,K)}{\partial L} &= W - \lambda L^{\alpha-1} \\
\frac{\partial \mathcal{L}(\lambda,L,K)}{\partial K} &= R - \lambda K^{\alpha-1}.
\end{aligned}$$

The first-order conditions yield

$$\begin{aligned}
Q^{\frac{\alpha}{\beta}} - \frac{(L^*)^\alpha}{\alpha} - \frac{(K^*)^\alpha}{\alpha} &= 0 \\
&\Longrightarrow Q^{\frac{\alpha}{\beta}} = \frac{(L^*)^\alpha}{\alpha} + \frac{(K^*)^\alpha}{\alpha} \\
W - \alpha\lambda^* (L^*)^{\alpha-1} &= 0 \\
&\Longrightarrow W = \lambda^* (L^*)^{\alpha-1} \\
R - \alpha\lambda^* (K^*)^{\alpha-1} &= 0 \\
&\Longrightarrow R = \lambda^* (K^*)^{\alpha-1}.
\end{aligned}$$

Dividing the last two results we have

$$\begin{aligned}
\frac{W}{R} &= \frac{\lambda^* (L^*)^{\alpha-1}}{\lambda^* (K^*)^{\alpha-1}} = \left(\frac{K^*}{L^*}\right)^{1-\alpha} \\
&\Longrightarrow K^* = W^{\frac{1}{1-\alpha}} R^{-\frac{1}{1-\alpha}} L^*
\end{aligned}$$

and so putting this in the first results we have

$$\begin{aligned}
Q^{\frac{\alpha}{\beta}} &= \frac{(L^*)^\alpha}{\alpha} + \frac{(K^*)^\alpha}{\alpha} = \frac{(L^*)^\alpha}{\alpha} + \frac{\left(W^{\frac{1}{1-\alpha}} R^{-\frac{1}{1-\alpha}} L^*\right)^\alpha}{\alpha} \\
&= \frac{(L^*)^\alpha}{\alpha}\left(1 + W^{\frac{\alpha}{1-\alpha}} R^{-\frac{\alpha}{1-\alpha}}\right) = \frac{(L^*)^\alpha}{\alpha} W^{\frac{\alpha}{1-\alpha}} \left(W^{-\frac{\alpha}{1-\alpha}} + R^{-\frac{\alpha}{1-\alpha}}\right) \\
&\Longrightarrow (L^*)^\alpha = \alpha Q^{\frac{\alpha}{\beta}} W^{-\frac{\alpha}{1-\alpha}} \left(W^{-\frac{\alpha}{1-\alpha}} + R^{-\frac{\alpha}{1-\alpha}}\right)^{-1} \\
&\Longrightarrow L^* = \alpha^{\frac{1}{\alpha}} Q^{\frac{1}{\beta}} W^{-\frac{1}{1-\alpha}} \left(W^{-\frac{\alpha}{1-\alpha}} + R^{-\frac{\alpha}{1-\alpha}}\right)^{-\frac{1}{\alpha}}
\end{aligned}$$

and

$$\begin{aligned}
K^* &= W^{\frac{1}{1-\alpha}} R^{-\frac{1}{1-\alpha}} L^* = W^{\frac{1}{1-\alpha}} R^{-\frac{1}{1-\alpha}} \times \alpha^{\frac{1}{\alpha}} Q^{\frac{1}{\beta}} W^{-\frac{1}{1-\alpha}} \left(W^{-\frac{\alpha}{1-\alpha}} + R^{-\frac{\alpha}{1-\alpha}}\right)^{-\frac{1}{\alpha}} \\
&= \alpha^{\frac{1}{\alpha}} Q^{\frac{1}{\beta}} R^{-\frac{1}{1-\alpha}} \left(W^{-\frac{\alpha}{1-\alpha}} + R^{-\frac{\alpha}{1-\alpha}}\right)^{-\frac{1}{\alpha}}.
\end{aligned}$$

The cost function is

$$\begin{aligned}
C^*(Q, W, R) &= WL^* + RK^* \\
&= W\alpha^{\frac{1}{\alpha}} Q^{\frac{1}{\beta}} W^{-\frac{1}{1-\alpha}} \left(W^{-\frac{\alpha}{1-\alpha}} + R^{-\frac{\alpha}{1-\alpha}}\right)^{-\frac{1}{\alpha}} \\
&\quad + R\alpha^{\frac{1}{\alpha}} Q^{\frac{1}{\beta}} R^{-\frac{1}{1-\alpha}} \left(W^{-\frac{\alpha}{1-\alpha}} + R^{-\frac{\alpha}{1-\alpha}}\right)^{-\frac{1}{\alpha}} \\
&= \alpha^{\frac{1}{\alpha}} Q^{\frac{1}{\beta}} W^{-\frac{\alpha}{1-\alpha}} \left(W^{-\frac{\alpha}{1-\alpha}} + R^{-\frac{\alpha}{1-\alpha}}\right)^{-\frac{1}{\alpha}} \\
&\quad + \alpha^{\frac{1}{\alpha}} Q^{\frac{1}{\beta}} R^{-\frac{\alpha}{1-\alpha}} \left(W^{-\frac{\alpha}{1-\alpha}} + R^{-\frac{\alpha}{1-\alpha}}\right)^{-\frac{1}{\alpha}} \\
&= \alpha^{\frac{1}{\alpha}} Q^{\frac{1}{\beta}} \left(W^{-\frac{\alpha}{1-\alpha}} + R^{-\frac{\alpha}{1-\alpha}}\right)^{-\frac{1}{\alpha}} \left(W^{-\frac{\alpha}{1-\alpha}} + R^{-\frac{\alpha}{1-\alpha}}\right) \\
&= \alpha^{\frac{1}{\alpha}} Q^{\frac{1}{\beta}} \left(W^{-\frac{\alpha}{1-\alpha}} + R^{-\frac{\alpha}{1-\alpha}}\right)^{1-\frac{1}{\alpha}} \\
&= \alpha^{\frac{1}{\alpha}} Q^{\frac{1}{\beta}} \left(W^{-\frac{\alpha}{1-\alpha}} + R^{-\frac{\alpha}{1-\alpha}}\right)^{-\frac{1-\alpha}{\alpha}}
\end{aligned}$$

and average costs are

$$\begin{aligned} AC^*(Q,W,R) &\equiv \frac{C^*(Q,W,R)}{Q} \\ &= \frac{Q^{\frac{1}{\beta}}\left(W^{-\frac{\alpha}{1-\alpha}}+R^{-\frac{\alpha}{1-\alpha}}\right)^{-\frac{1-\alpha}{\alpha}}}{Q} \\ &= Q^{\frac{1}{\beta}-1}\left(W^{-\frac{\alpha}{1-\alpha}}+R^{-\frac{\alpha}{1-\alpha}}\right)^{-\frac{1-\alpha}{\alpha}} \end{aligned}$$

and so

$$\begin{aligned} \frac{\partial AC^*(Q,W,R)}{\partial Q} &= \frac{\partial}{\partial Q}\left(Q^{\frac{1}{\beta}-1}\left(W^{-\frac{\alpha}{1-\alpha}}+R^{-\frac{\alpha}{1-\alpha}}\right)^{-\frac{1-\alpha}{\alpha}}\right) \\ &= \left(\frac{1}{\beta}-1\right)Q^{\frac{1}{\beta}-2}\left(W^{-\frac{\alpha}{1-\alpha}}+R^{-\frac{\alpha}{1-\alpha}}\right)^{-\frac{1-\alpha}{\alpha}} \\ &= \left(\frac{1-\beta}{\beta}\right)Q^{\frac{1}{\beta}-2}\left(W^{-\frac{\alpha}{1-\alpha}}+R^{-\frac{\alpha}{1-\alpha}}\right)^{-\frac{1-\alpha}{\alpha}} \end{aligned}$$

so that $\frac{\partial AC^*}{\partial Q} > 0$ (decreasing returns to scale) if $\beta < 1$, $\frac{\partial AC^*}{\partial Q} = 0$ (constant returns to scale) if $\beta = 1$, and $\frac{\partial AC^*}{\partial Q} < 0$ (increasing returns to scale) if $\beta > 1$.

4.3.4 The Hessian of the Lagrangian

Problem 4.54 *[B**] In Problem 4.32 on page 212 we showed that the problem of minimizing*

$$f(x_1,x_2) = x_1^2 + \frac{1}{2}x_2^2 - x_1x_2$$

subject to $x_1 + x_2 = 1$ *has a solution* $x_1^* = \frac{2}{5}, x_2^* = \frac{3}{5}, \lambda^* = \frac{1}{5}$. *Show that the Hessian of the Lagrangian is not positive definite, but that it can be used to show that the solution is a constrained local minimum.*

Answer: The Lagrangian is

$$\mathcal{L}(\lambda,x_1,x_2) = x_1^2 + \frac{1}{2}x_2^2 - x_1x_2 + \lambda(1-x_1-x_2).$$

Differentiating the Lagrangian with respect to λ, x_1, x_2 results in

$$\begin{aligned} \frac{\partial\mathcal{L}}{\partial\lambda} &= \frac{\partial}{\partial\lambda}\left(x_1^2+\frac{1}{2}x_2^2-x_1x_2+\lambda(1-x_1-x_2)\right) = 1-x_1-x_2 \\ \frac{\partial\mathcal{L}}{\partial x_1} &= \frac{\partial}{\partial x_1}\left(x_1^2+\frac{1}{2}x_2^2-x_1x_2+\lambda(1-x_1-x_2)\right) = 2x_1-x_2-\lambda \\ \frac{\partial\mathcal{L}}{\partial x_2} &= \frac{\partial}{\partial x_2}\left(x_1^2+\frac{1}{2}x_2^2-x_1x_2+\lambda(1-x_1-x_2)\right) = x_2-x_1-\lambda \end{aligned}$$

from which the Hessian of the Lagrangian can be calculated from

$$\begin{aligned}
\frac{\partial \mathcal{L}}{\partial \lambda} &= 1 - x_1 - x_2 \\
&\Longrightarrow \frac{\partial^2 \mathcal{L}}{\partial \lambda^2} = 0, \frac{\partial^2 \mathcal{L}}{\partial \lambda \partial x_1} = -1, \frac{\partial^2 \mathcal{L}}{\partial \lambda \partial x_2} = -1 \\
\frac{\partial \mathcal{L}}{\partial x_1} &= 2x_1 - x_2 - \lambda \\
&\Longrightarrow \frac{\partial^2 \mathcal{L}}{\partial \lambda \partial x_1} = -1, \frac{\partial^2 \mathcal{L}}{\partial x_1^2} = 2, \frac{\partial^2 \mathcal{L}}{\partial x_1 \partial x_2} = -1 \\
\frac{\partial \mathcal{L}}{\partial x_2} &= x_2 - x_1 - \lambda \\
&\Longrightarrow \frac{\partial^2 \mathcal{L}}{\partial \lambda \partial x_2} = -1, \frac{\partial^2 \mathcal{L}}{\partial x_1 \partial x_2} = -1, \frac{\partial^2 \mathcal{L}}{\partial x_2^2} = 1
\end{aligned}$$

as

$$H(\lambda, x_1, x_2) = \begin{bmatrix} 0 & -1 & -1 \\ -1 & 2 & -1 \\ -1 & -1 & 1.0 \end{bmatrix}.$$

The Hessian $H(\lambda, x_1, x_2)$ is neither positive nor negative definite since the 1, 1 element is 0. (Or the eigenvalues are $\lambda_1 = -1.21, \lambda_2 = 1.54, \lambda_3 = 2.68$ with $\lambda_1 < 0$.) But we have

$$M_3 = \det \begin{bmatrix} 0 & -1 & -1 \\ -1 & 2 & -1 \\ -1 & -1 & 1 \end{bmatrix} = -5 < 0$$

and so $x_1^* = \frac{2}{5}, x_2^* = \frac{3}{5}, \lambda^* = \frac{1}{5}$ is a constrained local minimizer.

Problem 4.55 *[B**] For a household with the utility function*

$$U(Q_1, Q_2) = 3Q_1^{\frac{1}{3}} + 3Q_2^{\frac{1}{3}}$$

the solution to the first-order conditions is

$$\lambda^* = Y^{-\frac{2}{3}} \left(P_1^{-\frac{1}{2}} + P_2^{-\frac{1}{2}} \right)^{\frac{2}{3}}, Q_1^* = \frac{Y P_1^{-\frac{3}{2}}}{P_1^{-\frac{1}{2}} + P_2^{-\frac{1}{2}}}, Q_2^* = \frac{Y P_2^{-\frac{3}{2}}}{P_1^{-\frac{1}{2}} + P_2^{-\frac{1}{2}}}.$$

Use the Lagrangian of the Hessian to show that the solution corresponds to a constrained local maximum.

Answer: We have

$$\mathcal{L}(\lambda, Q_1, Q_2) = 3Q_1^{\frac{1}{3}} + 3Q_2^{\frac{1}{3}} + \lambda (Y - P_1 Q_1 - P_2 Q_2)$$

so that

$$\begin{aligned}\frac{\partial \mathcal{L}(\lambda, Q_1, Q_2)}{\partial \lambda} &= Y - P_1 Q_1 - P_2 Q_2 \\ \frac{\partial \mathcal{L}(\lambda, Q_1, Q_2)}{\partial Q_1} &= Q_1^{-\frac{2}{3}} - \lambda P_1 \\ \frac{\partial \mathcal{L}(\lambda, Q_1, Q_2)}{\partial Q_2} &= Q_2^{-\frac{2}{3}} - \lambda P_2.\end{aligned}$$

The Hessian of the Lagrangian is found from

$$\begin{aligned}\frac{\partial^2 \mathcal{L}(\lambda, Q_1, Q_2)}{\partial \lambda^2} &= 0, \frac{\partial^2 \mathcal{L}(\lambda, Q_1, Q_2)}{\partial \lambda \partial Q_1} = -P_1, \frac{\partial^2 \mathcal{L}(\lambda, Q_1, Q_2)}{\partial \lambda \partial Q_2} = -P_2 \\ \frac{\partial^2 \mathcal{L}(\lambda, Q_1, Q_2)}{\partial Q_1^2} &= \frac{\partial}{\partial Q_1}\left(Q_1^{-\frac{2}{3}} - \lambda P_1\right) = -\frac{2}{3} Q_1^{-\frac{5}{3}} \\ \frac{\partial^2 \mathcal{L}(\lambda, Q_1, Q_2)}{\partial Q_1 \partial Q_2} &= \frac{\partial}{\partial Q_2}\left(Q_1^{-\frac{2}{3}} - \lambda P_1\right) = 0 \\ \frac{\partial^2 \mathcal{L}(\lambda, Q_1, Q_2)}{\partial Q_2^2} &= \frac{\partial}{\partial Q_2}\left(Q_2^{-\frac{2}{3}} - \lambda P_1\right) = -\frac{2}{3} Q_2^{-\frac{5}{3}}\end{aligned}$$

as

$$H(\lambda, Q_1, Q_2) = \begin{bmatrix} 0 & -P_1 & -P_2 \\ -P_1 & -\frac{2}{3} Q_1^{-5/3} & 0 \\ -P_2 & 0 & -\frac{2}{3} Q_2^{-5/3} \end{bmatrix}.$$

We have

$$\begin{aligned}\det[H(\lambda^*, Q_1^*, Q_2^*)] &= \det \begin{bmatrix} 0 & -P_1 & -P_2 \\ -P_1 & -\frac{2}{3}(Q_1^*)^{-\frac{5}{3}} & 0 \\ -P_2 & 0 & -\frac{2}{3}(Q_2^*)^{-\frac{5}{3}} \end{bmatrix} \\ &= \frac{2}{3} P_1^2 (Q_2^*)^{-\frac{5}{3}} + \frac{2}{3} P_2^2 (Q_1^*)^{-\frac{5}{3}} > 0\end{aligned}$$

so λ^*, Q_1^*, Q_2^* is a constrained local maximizer.

Problem 4.56 *[B**] For a household with utility*

$$U(Q_1, Q_2) = -Q_1^{-1} - Q_2^{-1}$$

use the Hessian of the Lagrangian to show that the solution to the first-order conditions λ^*, Q_1^*, Q_2^* corresponds to a constrained local maximum.

Answer: The Hessian of the Lagrangian

$$\mathcal{L}(\lambda, Q_1, Q_2) = -Q_1^{-1} - Q_2^{-1} + \lambda(Y - P_1 Q_1 - P_2 Q_2)$$

is

$$H(\lambda, Q_1, Q_2) = \begin{bmatrix} 0 & -P_1 & -P_2 \\ -P_1 & -2Q_1^{-3} & 0 \\ -P_2 & 0 & -2Q_2^{-3} \end{bmatrix}$$

so that

$$\begin{aligned}\det\left[H\left(\lambda^*, Q_1^*, Q_2^*\right)\right] &= \det\begin{bmatrix} 0 & -P_1 & -P_2 \\ -P_1 & -2\left(Q_1^*\right)^{-3} & 0 \\ -P_2 & 0 & -2\left(Q_2^*\right)^{-3} \end{bmatrix} \\ &= 2P_1^2\left(Q_2^*\right)^{-3} + 2P_2^2\left(Q_1^*\right)^{-3} > 0\end{aligned}$$

and the solution is a constrained local maximizer.

Problem 4.57 *[A **] Consider the problem of maximizing utility*

$$U\left(Q_1, Q_2\right) = U_1\left(Q_1\right) + U_2\left(Q_2\right)$$

where $U_1'\left(Q_1\right) > 0, U_2'\left(Q_2\right) > 0, U_1''\left(Q_1\right) < 0, U_2''\left(Q_2\right) < 0$ *with Lagrangian*

$$\mathcal{L}\left(\lambda, Q_1, Q_2\right) = U_1\left(Q_1\right) + U_2\left(Q_2\right) + \lambda\left(Y - P_1Q_1 - P_2Q_2\right).$$

Use the the Hessian of the Lagrangian to show that the solution to the first-order conditions λ^*, Q_1^*, Q_2^* *is a constrained local maximum.*

Answer: From

$$\begin{aligned}\frac{\partial\mathcal{L}}{\partial\lambda} &= Y - P_1Q_1 - P_2Q_2 \\ \frac{\partial\mathcal{L}}{\partial Q_1} &= U_1'\left(Q_1\right) - \lambda P_1 \\ \frac{\partial\mathcal{L}}{\partial Q_2} &= U_2'\left(Q_2\right) - \lambda P_2\end{aligned}$$

we have

$$\begin{aligned}\frac{\partial^2\mathcal{L}}{\partial\lambda^2} &= 0, \frac{\partial^2\mathcal{L}}{\partial\lambda\partial Q_1} = -P_1, \frac{\partial^2\mathcal{L}}{\partial\lambda\partial Q_2} = -P_2 \\ \frac{\partial^2\mathcal{L}}{\partial Q_1^2} &= U_1''\left(Q_1\right), \frac{\partial^2\mathcal{L}}{\partial Q_1\partial Q_2} = 0, \frac{\partial^2\mathcal{L}}{\partial Q_2^2} = U_2''\left(Q_2\right)\end{aligned}$$

so that the Hessian of the Lagrangian evaluated at λ^*, Q_1^*, Q_2^* is

$$H\left(\lambda^*, Q_1^*, Q_2^*\right) = \begin{bmatrix} 0 & -P_1 & -P_2 \\ -P_1 & U_1''\left(Q_1^*\right) & 0 \\ -P_2 & 0 & U_2''\left(Q_2^*\right) \end{bmatrix}$$

and so

$$\det\begin{bmatrix} 0 & -P_1 & -P_2 \\ -P_1 & U_1''\left(Q_1^*\right) & 0 \\ -P_2 & 0 & U_2''\left(Q_2^*\right) \end{bmatrix} = -P_1^2U_2''\left(Q_2^*\right) - P_2^2U_1''\left(Q_1^*\right) > 0$$

since $U_1''\left(Q_1^*\right) < 0, U_2''\left(Q_2^*\right) < 0$. It follows that λ^*, Q_1^*, Q_2^* is a constrained local maximizer.

Made in United States
Orlando, FL
11 November 2026

90609401R00142